Lecture Notes in Biomathematics

ctd. on inside back cover

Lecture Notes in Biomathematics

Managing Editor: S. Levin

81

C. Castillo-Chavez S.A. Levin
C.A. Shoemaker (Eds.)

Mathematical Approaches to Problems in Resource Management and Epidemiology

Proceedings of a Conference
held at Ithaca, NY, Oct. 28–30, 1987

Springer-Verlag

Berlin Heidelberg New York London Paris Tokyo Hong Kong

Mathematics Subject Classification (1980): 92A 15, 92A 17

ISBN 978-3-540-51820-4   ISBN 978-3-642-46693-9 (eBook)
DOI 10.1007/978-3-642-46693-9

2146/3140-543210 – Printed on acid-free paper

TABLE OF CONTENTS

PREFACE

Increasingly, mathematical methods are being used to advantage in addressing the problems facing humanity in managing its environment. Problems in resource management and epidemiology especially have demonstrated the utility of quantitative modeling. To explore these approaches, the Center of Applied Mathematics at Cornell University organized a conference in Fall, 1987, with the objective of surveying and assessing the state of the art. This volume records the proceedings of that conference.

Underlying virtually all of these studies are models of population growth, from individual cells to large vertebrates. Cell population growth presents the simplest of systems for study, and is of fundamental importance in its own right for a variety of medical and environmental applications. In Part I of this volume, Michael Shuler describes computer models of individual cells and cell populations, and Frank Hoppensteadt discusses the synchronization of bacterial culture growth. Together, these provide a valuable introduction to mathematical cell biology.

One of the most important works in all of mathematical biology was Volterra's investigations of the fluctuations of the Adriatic fisheries, and mathematical models remain a central component of resource management today. In Part II, a series of five papers by leading researchers describe mathematical, biological, and economic aspects of the subject aptly named mathematical bioeconomics. Colin Clark, one of the founders of this line of research, surveys the subject, tracing its history from the works of Faustmann, Schaefer and Gordon. Louis Gross discusses particular problems associated with models of plant populations, and Jon Conrad complements this biological analysis with a discussion of economic aspects of resource modeling. Christine Shoemaker and Sharon Johnson develop the methodology of stochastic nonlinear optimal control, discussing computational difficulties. Henry Wan concludes the section with an application to tree farming, relating his paper to those of Clark, Shoemaker, and Conrad, and developing the ideas of Faustmann mentioned earlier by Clark.

The mathematical tradition also has deep roots in the study of the epidemiology of infectious diseases, tracing back to the work of En'ko, Hamer, Ross, and others. Parts III and IV discuss current research problems in this area, beginning with Wei-min Liu's survey of diseases that involve several host species. Fred Brauer discusses the implications of relaxing the common assumption that population size does not vary much during disease cycles, an assumption that is often appropriate for human populations, but not for other hosts. Stavros Busenberg, Kenneth Cooke, and Mimmo Iannelli examine the threshold problem for age-structured populations, extending classical results; Viggo Andreasen complements this work with further studies of age-structured models, extending these ideas to situations in which two strains of a virus are cocirculating in a population. Finally, Fred Adler, Lincoln Smith and Carlos Castillo-Chavez show how epidemiological approaches can be applied to parasitoid-host systems.

Part IV surveys current work on AIDS, the most timely of epidemiological problems. A fuller treatment of this topic will be presented in a forthcoming volume in this series, edited by Carlos Castillo-Chavez, and concerned exclusively with AIDS. Nonetheless, the four papers presented in this section, involving the work of ten researchers, provide a fairly complete introduction to the state of the art. Four aspects are considered: the outline of a simulation model presently being tested by the Center for Disease Control in Atlanta, the role of long periods of infectiousness, the effects of social mixing, and the possible demographic consequences of the AIDS epidemic.

Applied mathematical ecology involves model construction, fitting models to particular situations, and mathematical exploration of model properties. Too often, the latter two aspects are given short shrift in the mathematical ecology literature. Part V provides a discussion of statistical problems associated with fitting models to data. David Ruppert begins with a review of recent developments in fitting functional relationships, and H.T. Banks and B.G. Fitzpatrick discuss the parameter estimation problem for partial differential equations. Don Ludwig concludes this section with a discussion of the problem of selecting management models, pointing out that mindless attention to detail may lead one down the garden path—and into the fish pond.

The last section discusses general aspects of dynamic modeling, including methodologies from dynamical systems theory, interacting particle systems, and differential geometry. Stephen Ellner discusses the application of the methods of Takens, and the problem of chaos. Richard Durrett discusses the application of stochastic models to describing the spatial spread of populations and epidemics. Ethan Akin makes a pitch for the use of differential geometric methods, and John Guckenheimer concludes the volume with a balanced discussion of the difficulties in applying dynamical systems theory.

For this volume, heterogeneity is the rule, but heterogeneity is expressed through a single language—mathematics. The approaches discussed include optimization and numerical methods, dynamical systems theory, aggregation and simplification, and stochastic methods. The problems attacked include ones selected from theoretical epidemiology, biological control, cell growth, and resource management. Several of the articles review important past contributions in the area, identify critical problems still in need of attention and suggest new directions for research. Invited participants were drawn from a variety of fields, including theoretical biology, statistics, economics, and pure and applied mathematics. The organization of the volume reflects this diversity of interests.

The experience of organizing the conference and volume has been very stimulating, and the symposium provided the unique opportunity to bring together researchers of diverse backgrounds. We are grateful for the support of the Center for Applied Mathematics, and particularly its directors, Michael Todd and Sid Leibovich, for their moral and financial support. We acknowledge happily the outstanding organizational efforts of Dolores Pendell. We also thank the Center for Environmental Research for its partial support of the special year.

Finally, we owe a debt of gratitude to all the expert scientists who reviewed the manuscripts included in this volume, and especially to the authors for their excellent papers. Above all, we express our deep gratitude to Ilka Lee, our editorial assistant, for her work in making a silk purse out of a sow's ear.

C. Castillo-Chavez
S.A. Levin
C. Shoemaker
Ithaca, N.Y.

Part I. Cell Population Dynamics

COMPUTER MODELS OF INDIVIDUAL LIVING CELLS IN CELL POPULATIONS

Michael L. Shuler
School of Chemical Engineering
Cornell University
Ithaca, New York 14853-5201

Abstract

The biosynthetic capacity of an individual cell is dependent on its structure. The response of large population of cells reflects the aggregated response of individual cells. Individual cell's differ from one and another. The use of population balance equations to describe the dynamic response of populations to perturbations in their environment is computationally difficult when both the structure of individual cells and their distribution within the population are important. We circumvent these computational problems by building highly structured models of individual cells and then using a finite-representation technique to model the whole population. Application of this technique to predicting protein production from recombinant DNA is described.

1. Introduction

How do cells grow? How do they regulate their response to changes in their external environment? How can we manipulate the extracellular environment to force the cells to make a given metabolic product?

These questions are representative of a wide group of questions we seek to answer. A bioreaction engineer must construct a large macroscopic reactor and its operating strategy in such a way as to maximize the formation of a product from a population of living cells. Each cell is a chemical reactor - a complex and highly regulated reactor. The regulatory systems of such cells are not fully understood. They are, however, far more sophisticated than any man-developed process control system for chemical reactors. Any macroscopic reactor will contain "billions and billions" of these little cellular reactors. Each cellular reactor is distinct; it is an individual with a unique physiological and, hence, biochemical state. Ideally the engineer wants to quantitatively predict the aggregated response of such a population to any change in the extracellular environment (e.g. nutrient levels, pH, temperature, etc.). To accomplish this objective the engineer must be able to predict *a priori* the response of individual cells to changes in its environment. This problem is essentially the same one that an ecologist faces when attempting to learn how a particular ecosystem will respond to environmental perturbations.

This objective leads to the modeling approach I will describe. Unlike many of the other papers in this volume the appropriate model is very detailed. The model attempts to achieve a rather literal translation of the physical reality of a cell. Such an approach yields a model which does not admit an analytical solution, but rather demands a numerical solution. Thus, the models described in this chapter are rather different from many of the approaches suggested elsewhere in this volume.

2. Some Basic Modeling Concepts for Cells

Process models for bioreactors and models for the response of biota in a natural environment can be described in the same terms. The main difference is that bioreactors contain usually single-species while the natural environment contains many interacting species. I will begin by considering single species or "pure" populations.

The intellectual framework for modeling such populations was developed in an important review article by Tsuchiya, Fredrickson, and Aris (1966). The main characteristics of this framework were based on the definitions of structure and segregation. A model is said to be structured if two or more sub-components must be specified to fix the physiological state of the culture. A subcomponent could be a distinct cellular organelle or structure (e.g. cell wall, starch granules, vacuoles, etc.) or a separate chemical specie (e.g. RNA, protein, DNA, polysaccharides, etc.).

Most often models are chemically structured and a structured model might lump cellular components into groups such as RNA, cytoplasmic protein, DNA, cell membrane lipids, etc. An unstructured model assumes that a single variable is adequate to describe the population. Typically this single variable is related to the quantity of biomass.

Although unstructured models are attractive due to their simplicity, they are also very limited. Consider that a cell's regulatory system functions by altering its metabolic abilities in response to changing environmental conditions. Formation of new enzymes might be induced while synthesis of other enzymes are repressed. Thus, the internal composition of the cell changes as well as its biosynthetic capabilities. An unstructured model simply does not allow for such changes. The biosynthetic capability of a culture to perform a certain chemical transformation does not necessarily double if the cell mass doubles while environmental changes are occurring. It can be mathematically demonstrated that unstructured models cannot predict the dynamic response of a culture to perturbations in the extracellular environment (Fredrickson, Ramkrishna, and Tsuchiya, 1971).

In addition to chemical structure the property of segregation can be important. Segregation recognizes that a population consists of many different distinct individuals. The distribution of properties among individuals in the population can be extremely important to the future productivity of a culture (particularly cells with plasmids encoding high levels of protein production). A non-segregated model ignores differences among individuals. Non-segregated models assume that a population-averaged value of a property is totally sufficient to predict the ultimate response of the population. The assumption of uniformly identical cells results in reasonable predictions in many cases but fails in some very important cases.

For example, consider a population where copies of a particular gene are amplified with 80% of the population having 5 copies and the other 20% having 100 copies. In many cases the amount of protein made per gene is non-linear with the number of gene copies. Thus, a cell with 100 copies might make eight times as much gene product as one with 5 copies and twice as much as one with 24 copies. If we normalize gene productivity with that of low copy number cells (5 copies or less), one would predict production of 1200 units of gene product using the actual distribution of gene copies while using a population average value of 24 copies per cell one would predict 2000 units of gene product. With the advent of genetic engineering gene amplification to several hundred copies per cell is readily achievable but there is a great deal of variation from cell to cell.

Thus, many situations of practical significance demand models with characteristics of both structure and segregation. Models that contain a significant level of structure and segregation have proved to be numerically intractable when approached from the population balance point of view (J.E. Bailey, personal communication). Models derived from a population-balance approach result in a set with a large number of integral-differential equations due to the need for integration of frequency functions. Because of mathematical complexities, population models are written typically for distributions in terms of only a single variable (e.g. cell age or size). These models, while segregated, do not explicitly contain structure. A good review of such moels and how to develop the appropriate population balance equation is given by Ramkrishna (1979). An attempt to include structure in population models is described by Nishimura and Bailey (1980); their approach combines elements of the population balance approach and our approach (Domach and Shuler, 1984b) of using a finite-representation technique based on single-cell models. Nonetheless, no one has written a model with both structure and segregation by viewing the population as a quasicontinuum and using frequency functions.

3. Why Single Cell Models?

Our approach to circumvent this problem was to mimic nature. Highly structured single-cell models are first developed and population models are created by a finite-representation technique. This approach results in a large number of non-linear ordinary differential equations. This set of equations can be integrated forward in time with a predictor-corrector technique with variable step size. The equations exhibit stiffness during any portion of the cell cycle where there is a large flux of material through a small metabolic pool (see Park, 1974). Nonetheless, models of cellular populations formulated in this manner are a good deal more tractable than highly structured-segregated models written from a population balance perspective.

The use of the single-cell approach is also quite attractive from the biological perspective. The cell is the essential element in biology as molecules are in chemistry. Biologists and the biological literature are constructed in an intellectual framework focused on cellular rather than population behavior. Single-cell models facilitate communication between modelers and experimentalists, particularly in terms of cell division cycle events (e.g. initiation of chromosome replication, septum formation, and cell division). Also, single cell models promote the explicit recognition of cell size and shape changes with respect to cell physiology.

4. Description of Cornell Single Cell Model

Fig. 1 depicts the current base Cornell Single Cell Model. This model is described in detail elsewhere (Shuler, Leung, and Dick, 1979; Shuler and Domach, 1983; Domach, et al., 1984; Shuler, 1985). The cell responds explicitly to changes in either glucose or ammonium ion medium concentration. The ability to respond explicitly to changes in environmental parameters is critical in using models for predicting bioreactor response to various operating strategies. Many other bacterial models depend on the growth rate as an input parameter. In our model growth rate is predicted by the model from the specification of nutrient concentrations.

In the base model all of the cells 2000 to 3000 components are lumped into 20 model components. The choice of model components is based on the judgement of the modeler and requires a satisfactory knowledge of microbial physiology. However, these choices of model components may not be the best minimum set of components for all potential uses of the model. These choices do form a realistic and reasonably robust formulation that can be easily expanded.

Often the biological scientist will have developed a fairly extensive understanding of a cellular subsystem. However, the complex non-linear nature of the cell makes it difficult to make quantitative and even sometimes qualitative predictions of how a subcellular system interacts with whole cell physiology and changes in nutrient concentration. The model, however, provides a tool to relate the details of cellular subsystem to the overall cellular physiology. To use the model in this way the understanding of the biological system must be converted into mathematical statements usually with the addition of explicit components. For example, a more detailed understanding of amino acid metabolism would require breaking the P_1 pool into different types of amino acid pools. These additional pools would still interact with the rest of the model cell with only minor modifications to the rest of the model. The model with an expanded description of amino acid metabolism could probe the response of auxotrophic mutants, effects of gene amplification in these metabolic pathways, etc. The point is that the base model and its choice of model components facilitates model expansion when desirable without the necessity of writing completely new models. The single cell model is "modular" in at least some sense.

The base model and any modifications require mass balance equations for each component. These equations must be written recognizing that the cell is an expanding reactor. For example, the basic equation for precursors is:

Rate of Change of Amount = of P_i in the cell	Maximum Rate of Formation per Unit Cell Volume	Term for Feedback Repression or Inhibition	Terms for Dependence on Availability of Reactants

	Cell Volume	Rate of Consumption of P_i to Make Other Cellular Products	Rate of Degradation of P_i

$$\qquad \text{(1)}$$

For macromolecules the basic equation is:

Rate of Change of Amount = of M_i in the cell	Maximum Rate of Formation per Unit of Template Molecules	Terms for Dependence on Availability of Precursors, Energy, and Other Small Molecules	Amount of Template

	- Rate of Degradation		

$$\qquad \text{(2)}$$

Typically terms use saturation-type kinetics. For example:

$$\left(\frac{A_2/V}{K_{P_iA_2}+A_2/V} \right) \qquad (3)$$

or for feedback control terms

$$\left(\frac{K_{IP_i}}{K_{IP_i}+P_i/V} \right) \qquad (4)$$

As specific examples, consider the equations for formation of amino acids and for proteins (polymer of amino acids).

$$\left(\frac{dP_1}{dt}\right) = k_1 \left(\frac{K_{P_1}}{K_{P_1}+P_1/V}\right)\left(\frac{A_1/V}{K_{P_1A_1}+A_1/V}\right)\left(\frac{A_2/V}{K_{P_1A_2}+A_2/V}\right)V - k_{TP_1}\frac{K_{TP_1}}{K_{TP_1}+A_2/V}P_1 - \gamma_1\left(\frac{dM_1}{dt}\right)$$

$$- \varepsilon_2\left(\frac{dP_2}{dt} + \gamma_2\frac{dM_2}{dt} + \varepsilon_3\frac{dP_3}{dt} + \varepsilon_3\gamma_3\frac{dM_3}{dt}\right) - \varepsilon_4\left(\frac{dP_4}{dt} + \gamma_4\frac{dM_4}{dt}\right) \qquad (5)$$

$$\frac{dM_1}{dt} = \mu_1\left(\frac{A_2/V}{K_{M_1A_2}+A_2/V}\right)\left(\frac{P_1/V}{K_{M_1P_1}+P_1/V}\right)0.85M_{2RTM} - k'_{TM_1}M_1 - k_{TM_1}\left(\frac{K_{TM_1}}{K_{TM_1}+A_2/V}\right)M_1$$

$$(6)$$

The symbols for A_1, A_2, P_1, P_2, P_3, P_4, M_1, M_2, M_{2RTM}, and M_4 are defined in Fig. 1 and in these equations represent the amount of material per cell. Also, V is cellular volume and t is time. The constant, k_1, is the maximum rate of formation of amino acids per unit volume of cell volume and μ_3 is the maximum rate of protein synthesis per unit amount of ribosomal RNA. Kinetic parameters for amino and protein degradation are given by k_{TP_1}, k'_{TM1}, and k_{TM_1}. All constants of the form K_i or K_{ij} are saturation parameters. Stoichiometric parameters are given by ε_i or γ_i.

In addition to this set of kinetic equations describing the chemical conversions within the cell the model contains three auxiliary sets of calculations: (1) cell size and shape, (2) cell division, and (3) control of initiation of chromosome replication.

Calculation of cell size and shape depends upon the assumption that an *E. coli* cell is a cylinder with two hemispherical caps. The model equations generate values for total amount of cell envelope and total cell mass. If the cell envelope thickness is assumed constant, then the total cell surface area is known. If the cytoplasmic density is known and assumed constant (a good assumption for gram negative bacteria like *E. coli*), then knowing the cell mass allows calculation of total cell volume. Knowing the total cell volume and surface area coupled with the assumption that a cell is a cylinder with hemispherical caps, leads to calculation of cell length and width (e.g. two equations and two unknowns). The calculation is done in the same manner during the period of septation when two partially formed hemispherical caps form the cross-wall.

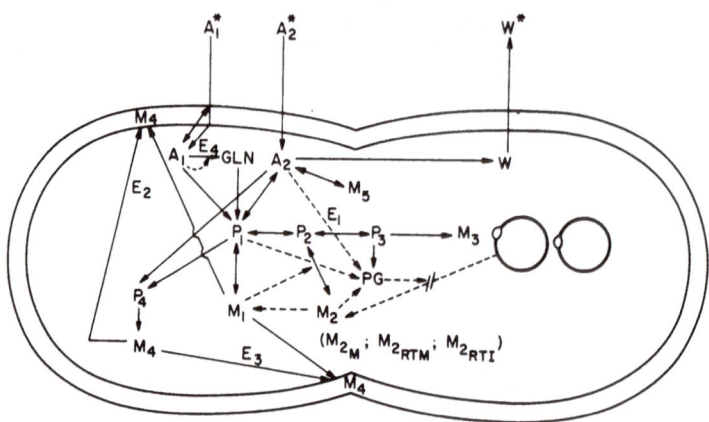

FIGURE 1. An idealized sketch of the model of *E. coli* B/rA growing in a glucose-ammonium salts medium with glucose or ammonia as the limiting nutrient. At the time shown the cell has just completed a round of DNA replication and initiated cross-wall formation and a new round of DNA replication. Solid lines indicate the flow of material, while dashed lines indicate flow of information. Reproduced with permission from Shuler and Domach, 1983.

A_1 = ammonium ion

A_2 = glucose (and associated compounds in the cell

W = waste products (CO_2, H_2O, and acetate) formed from energy metabolism during aerobic growth

P_1 = amino acids

P_2 = ribonucleotides

P_3 = deoxyribonucleotides

P_4 = cell envelope precursors

M_1 = protein (both cytoplasmic and envelope)

M_{2RTI} = immature "stable" RNA

M_{2RTM} = mature "stable" RNA (r-RNA and r-RNA - assume 85% r-RNA through-out)

M_{2M} = messenger RNA

M_3 = DNA

M_4 = non-protein part of cell envelope (assume 16.7% peptidoglycan, 47.6% lipid, and 35.7% polysaccharide)

M_5 = glycogen

PG = ppGPP

E_2, E_3 = molecules involved in directing cross-wall formation and cell envelope synthesis - the approach used in the prototype model was used here but more recent experimental support is available

GLN = glutamine

E_4 = glutamine synthetase

* = the material is present in the external environment.

The model also requires that crosswall formation begins at the time of termination of chromosome replication due to the activation of enzymes associated in crosswall formation.

The mechanism for control of initiation of chromosome replication has been adapted from the observations of Messer, et al. (1975) and Womack and Messer (1978). The model generates a burst of synthesis of a protein from the dnaA gene at the time that point of the chromosome is replicated. The dnaA gene product inhibits a necessary step for initiation of chromosome replication. The model also allows the production of an anti-repressor at a rate proportional to the rate of cell envelope synthesis. The anti-repressor can combine with the dnaA gene product and inactivate it. When the concentration of the dnaA gene product is reduced to a sufficiently low level due to titration by antirepressor and dilution by cell growth, then chromosome replication is allowed to begin.

These decision-making, auxiliary calculations are all directed by the chemical changes calculated from the mass balance equations described earlier. Unlike many other models, decisions on cell size and shape changes and timing of cell division are natural consequences of changes in cellular composition.

The kinetic and stoichiometric parameters in the twenty-component model is over 100. Almost all of these parameters can be evaluated from data in the literature using in vitro experiments on specific enzymes or sub-cellular components, growth and composition data coupled with mass balance calculations, or heuristic rules for saturation constants based on measured values of the normal intracellular concentrations of key small molecular weight molecules. Four parameters associated with crosswall formation were adjusted using computer runs for model cells growing under glucose-limitation and aerobic conditions. Two computer experiments were necessary: one at maximum growth rate in a minimal medium and one with the glucose concentration set to give one-half maximal growth rate. Model predictions at any other conditions were free of adjustable parameters.

Other conditions include nitrogen-limited growth (Shuler and Domach, 1983), anaerobic conditions (Ataai and Shuler, 1985a), dual nutrient limitation (Lee, et al., 1984), and containing plasmids encoding foreign protein synthesis (Ataai and Shuler, 1986, 1987). Under all of these circumstances the model makes predictions which compare well with actual experimental data on a quantitative basis. Extension of the model to each of these conditions did not require adjustment of parameters but only the insertion of a more detailed description of some subcellular system.

An interesting example of the use of the single-cell model is the prediction and our experimental confirmation of multiple steady states in ammonium-limited cultures at high dilution rates in a chemostat (Shu, et al., 1987). This bistability is due to the presence and interaction of dual uptake systems for ammonium ion and two intracellular routes for conversion of ammonium ion into the amino acid, glutamate. Bistability at high dilution rates has not been reported previously. The model also makes the unexpected prediction that the maximum growth rate in a chemostat can exceed that in batch culture. Our experimental data is consistent with this prediction. Thus, the model is sufficiently robust to predict unexpected and previously unreported experimental results.

In all of the above cases the single-cell model has functioned as a structured but non-segregated model. Models with segregation become more necessary as we probe the dynamic behavior of continuous flow bioreactors.

5.   Population Models Using a Finite-Representation Technique

The first literature reports on the use of a single-cell model to construct a population model were given by Nishimura and Bailey (1980, 1981). The single-cell model invoked was very simple and semi-empirical. However, this model did establish the feasibility of such an approach. To use a more complex single cell model we must try to ascertain mechanisms for cell to cell variations in populations. Although synchronously dividing cultures can be generated, synchronory cannot be maintained for more than three or four generations. Apparently identical cells exhibit variations in cell cycle division times and cell size at division. Many causes for this asynchrony are possible and probably operative. However, using the single-cell model and allowing a random variation in the amount of enzyme involved in cross-wall formation is sufficient to predict observed variations in cycle time and division size, as well as the negative correlation in these parameters between parent and progeny (Domach and Shuler, 1984a).

Having a workable description of the cause of asynchrony allows the construction of a population model using a finite-representation technique (Domach and Shuler, 1984b; Ataai and Shuler, 1985b). Currently, the population is divided into 225 sub-populations with each represented by a single-cell model. All cells within that subgroup are required to behave identically with the model cell for that group. Start-up procedures are facilitated by assuming an ideal cell distribution initially. Thus, the fraction of the population in the cell compartment with the smallest size is initially twice that of the cell compartment at the verge of division.

The population model structure is based on a continuous flow stirred tank reactor (CFSTR) (or chemostat). A cell-free stream with known levels of nutrient enters the reactor at a specified flow rate. The reactor is of known and constant volume.

The model does a balance on each cell group and on substrate concentration at specific intervals (e.g. 0.01h). Hydraulic washout, nutrient consumption, growth in terms of increase of each cellular component, and cell division are determined; and this information is used to reinitialize the main array. Between intervals each model cell is "run" assuming a constant substrate concentration as determined from the previous mass balance calculation. A summary of the procedure is given in Figure 2.

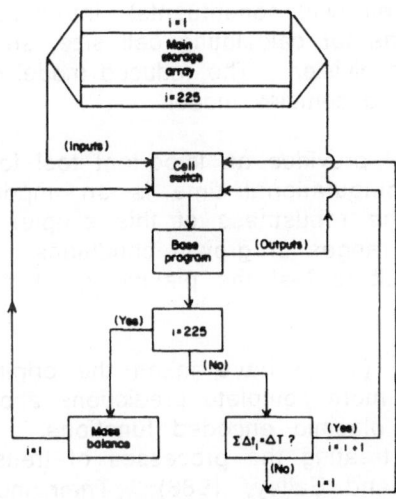

FIGURE 2. A flow sheet of the operation of the population routine is shown. The approach is based on a finite representation technique and each computer cell (i = 1 - 225) is "turned on" for a set on time (ΔT = 0.01 h). Remaining details are covered in the text. Reprinted with permission, Domach and Shuler, 1984b.

The population model has been used to predict dynamic responses of both aerobic reactors (Domach and Shuler, 1984b) and anaerobic reactors (Ataai and Shuler, 1985b). We currently are extending this approach to examine populations containing plasmids. The primary randomizing force in such a population is the unequal, random partitioning of plasmid upon cell division. Different plasmid contents result in differential growth rates among cells; cells with fewer plasmids devote fewer cellular resources to non-growth related functions and thus grow more quickly than cells with large numbers of plasmids. A population model mimicking the correct distribution of plasmids should also mimic the distribution of growth rates and productivity with respect to plasmid encoded gene products. We have constructed a preliminary population model which explicitly includes a mechanism for unequal plasmid-partitioning (Kim, et al., 1988).

6. Other Examples of Use of Single-Cell Models

The Cornell single cell model is rather complex. Joshi and Paulson (1988) have applied a technique of model analysis to the single-cell model to identify which cellular pools respond on a time-scale similar to that for growth (e.g.

minutes). Components that respond much more quickly or slowly can be considered to be in a quasi-steady state. Thus, Joshi and Paulson (1988) reduced the 18 component Cornell model to one with three main pools (protein, nucleic acids, and cell wall constituents), intracellular glucose concentration, and auxiliary equations for calculating cell size, shape, chromosome initiation, septum formation, and division. The reduced model displayed dynamic responds very similar to the more complex model.

This simpler model provides an important tool for predicting growth related phenomena when computational time is an important consideration (as in process control). The robustness of this simpler formulation has not been tested over extreme ranges of growth conditions. The simplified model also makes it more difficult to test the plausibility of hypotheses concerning sub-cellular mechanisms.

Peretti and Bailey (1987) have taken the original single cell model and extended it to make more complete predictions about the interactions of host cell metabolism and plasmid encoded functions. They modified the Cornell single-cell model by treating the processes of transcription and translation in more detail (Peretti and Bailey, 1986). Their model for example explicitly accounted for the competition of RNA polymerase. Their model would predict the implications of changes of plasmid copy number and of the strengths of the promoter and ribosome binding site on plasmid-encoded protein synthesis and cellular growth rates. Their results suggest that cloned-gene expression is limited by the cell's capacity for transcription. They also suggest that augmenting the strength of the ribosome-binding site will increase target protein production more than a similar increase in promoter strength. Thus, this rather detailed model can provide direct guidance to the experimentalist in terms of what molecular-level manipulations are likely to be most successful.

7. Summary

The single-cell model approach and population models constructed using an ensemble of single-cell models provide an excellent means to simulate a wide variety of cellular responses to perturbations in the extracellular environment. The model's structure facilitates the insertion of new information on biological mechanisms and allows the direct interpretation of model predictions in terms of biological mechanisms. As the model becomes more realistic (and complex), it becomes increasingly more computationally intensive; the retention of computational accessibility will require greater attention to the use of more computationally efficient algorithms and the elimination of terms not relevant to the time-scale of interest of the experimenter.

Acknowledgement

The work described in this chapter has been supported previously by the National Science Foundation and currently by the Office of Naval Research. Many of the computer simulations were conducted using the Cornell National Super-computer Facility, a resource of the Center for Theory and Simulations in Science and Engineering (Cornell Theory Center, which receives major funding from the National Science Foundation and IBM Corporation, with additional support from New York State and members of the CORPORATE RESEARCH INSTITUTE).

References

Ataai, M.M. and M.L. Shuler. 1985a. Simulation of the growth pattern of a single cell of *Escherichia coli* under anaerobic conditions. *Biotechnol. Bioeng.* *27*(7):1026-1035.

Ataai, M.M. and M.L. Shuler. 1985b. Simulation of CFSTR through development of a mathematical model for anaerobic growth of *Escherichia coli* cell population. *Biotechnol. Bioeng. 27*(7):1051-1055.

Ataai, M.M. and M.L. Shuler. 1986. Mathematical model for the control of ColE1 type plasmid replication. *Plasmid 16*:204-212.

Ataai, M.M. and M.L. Shuler. 1987. A mathematical model for prediction of plasmid copy number and genetic stability in *Escherichia coli. Biotechnol. Bioeng. 30*:389-397.

Domach, M.M. and M.L. Shuler. 1984a. Testing of a potential mechanism for *E. coli* temporal cycle imprecision with a structured model. *J. Theor. Biol. 106*:577-585.

Domach, M.M. and M.L. Shuler. 1984b. A finite representation model for an asynchronous culture of *E. coli. Biotechnol. Bioeng. 26*:877-884.

Domach, M.M., S.K. Leung, R.E. Cahn, G.G. Cocks, and M.L. Shuler. 1984. Computer model for glucose-limited growth of a single cell of *Escherichia coli* B/r A. *Biotechnol. Bioeng. 26*(3):203-216.

Fredrickson, A.G., D. Ramkrishna, and H.M. Tsuchiya. 1971. The necessity of including structure in mathematical models of unbalanced microbial growth. *Chem. Eng. Symp. Series 67*(108):53.

Joshi, A. and B.O. Palsson. 1988. *Escherichia coli* growth dynamics: A three-pool biochemically based description. *Biotechnol. Bioeng. 31*:102-116.

Kim, B.G., J. Shu, L.A. Laffend, and M.L. Shuler. 1988. On predicting protein production from recombinant DNA in bacteria. *Forefronts*. (In Press)

Lee, A.L., M.M. Ataai, and M.L. Shuler. 1984. Double substrate limited growth of *Escherichia coli. Biotechnol. Bioeng. 26*:1398-1401.

Messer, W., L. Dankworth, R. Tippe-Schindler, J.E. Womack, and G. Zahn. 1975. Regulation of the initiation of DNA replication in *E. coli*: Isolation of I-RNA and the control of I-RNA synthesis. In *DNA Synthesis and Its Regulation*, M. Goulian and P. Hanawalt, Eds. (Benjamin, Menlo Park, CA), p. 602.

Nishimura, Y. and J.E. Bailey. 1980. On the dynamics of Cooper-Helmstetter-Donachie procaryote populations. *Math. Biosci. 51*:305.

Nishimura, Y. and J.E. Bailey. 1981. Bacterial population dynamics in batch and continuous-flow microbial reactors. *AIChE J. 27*:73.

Park, D.J.M. 1974. The hierarchial structure of metabolic networks and the construction of efficient metabolic simulators. *J. Theor. Biol. 46*:31-74.

Peretti, S.W. and J.E. Bailey. 1986. A mechanistically detailed model of cellular metabolism for glucose-limited growth of *Escherichia coli* B/r. *Biotechnol. Bioeng. 28*:1672.

Peretti, S.W. and J.E. Bailey. 1987. Simulations of host-plasmid interactions in *Escherichia coli*: Copy number, promoter strength, and ribosome binding site strength effects on metabolic activity and plasmid gene expression. *Biotechnol. Bioeng. 29*:316-328.

Ramkrishna, D. 1979. Statistical models of cell populations. *Adv. in Biochem. Eng. 11*:1-49.

Shu, J., P. Wu, and M.L. Shuler. 1987. Bistability in ammonium-limited cultures of *Escherichia coli* B/r. *Chemical Eng. Comm. 58*:185-194. (Special issue in honor of Neal Amundson).

Shuler, M.L. 1985. On the use of chemically structured models for bioreactors. *Chemical Eng. Communications 36*:161-189.

Shuler, M.L., S. Leung, and C.C. Dick. 1979. A mathematical model for the growth of a single bacterial cell. *Ann. N.Y. Acad. Sci. 326*:35-56.

Shuler, M.L. and M.M. Domach. 1983. Mathematical models of the growth of individual cells. Tools for testing biochemical mechanisms. In *Foundations of Biochemical Engineering: Kinetics and Thermodynamics in Biological Systems*. Ed., H.W. Blanch, E.P. Papoutsakis, and G. Stephanapoulos. ACS Symp. Series *207*, Am. Chem. Soc., Washington, D.C., pp. 93-133.

Tsuchiya, H.M., A.G. Fredrickson, and R. Aris. 1966. Dynamics of mirobial cell populations. *Adv. Chem. Eng. 6*:125.

Womack, J.E. and W. Messer. 1978. Stability of origin-RNA and its implications on the structure of the origin of replication in *E. coli*. In *DNA Synthesis-Present and Future*. I. Molineaux and M. Kohiyama, Eds, (Plenum, New York), pp. 41-48.

SYNCHRONIZATION OF BACTERIAL CULTURE GROWTH

F. C. Hoppensteadt
Department of Mathematics, Michigan State University
East Lansing, Michigan 48824

Abstract

Synchronization of cell doubling times due to alternating starvation-nutrition cycles is studied here using a method based on nonlinear Lexis diagrams and the assumption that the cell cycle has three phases, pre-replication, replication and post-replication, the middle of which is always of fixed length once started.

1. Introduction.

Synchronization of cell doubling times in bacterial cell cultures can be induced by mechanical, chemical or thermal shock or by an alternate regimen of starvation and nutrition.

These possibilities are investigated here using a theory of nonlinear Lexis diagrams. These diagrams plot cell cycle phase (i.e., maturation) against time. The result is a diagram that shows

    1) how batch cultures approach stationary phase;

    2) how alternate starvation-nutrition cycles can induce culture synchronization when applied with appropriate frequency; and

    3) how shock methods can lead to a weak form of synchronization.

Synchronization can occur in experimental settings. The models considered here account for the three developmental periods of cell growth (pre-replication, replication and post-replication) where the replication period has fixed length while the other two can be modulated by shock or by the availability of nutrients. Shocks significant to disrupt the replication cycle pose too great a threat to the culture's viability for experimentation. Nutrient modulation enables an experimenter to alter a cell's aging rate in either of the two non-replication phases with minimal disturbance of the cells.

Once replication begins a constant time is required to finish. We propose that this fact can explain synchronization by starvation-nutrition cycles. This work was motivated by the work of Kepes, et. al., [1 - 8] where machines for automatic cell synchronization are derived based on nutrition-starvation cycles.

2. Model of cell aging.

Let the variable p denote the phase of a cell. It is a time-like variable that measures the extent of cell maturation. We suppose that there are three distinct developmental periods for a cell

    I. Pre-replication, or accumulation period.

    II. Replication period.

    III. Post-replication.

The phase variable is normalized so that $p = 1$ corresponds to division phase, at which the cell divides into two daughters. We say that developmental period I is the set of phases $0 < p < p_I$, the second one is $p_I < p < p_{II}$, and the third is

$p_{II} < p < 1$.

    The number of cells at developmental phase p at (chronological) time t is denoted by

    $u(p,t)$

The rate at which p changes with respect to chronological time describes the aging rate of the cell, and this depends on parameters describing growth media and on the phase at which a cell finds itself since, for example, various chemicals are taken up by the cell during various phases of the cycle. There is feedback of the cell population size to growth dynamics because nutrients are depleted by uptake kinetics and metabolic by-products that inhibit cell growth are produced (e.g., lactic acid). Therefore, we describe cell aging by the equation

    $dp/dt = F(t,p,U)$

where

    $U(t) = \int_o^1 u(p,t)\, dp$

is the total cell population at time t. The function F describes the growth rate of a single cell at time t that has phase p and is in a population of size U. The assumption of growth rate depending only on t, p and U is special, but it makes the point of how aging dynamics interact with a culture's age structure to enable synchronization.

    The culture's structure is described by the function $u(p,t)$. The change in this over a short time interval, say of length h, is

    $u(p(t+h),t+h) - u(p,t) = - h\, d(p,t)\, u(p,t)$

where d is the instantaneous death rate. Expanding the left side of this equation using Taylor's formula, dividing the equation by h and setting $h = 0$ in the result leads to the equation

$$\frac{\partial u}{\partial t} + \left(\frac{dp}{dt}\right)\frac{\partial u}{\partial p} = - d(p,t)\, u$$

Or,

$$\frac{\partial u}{\partial t} + F(p,t,U)\frac{\partial u}{\partial p} = - d(p,t)\, u$$

This is a nonlinear equation for the population distribution u, and it is similar to ones derived by M'Kendrick, von Foerster, Rubinow and Monod.

    The growth rates during the three phases are denoted in the three equations. For example, if t is measured in hours, $c = 2/3$; if it is measured in minutes, then $c = 40$. Under optimal growth conditions, both F_I and F_{II} are significantly greater than c.

$$\frac{dp}{dt} = F_I(p,t,U) \qquad\qquad 0 < p < p_I$$

$$\frac{dp}{dt} = c \qquad\qquad p_I < p < p_{II}$$

$$\frac{dp}{dt} = F_{II}(p,t,U) \qquad\qquad p_{II} < p < 1$$

An alternate formulation of this model would allow the values of p_I and p_{II} to vary with time, but with their difference being fixed. Changing the aging rates through the I and III intervals includes this and facilitates plotting the results.

3. Synchronization of cultures.

The fixed length of the replication period, no matter how unfavorable the growth conditions, provides the key to synchronization. If the culture is deprived of nutrients, those cells in the replication period will continue aging until they reach phase p_{II}. When nutrition becomes available again, this newly formed single-age group, or cohort, continues to age until division. The daughters of this culture then have the same age. Repeating this process forces more cell lines into this cohort. Figure 1 depicts a typical experiment.

Actual growth rates in various experimental settings involve changes in c, F_I and F_{II} [9,10]. Once these are known, the population distributions can be determined by solving the equations

$$\frac{\partial u}{\partial t} + F_I(p,t,U)\frac{\partial u}{\partial p} = 0 \qquad \text{for } 0 < p < p_I$$

$$\frac{\partial u}{\partial t} + c\frac{\partial u}{\partial p} = 0 \qquad \text{for } p_I < p < p_{II}$$

$$\frac{\partial u}{\partial t} + F_{II}(p,t,U)\frac{\partial u}{\partial p} = 0 \qquad \text{for } p_{II} < p < 1$$

simultaneously with the equations

$$U(t) = \int_0^1 u(p,t)\,dp \qquad \text{(Total population size)}$$
$$u(p,0) = u_0(p) \qquad \text{(Initial population size)}$$
$$u(0,t) = 2\,u(1,t) \qquad \text{(Cell fission)}$$

This problem can be solved in a straightforward way using the method of characteristics.

Optimal synchronization occurs if the starvation is timed to occur when the major cohort is in its replication phase and to last longer than the replication phase. Then the major cohort will hold at phase p_{II} while others accumulate.

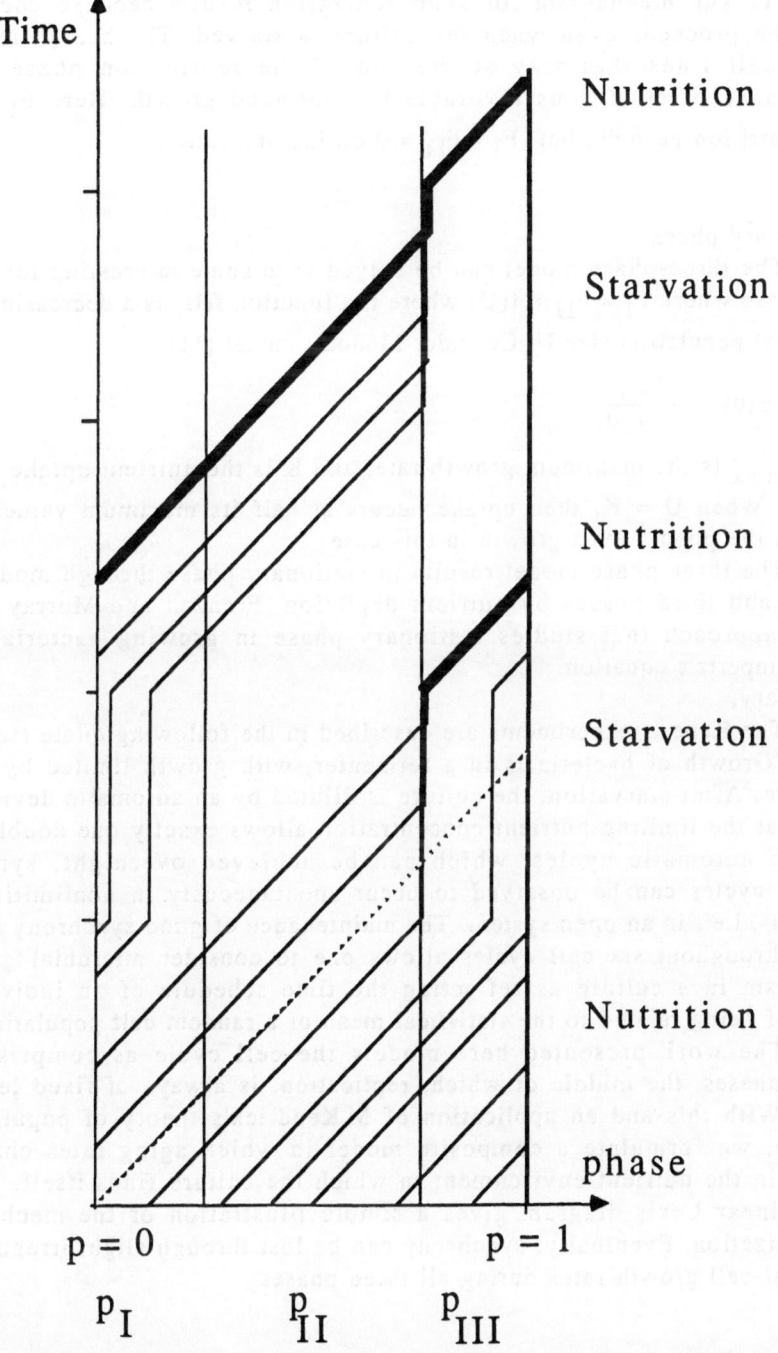

Time

Nutrition

Starvation

Nutrition

Starvation

Nutrition

phase

$p = 0$ $p = 1$

p_I p_{II} p_{III}

Figure 1: The mechanism for synchronization results because once started, replication proceeds even when the culture is starved. The heavy line denotes several cell lines that stay at the end of the replication phase until the reappearance of conditions favorable to continued growth. Here $F_I = F_{II} = c$ during nutrition periods, but $F_I = F_{II} = 0$ during starvation.

4. Stationary phase.

The three-phase model can be solved with some interesting results in the special case where $F_I = F_{II} = f(U)$ where the function $f(U)$ is a decreasing function of the total population size U. Consider Monod's model [11]

$$f(U) = \frac{V_{max}}{K + U}$$

where V_{max} is the maximum growth rate, and K is the nutrient uptake saturation constant. When $U = K$, then uptake occurs at half its maximum value. Figure 2 describes the population's growth in this case.

The three phase model results in stationary phase through modulation of the first and third phases by nutrient depletion. Frenzen and Murray [13] took another approach that studies stationary phase in growing bacterial cultures using Gompertz's equation.

5. Summary.

The Kepes's experiments are described in the following quote (see [1]).

"Growth of bacteria is in a fermenter, with growth limited by inorganic phosphate. After starvation, the culture is diluted by an automatic device in such a way that the limiting nutrient concentration allows exactly one doubling. After 10 to 16 automatic cycles, which can be achieved overnight, synchronous bacterial cycles can be observed to occur spontaneously in nonlimiting culture conditions, i.e., in an open system. The maintenance of good synchrony in an open system throughout six cell cycles allows one to consider microbial growth and metabolism in a culture as reflecting the time schedule of an individual cell instead of being related to the statistical mean of a random cell population."

The work presented here models the cell cycle as comprising three distinct phases, the middle of which, replication, is always of fixed length once started. With this and an application of M'Kendrick's theory of population age-structure, we formulate a composite model in which aging rates change with changes in the nutrient environment in which the culture finds itself. With this, the nonlinear Lexis diagram gives a simple illustration of the mechanism for synchronization. Eventually, synchrony can be lost through slight irregularities in individual cell growth rates during all three phases.

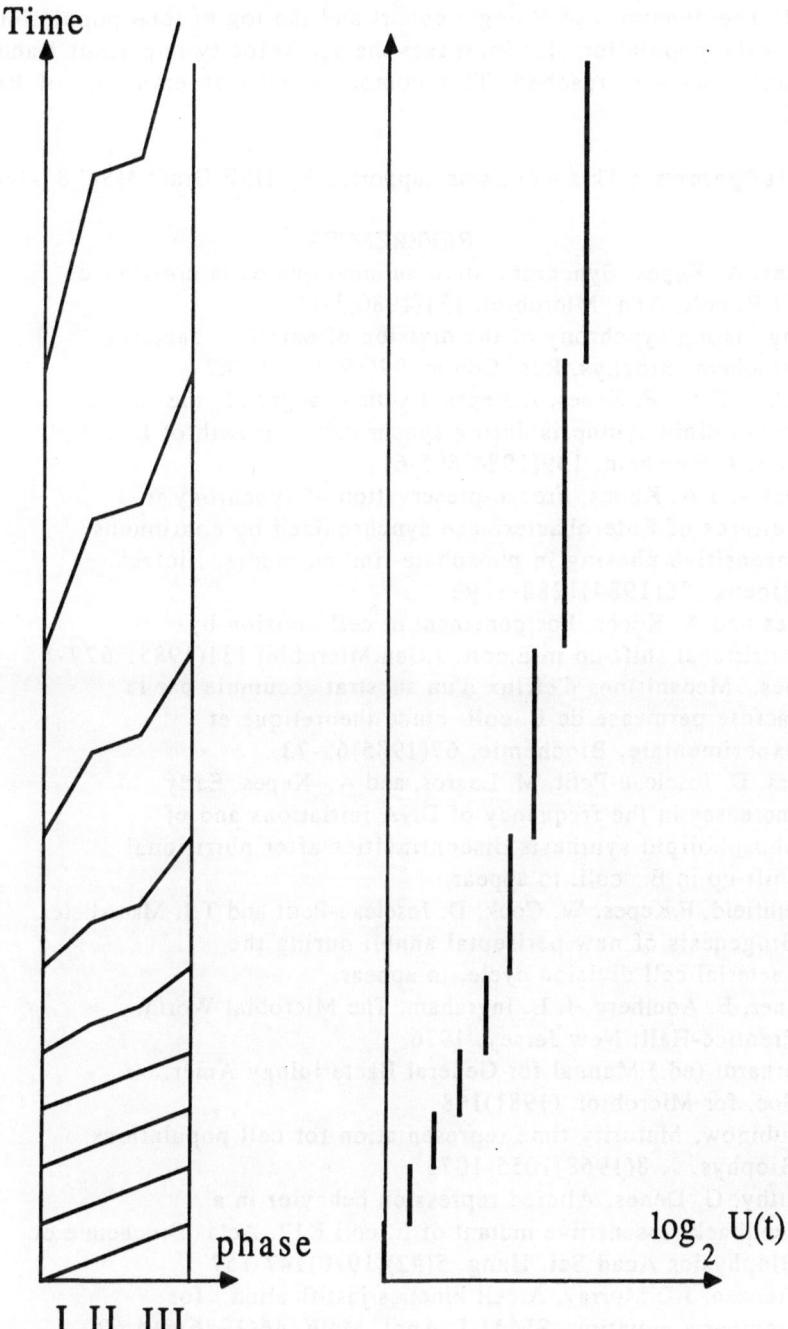

Figure 2: The dynamics of a single cohort and the log of total population size are plotted. As the population size increases the age velocity in periods I and III slow as stationary phase is reached. This compares with observations of Patthy and Denes [12].

Acknowledgement: This work was supported by NSF Grant MSC 85-14000.

REFERENCES
1. F. Kepes, A. Kepes, Synchronisation automatique de la croissance de E. coli, Ann. Microbiol. 131(1980)3-16
2. --, Long-lasting synchrony of the division of enteric bacteria, Biochem. Biophys. Res. Comm. 99(1981)761-767
3. D. Joseleau-Petit, F. Kepes, A.Kepes, Cyclic changes of the rate of phospolipid synthesis during synchronous growth of E. coli, Eur. J. Biochem. 139(1984)605-611
4. F. Kepes and A. Kepes, Freeze-preservation of synchrony in cultures of Enterobacteriacae synchronized by continuous insensitive phasing in phosphate-limited media. Biotech. Bioeng. 26(1984)1288-1293
5. F. Kepes and A. Kepes, Postponement of cell division by nutritional shift-up in E.coli. J.Gen.Microbiol.131(1985) 677-85
6. F. Kepes, Mecanismes d'efflux d'un substrat accumule par la lactose permease de E. coli: etude theoretique et experimentale, Biochemie, 67(1985)69-73
7. F. Kepes, D. Joseleau-Petit, M. Legros, and A. Kepes. Early increases in the frequency of DNA initiations and of phospholipid synthesis discontinuities after nutritional shift-up in E. coli. to appear.
8. L.I. Rothfield, F.Kepes, W. Cook, D. Joseleau-Petit and T.J. MacAlister, Biogenesis of new periseptal annuli during the bacterial cell division cycle, to appear.
9. R. Stainer, E. Adelberg, J. L. Ingraham, The Microbial World, Prentice-Hall, New Jersey, 1976.
10. P. Gerhardt (ed.) Manual for General Bacteriology Amer. Soc. for Microbiol. (1981)168
11. S.I.Rubinow, Maturity time representation for cell populations, Biophys. J. 8(1968)1055-1073
12. L. Patthy, G. Denes, Altered repression behavior in a feedback insensitive mutant of E. coli K12. Acta Biochemie et Biophysics Acad Sci. Hung. 5(#2)(1970)147-157
13. C.L.Frenzen, J.D.Murray, A cell kinetics justification for Gompertz equation, SIAM J. Appl. Math. 46(1986)614-629.

Part II. Resource Management

BIOLOGICAL RESOURCE MODELING—A BRIEF SURVEY

Colin W. Clark

Institute of Applied Mathematics

University of British Columbia

Vancouver, B.C., Canada V6T 1Y4

Abstract

The use of mathematical models in natural resource management has increased greatly over the past two or three decades, both in response to the perceived need for a more careful shepherding of the renewable resource base upon which life itself depends, and also as a result of improvements in analytical and computational techniques. In this review I first describe two classical resource models, one designed for fisheries and the other for forestry. In subsequent sections I describe some of the major directions in which these models have been extended to become more realistic and more useful. I conclude with a call to resource modelers to extend their outlook to encompass global resource issues.

1. Two Basic Models

Fisheries. The following model of optimal fishery exploitation derives from the work of a biologist, M.B. Schaefer (1954), and an economist, H.S. Gordon (1954). The simple dynamic version was developed by Clark (1973), a mathematician.

Resource dynamics are modeled by the differential equation

$$\frac{dx}{dt} = F(x) - h(t); \qquad x(0) = x_o \tag{1.1}$$

where $x(t)$ denotes resource biomass at time t, $F(x)$ is the natural replenishment rate, and $h(t)$ is the rate of harvest. The assumed objective of management is to maximize

$$PV = \int_0^\infty e^{-\delta t} \pi(x, h)\, dt \tag{1.2}$$

where $\pi(x, h)$ denotes the net benefit flow resulting from harvest rate h at stock level x, and δ is the discount rate.

It follows from optimal control theory that, subject to the usual provisos, the above optimization problem possesses an optimal equilibrium solution x^*, h^*, determined by the equation

$$F'(x) + \frac{\partial \pi / \partial x}{\partial \pi / \partial h} = \delta, \qquad h = F(x) \tag{1.3}$$

The optimal approach to this equilibrium is monotonic, $x(t) \rightarrow x^*$ monotonically. In general the limit is approached asymptotically, but in the case where π is linear in h, a bang-bang approach is optimal. Complete details, and discussion of the implications, appear in my book (Clark 1976).

A practically important alternative to the centralized management tacitly assumed in the objective (1.2) is the case of unregulated, decentralized exploitation (Gordon 1954). A full treatment requires explicit consideration of fishermen's behavioral motives (Anderson 1977, Clark 1980), but the general rule is that such uncontrolled exploitation corresponds to discounting the future utterly, i.e. to the case $\delta \rightarrow \infty$ in Eq. (1.3). Thus unregulated exploitation of common-property resource stocks leads to overexploitation and also to the dissipation of economic rents. The history of fisheries exploitation and management attests all too vividly to the validity of this prediction (which is by no means limited to fishery resources). Unfortunately, many well-intended programs designed to overcome these problems have been less successful than desired, particularly from the economic viewpoint. Resource issues are complex biologically, economically, and politically, and simplistic solutions which "look good on paper" are seldom adequate. Mathematical models of increasing levels of sophistication can help to understand these difficulties, and to suggest appropriate management strategies.

Forestry. The basic model in forest management goes back to a German forester, Faustmann (1849). Let $V(t)$ denote the net revenue derived from logging a stand of trees of age t. If the stand is to be logged at age T, and a new stand initiated, then the present value of the sequence of rotations equals

$$V(T) \sum_{k=1}^{\infty} e^{-k\delta T} = \frac{V(T)}{e^{\delta T} - 1}$$

Maximization of this requires that

$$\frac{V'(T)}{V(T)} = \frac{\delta}{1 - e^{-\delta T}} \qquad (1.4)$$

which is Faustmann's formula.

Because the growth rate of trees $V'(T)$ is typically low (a few percent per year), the impact of the discount rate on the optimal rotation age, and on the average annual yield $V(T)/T$, can be very strong (Clark 1976, p. 261). Expenditures on reforestation and other silvicultural practices are also subject to the discount rule; unless discount rates are low, such "investments" will not be considered to be desirable. Over 50% of the logged-over forest land in British Columbia has never been replanted for this reason.

The Schaefer and Faustmann models may seem unrelated, but both prescribe a certain "intensity" of harvesting for a renewable resource. In the fishery model, high discount rates imply a low stock biomass and low yield at equilibrium (provided that the resource is sufficiently valuable). In the forest model, high discount rates imply a low rotation age, and here also low forest biomass and yield from an evenly age-distributed forest.

A rigorous analytic solution of the above two-state-variable, two-control-variable optimal control problem appears in Clark et al. (1979). The solution is slightly complicated, but has a clear economic interpretation. It turns out that there is a unique long-term equilibrium solution, given by a version of Eq. (1.3) that includes costs of interest and depreciation. The approach to this equilibrium from above is not monotonic, however, but involves a single boom-and-bust cycle, with temporary initial "overinvestment," followed by a period of temporary "overexploitation."

The above model has been applied to the Antarctic whaling industry, 1926–1965, by Clark and Lamberson (1982). The post-war (1946–65) expansion predicted by the model is 9 factory fleets, but some 26 were actually constucted. (Only one factory fleet is needed to harvest the sustainable yield.) This is not surprising in view of the fact that Antarctic whaling was carried out competitively by fleets from several nations. A competitive version of the dynamic investment model developed by McKelvey (1984) agrees more closely with the observations.

Investment phenomena associated with forestry have been modeled by Allard et al. (1988).

3. Theory of Resource Regulation

The early models of renewable resource exploitation described in Sec. 1 clarified the economic factors underlying overexploitation. As noted in Sec. 2, these models were later extended to include overinvestment and optimal investment. Models capable of predicting the effects of re-source regulation, and identifying optimal regulatory methods, were developed by Clark (1980). Using the basic (specifically, dynamic) modeling framework described above, it is possible to de-velop useful models of fishery regulation that cover such instruments as Total Allowable Catch quotas (TAC's), closed fishing seasons, license limitation, gear restriction, as well as taxes and individual transferable quotas (ITQ's). Investment phenomena, market responses, and multi-sector industrial structures can also be included (see Clark 1985). Here I will merely describe a model of ITQ's, which clarifies the well-known economic dictum that quotas and taxes are equivalent, in a certain sense.

Assume that each individual fishing "unit" (e.g. vessel, but I will use the term "fisherman") can control its input of effort $E(t)$ over time. Individual harvest rate is $h(t) = qE(t)x(t)$, and cost is $c(E(t))$, a nonlinear function of the usual kind. In the absence of regulation the competitive fisherman is motivated to maximize his net revenue flow $\pi = pqxE - c(E)$, so that

$$c'(E) = pqx \tag{3.1}$$

In contrast, an optimally (cooperatively) controlled fishery would maximize total discounted net revenue

$$TPV = \int_0^\infty N\pi(x, E)\, dt \tag{3.2}$$

where N (also a control variable) denotes the number of (identical) vessels. The maximum principle leads to the necessary condition (among others) that

$$c'(E) = (p - \lambda)qx \qquad (3.3)$$

where $\lambda = \lambda(t)$ is the adjoint variable, representing the "shadow price" of the resource stock, a price that is not realized in any market. If $\pi(x, E) > 0$ then also $\lambda > 0$, and a comparison of Eqs. (3.1) and (3.3) shows that the unregulated fisherman exerts excessive effort. It also follows that $N(t)$ will be excessive in the unregulated fishery.

One conceivable approach to achieving optimal control of a competitive fishery is to impose a tax τ on the catch. The fisherman's net revenue then becomes $\pi(x, E; \tau) = (p - \tau)qxE - c(E)$, and maximization yields

$$c'(E) = (p - \tau)qx \qquad (3.4)$$

The shadow price λ can thus be realized by a tax τ; all that the government has to do is compute λ and collect the taxes! Neither of these may be feasible in practice.

Suppose, however, that the government awards each fisherman a quota Q:

$$h(t) \le Q \qquad (3.5)$$

and allows fishermen to buy and sell quotas among themselves. Given Q, the fisherman's optimization problem becomes

$$\text{maximize } \pi(x, E) \qquad \text{subject to } H = qxE \le Q \qquad (3.6)$$

But suppose the fisherman can buy (or sell) quota units at the quota price m. It is then easy to see that the equation

$$\frac{\partial \pi}{\partial h} = m$$

determines the fisherman's demand function for quota units. This equation reduces to

$$c'(E) = (p - m)qx \qquad (3.7)$$

and this is the sense in which quotas and taxes (Eq. 3.4) are economically equivalent.

What happens in practice is that the government determines the total quota NQ, allocates it to N fishermen, and allows a quota market to operate. The quota market throws up a quota price m, which the fisherman sees as an opportunity cost, and which therefore plays exactly the same role as a tax, or a shadow price.

If the government worries about the corruption involved in quota giveaways, it can combine taxes with quotas. In this case our equation becomes

$$c'(E) = (p - m_1 - \tau_1)qx \qquad (3.8)$$

where τ_1 is the tax and m_1 is the new (reduced) quota price. (This is a tricky way to compute shadow prices: vary τ_1 until $m_1 = 0$, then $\tau_1 = \lambda$.)

The modeling approach can be used to predict the reaction of the fishing industry to forms of regulation such as TAC's, seasonal closures, and license limitation. These methods are incapable of achieving economic rationalization of the fishery, and may actually cause increased overcapacity—a frequently observed phenomenon.

The theme of this section has countless variations, most of which have not yet been investigated. The most important doubtlessly pertain to fluctuations and uncertainties: Quotas are not equivalent to taxes under these circumstances.

4. Stochastic Resource Models

The recent literature in resource modeling has begun to address some of the many fascinating and important questions that arise when one considers stochastic fluctuations of biological resource stocks, and the inevitable parameter and structural uncertainty characteristics of natural resources. There appears to be no simple unifying principle in the subject; here I can only mention some leading themes. (See Mangel 1985 for technical details.)

Investment under uncertainty. The problem of optimal investment with fluctuating resource stocks has been addressed by Charles (1983). The results are qualitatively similar to the deterministic case, with the additional feature that investment is more strongly indicated when resource stocks are abundant than when they are rare. A Bayesian approach to optimal investment in an undeveloped (and hence poorly quantified) fishery is discussed by Clark et al. (1985). Not surprisingly, the prescription is to invest less than indicated by a certainty-equivalent model, on the grounds that initial surveys may be inaccurate, and mistakes on the under-investment side are more cheaply rectified than are overinvestments.

Parameter and stock uncertainty. The triumvirate of Ludwig, Walters, and Hilborn (see references) has spent many years developing adaptive control models for fishery regulation in the presence of highly uncertain parameter values. Much of the "received wisdom" of fishery management turns out to be ill-conceived under these circumstances. For example, the standard practice of attempting to maintain a fixed breeding stock, if successful, provides a minimum of information about the stock-recruitment relationship. Similarly, attempts to construct complex ecosystem models for multispecies fisheries may founder because of the statistical problems associated with multiple parameter estimation (see Ludwig, this conference).

Another problem of considerable economic importance in fisheries is that of stock estimation. How much increase in long-term yield can be expected from improved stock estimates, and what is the optimal expenditure on stock assessment? Clark and Kirkwood (1986) address this issue in a simplistic model, but much more work is needed. It seems likely that the political system encourages overexpenditure on stock assessment.

Forest fire risk. Reed and his coworkers (see references) have published an important series of papers addressed to the problem of including the risk of fires in forest management models. For the simplistic case, suppose there is a fixed probability λ that the forest will be destroyed by fire in any given year (λ is easily estimated from fire data). Reed (1984) shows that in this case the optimal rotation period T is given by

$$\frac{V'(T)}{V(T)} = \frac{\delta + \lambda}{1 - e^{-(\delta + \lambda)T}} \tag{4.1}$$

In other words, the optimal rotation period is obtained by adding the annual forest fire risk λ to the discount rate δ. Since λ is a "mortality rate," this is similar to Fisher's (1930) use of discounting in life history models.

As explained above, the discount rate can be a highly sensitive parameter in forest management. Reed's results show that forest fire risk is equally significant. As far as I am aware, all government forest services in North America implicitly use a zero discount rate in setting cutting schedules. Forest fire risk is either ignored, or is handled by some inappropriate rule of thumb. As a result, sustainable yields may have been vastly overestimated [leading to the same outcome as a high discount rate!], and the value of fire-suppression may have been underestimated. It is hard to imagine these important results being ignored by management authorities much longer, but rationality has long been a rare commodity in resource management.

5. Future Directions

It appears that resource modeling has come of age in the 1980's. Several journals are now devoted largely or exclusively to this field. Many management agencies now regularly recruit quantitatively trained personnel capable of employing modern techniques in modeling, statistics, and computation. Most such agencies, however, are primarily concerned with regional or national resource issues.

In contrast, few international resource-oriented agencies exist—indeed, none have jurisdictional authority. But resource conservation in its broadest sense is now beginning to assume fright-ening global dimensions. Among the potential resource-oriented problems which may have a serious impact on the future of mankind are: atmospheric pollution (acid rain, ozone redistribution, greenhouse gas buildup, nuclear aerosols), desertification, tropical forest destruction, mangrove destruction, and loss of genetic resources. Academic scientists may now be much more useful to humanity if they can devote their talents to these under-studied issues, rather than continuing to act as well-paid consultants to national agencies. If we don't do it, it won't get done.

REFERENCES

Allard, J., D. Errico and W.J. Reed. 1988. Irreversible investment and optimal forest exploitation. *Nat. Res. Modeling* (in press).

Anderson, L.G. 1977. *The Economics of Fisheries Management.* Johns Hopkins Univ. Press, Baltimore, MD.

Beverton, R.J.H. and S.J. Holt. 1957. *On the Dynamics of Exploited Fish Populations.* Ministry of Agriculture, Fisheries and Food (London), Fish.Invest. Ser. 2(19).

Charles, A. 1983. Optimal fisheries investment under uncertainty. *Can. J. Fish. Aquat. Sci.* 40: 2080–2091.

Clark, C.W. 1973. The economics of overexploitation. *Science* 181: 630–634.

Clark, C.W. 1976. *Mathematical Bioeconomics: The Optimal Management of Renewable Resources.* Wiley–Interscience, New York.

Clark, C.W. 1980. Towards a predictive model for the economic regulation of commercial fisheries. *Can. J. Fish. Aquat. Sci.* 37: 1111–1129.

Clark, C.W. 1985. *Bioeconomic Modelling and Fisheries Management.* Wiley–Interscience, New York.

Clark, C.W. and G.P. Kirkwood. 1986. Optimal harvesting of an uncertain resource stock and the value of stock surveys. *J. Environ. Econ. Manag.* 13: 235–244.

Clark, C.W. and R.H. Lamberson. 1982. An economic history and analysis of pelagic whaling. *Marine Policy* 6: 103–120.

Clark, C.W., A. Charles, J.R. Beddington and M. Mangel. 1985. Optimal capacity decisions in a developing fishery. *Marine Res. Econ.* 1: 25–54.

Clark, C.W., F.H. Clarke and G.R. Munro. 1979. The optimal exploitation of renewable resource stocks: problems of irreversible investment. *Econometrica* 47: 25–49.

Faustmann, M. 1849. Berechnung des Werthes, welchen Waldboden sowie nach nicht haubare Holzbestande für die Weldwirtschaft besitzen. *Allgemeine Forst und Jagd Zeit.* 25: 441.

Gordon, H.S. 1954. The economic theory of a common-property resource: the fishery. *J. Polit. Econ.* 62: 124–142.

Ludwig, D.A. 1982. Harvesting strategies for a randomly fluctuating population. *J. Cons. Intern. Expl. Mer.* 39: 168–174.

Ludwig, D.A. and R. Hilborn. 1982. Management of possibly overexploited stocks on the basis of catch and effort data. Univ. of Brit. Col., Vancouver, Inst. Appl. Math. Stat. Tech. Rep.No. 82-3.

Ludwig, D.A. and C.J. Walters. 1981. Measurement errors and uncertainty in parameter estimates for stock and recruitment. *Can. J. Fish. Aquat. Sci.* 38: 711–720.

Ludwig, D. A. and C. Walters. 1982. Optimal harvesting with imprecise parameter estimates. *Ecol. Modelling* 14: 273–292.

Mangel, M. 1985. *Decision and Control in Uncertain Resource Systems.* Academic Press, New York.

McKelvey, R.W. 1984. The dynamics of open-access exploitation of a renewable resource: the case of irreversible investment. Univ. Montana Math. Dept. Interdisc. Ser. Rep. No. 24.

Reed, W.J. 1984. The effects of the risk of fire on the optimal rotation of a forest. *J. Env. Econ. and Mgt.* 11: 180–190.

Reed, W.J. 1987. Protecting a forest against fire: optimal protection patterns and harvest policies. *Nat. Res. Modeling* 2: 23–54.

Reed, W.J. and D. Errico, 1985. Assessing the long-run yield of a forest stand subject to the risk of fire. *Can. J. For. Res.* 15: 680–687.

Schaefer, M.B. 1954. Some aspects of the dynamics of populations important to the management of commercial marine fisheries. *Bull. Inter-Amer. Trop. Tuna Comm.* 1: 25–56.

Walters, C.J. 1975. Optimal harvest strategies for salmon in relation to environmental variability and uncertain production parameters. *J. Fish. Res. Board Canada* 32: 1777–1784.

Walters, C.J. 1981. Optimum escapements in the face of alternative recruitment hypotheses. *Can. J. Fish. Aquat. Sci.* 38: 678–689.

Walters, C.J. 1984. Methods of managing fisheries under uncertainty. In R.M. May (Ed.) *Exploitation of Marine Communities.* Dahlem Konferenzen, Springer-Verlag, Berlin. pp. 263–274.

Walters, C.J. and R. Hilborn. 1976. Adaptive control of fishing systems. *J. Fish. Res. Board Canada* 33: 145–159.

MATHEMATICAL MODELING IN PLANT BIOLOGY:
IMPLICATIONS OF PHYSIOLOGICAL APPROACHES FOR RESOURCE MANAGEMENT

Louis J. Gross
Department of Mathematics and
Graduate Programs in Ecology and
Plant Physiology and Genetics
University of Tennessee
Knoxville, TN 37996-1300

Abstract

I provide a brief overview of mathematical models that have been developed for particular plant physiological processes, with emphasis on the difficulties involved in taking these upscale to deal with whole plant, crop and forest growth analyses. As the issues addressed by physiologists are often highly reductionist in nature, I point out the gap which has developed between our detailed knowledge of certain physiological processes and our general ignorance of appropriate ways to integrate these processes over whole plant or canopy scales. The importance of accurate integration for crop and forest management is discussed. Finally, I review models for the spread of plant pathogens, and indicate how these may be modified to take account of the spatial nature of plant infection and the continuum of resistance types within a natural population.

1. Introduction

Mathematical models have been applied to a wide variety of topics in plant physiology (Thornley, 1976). The majority of these focus on processes that are modeled independently such as photosynthesis, fluid transport, respiration, transpiration and stomatal response. The general goal of these models is to predict the effect of a variety of environmental factors, including radiation input, humidity, wind, CO_2 concentration and temperature, on the process rates. The models tend to be more general and realistic than precise, though all are empirical at some level. Modelers have an advantage in the wealth of physiological data which has been collected on these processes, but at the same time this is often too much of a good thing due to a general lack of accepted theory in the subject. Thus it is typically not clear what in the plethora of detailed biochemical knowledge about specific processes should be included and what can be safely ignored or lumped in with other components of the process under consideration.

The uncertainty of appropriate components to include in these models becomes even more critical when dealing with questions on longer time or larger spatial scales than those over which typical physiological approaches operate. Much of the detailed physiological models deal with processes on a cellular or organ (such as leaf) spatial scale and the natural time scales associated with these, normally from seconds to daily. I have argued elsewhere (Gross, 1986) that in dealing with plant processes, a natural breakdown of time scales would be physiological (within a day or so), acclimation (within genet lifespan), and evolutionary (over many generations). The appropriate time scale for many management applications is the acclimation one, and there has been much less work at this scale than at the physiological one. My purpose here is in part to indicate the variety of modeling approaches which have been undertaken at the physiological scale, and try to indicate a number of open areas regarding how to appropriately go to acclimation scales from these. A similar problem regarding scaling is discussed in the last section, namely how to make more realistic models for the spread of plant diseases which take into account the spatial interactions which are necessary for the disease to spread.

2. Physiological Models

The aim of most physiological models is to express the equilibrium rate of certain physiological processes as a function of environmental inputs and the current physiological state of the plant. A general form for this would be

$$R_1 = f_1(E_1,\ldots,E_k,P_1,\ldots,P_m,R_2,\ldots,R_n)$$
$$\vdots \tag{1}$$
$$R_n = f_n(E_1,\ldots,E_k,P_1,\ldots,P_m,R_1,\ldots,R_{n-1})$$

where $R_1,\ldots,R_n$ represent equilibrium rates of various physiological processes, $E_1,\ldots,E_k$ represent environmental factors, and $P_1,\ldots,P_m$ represent various physiological states. The physiological states could be viewed as independent of the R_i's but more realistically are controlled by the R_i's on a longer time scale than this equilibrium approach is meant to handle. For example, photosynthetic rate is affected by the chlorophyll concentration within a leaf, and this may change over time scales of days due to changes in a variety of reaction rates over the recent past. Thus the R_i's operate on the above mentioned physiological time scale, while the P_i's change on an acclimation time scale.

The form of (1) really is derived as the equilibrium solution of a system of differential equations which would track the concentrations of biochemical components of the associated reactions. It is somewhat rare to see a model actually applied in this form however, due to a general lack of knowledge about the dynamics of the component reactions. In fact, the vast majority of models of the form (1) are really derived as empirical fits to experimental data. In reality of course, both the physiological state and the environmental inputs are time-varying. A typical assumption is that these change on a slow time scale relative to that of the reaction rates, so that the equilibrium solution (1) just continuously tracks these changes. For a variety of processes, including stomatal conductance and photosynthetic rates, this assumption is not justified (Gross, 1986; Kirschbaum et al., 1988).

In addition to the dynamic assumptions inherent in (1), the approach is also inherently a local one, for it is assumed that the reactions guiding the process have no spatial component. That is, all the variables of the model should be viewed as spatial averages over whatever scale the process is assumed to be operating. Thus photosynthesis models at the leaf level based on biochemical reactions typically assume uniform concentrations of the various biochemical components of the process over the entire leaf. The nature of the averaging which is implied by this is non-trivial due to the nonlinearities in the models governing the process as a function of the reactants. Another assumption of these models concerns the simplifications of the biochemistry which are used in order to reduce the number of model parameters to tractable levels and to allow them to be estimated from available data. Many of the sub-reactions which determine the R_i's are extremely complicated, and though some aspects of them may be known in detail, typically one rapidly reaches the current level of ignorance when constructing a model. Thus it becomes necessary to lump component reactions.

Before leaving the general topic of physiological modeling, it should be emphasized that biophysical approaches are the basis for investigating the effects of most environmental factors in these models (Nobel, 1974). This is true of water relations, in which the volume changes of cells is related to external osmotic pressure through the Boyle-Van't Hoff relation and flux of water between various plant parts is governed by a transport equation based on the chemical potential of water in the parts. Similarly, mass and heat transfer between the plant and surrounding environment is derived from micrometeorological approaches (Montieth, 1975).

3. Whole Plant Processes

How are physiological processes scaled up to the whole plant? Though there are a variety of approaches, none are entirely satisfactory. One method is to use a macrodescriptor which is essentially empirical. For example, a frequently used estimate of whole plant respiration over a day is

$$R = kP + cW$$

where R is respiration, P is gross photosynthesis during the light period, W is plant dry weight, and c and k are constants. This was originally obtained from data on clover (McCree, 1970), but has since been applied in many simple growth models. Similar whole plant descriptors have been used for photosynthesis as a function of incident radiation and temperature, for transpiration as a function of these along with wind speed and humidity, and for a variety of other plant processes. These will sometimes have a mechanistic basis, but more typically use a regression approach to determine the interactions of the variety of factors affecting any of the processes of concern.

There are hosts of plant growth curves derived along these empirical lines, viewed as describing the time course of whole plant or community growth (Hunt, 1982). The difficulty is that, without the data to decide which curve is appropriate, one must proceed in an ad hoc manner in choosing a curve. If the data are available, then one really doesn't learn much new from the curve anyway, the typical use being interpolation. It also isn't clear how the growth curve should be modified by considering a plant in somewhat different conditions than those for the data from which the curve came. When large amounts of data are available, for example on crop varieties that have been planted in many different conditions for many years, these regression approaches work extremely well in predicting harvests. In fact, they are much better predictors than very complex mechanistic models such as those mentioned below. This is just a special case of the rule that, conditional on the availability of adequate data and ignorance of the exact mechanism of a process, statistical techniques provide far more accurate process prediction within the range of conditions included in the data base than mechanistic approaches with poorly understood functional forms or parameter values.

Alternative to descriptive approaches, one can build up to the whole plant level using brute force, meaning that one simply integrates the process of concern over the entire plant surface. For example, to estimate whole plant photosynthetic rate from a model for the rate of individual leaves

$$P = F(E_1, \ldots, E_k)$$

where P is photosynthetic rate per unit leaf area and the E_i's are environmental factors, one merely sums

$$P_{tot} = \int F(E_1(x), \ldots, E_k(x)) \, A(x) \, dx \qquad (2)$$

where P_{tot} is the photosynthetic rate for the whole plant, $E_i(x)$ is the value of environmental factor i at position x, and A(x) gives the amount of leaf area at position x. Here x will in general be in three space, and it may be very difficult to predict the spatial variation of environmental factors throughout the plant. In fact there are very sophisticated models to describe the spatial patterns of radiation throughout plant canopies, based on the architecture of the canopy, transmittance and reflectance of the leaves, quantity of branches, etc.(Ross, 1981).

Typically (2) is solved in a discrete manner, by breaking a canopy into layers and simply considering the fraction of leaf area in each layer subject to direct beam versus diffuse radiation. Although tests have been done of the radiation penetration portions of these (Baldocchi et al., 1985), the photosynthetic rate predicted from (2) hasn't been adequately tested yet. This is due to the difficulty of measuring whole canopy photosynthesis in field conditions. Note that (2) is really a simplification because the $E_i(x)$'s are time-dependent and the functional form of F will change with position in canopy.

Although approaches similar to (2) can be carried out for many physiological components of plant growth, the vast amount of variation of both environmental factors and physiological state throughout a plant canopy (I'm speaking here of commercially important crop and forest plants) limits the technique. One alternative is to use a highly simplified form for the variation in these factors, and derive from (2) a general relationship which might indicate how changes in basic parameters affect the process. For example in the photosynthetic case, the Monsi-Saeki theory says that light extinction in a canopy may be approximated by an exponential decay with depth (measured in units of leaf area per unit ground area, the leaf area index L) from the top of the canopy. If F(I) gives photosynthetic rate at irradiance I, then (2) becomes

$$P_{tot} = \int_0^{L_T} F(K \exp(-cL)) \, dL \qquad (3)$$

where c is an extinction coefficient, L_T is the total leaf area index of the canopy and K is a species-specific constant which depends on the transmission of a leaf as well as the irradiance at the top of the canopy (see France and Thornley, 1984, chap. 7, for more details). One advantage of this method is that it allows one to derive at least qualitative conclusions as to how changes in basic plant characteristics will affect the process. It provides a more mechanistically-based descriptor of whole plant processes than a purely empirical approach.

4. Growth and Yield Models

Just as there are several methods to move upscale from cellular level phenomena to the whole plant, there are a variety of techniques to model growth at the plant, canopy and community level. Descriptive approaches were already mentioned above, but it should be pointed out that some of these seem to be fairly general. For example there is strong evidence from both crop and forest data that annual biomass production for a crop or stand is given by

$$B = \alpha Q \qquad (4)$$

where Q is the annual interception of photosynthetically active radiation by the crop (McMurtrie, 1985). The constant of proportionality α will of course vary with species, nutrient and water conditions at the site, and probably stand history as well. However, estimates of the constants c and K in (3) are available for a wide variety of forests (Jarvis and Leverenz, 1983) from which it is possible to estimate Q. If a time series of data are available to estimate how α varies with time, it is possible to iterate (4) on a time scale shorter than that over which α changes significantly to estimate the time course of B, and this has been done for a number of crops (Charles-Edwards, 1982). This is an illustration of a top-down approach to growth modeling, in which the details of the physiology are lumped into a single parameter (here α). One could proceed from here to derive a physiologically based model for how α depends on environmental factors, a method which is utilized in some way in many growth models (Landsberg and McMurtrie, 1985).

Alternative to approaches based on (4), compartment models for whole plant growth are quite common (Thornley, 1976; France and Thornley, 1984). These models break up a plant or crop into compartments such as shoot and root, tracking perhaps several components of each, such as structural and storage dry weight and carbon and nitrogen concentration. They may be coupled to models for photosynthesis, soil nutrient uptake, and other possible inputs. Then differential equations are written to describe changes in each compartment, with respiratory losses typically being taken as proportional to the dry weight in a compartment.

An area of controversy in this approach concerns the nature of partitioning of new substrates (usually just taken to be carbon and nitrogen) among the various compartments. One approach is to simply assume that transport of these substrates follows a Fick's law of diffusion, so for example

$$J_c = \frac{\beta(C_s - C_r)}{r_c} \qquad (5)$$

where J_c is the flux of carbon from shoot to root, C_s and C_r are the carbon concentrations in the shoot and root respectively, β is a scaling factor, and r_c is a resistance to movement of carbon (France and Thornley, 1984). An alternative to this is to prescribe the partitioning of nutrients in a "goal-seeking" manner such that either a fixed carbon-to-nitrogen ratio is set and the dynamic behavior of the model is forced to seek this ratio (Reynolds and Thornley, 1982) or else this ratio is set in a way that depends on the root-to-shoot ratio (Johnson, 1985). Still another approach to partitioning is to assume that there are organizing principles of evolutionary origin which specify the partitioning of nutrients so as to maximize some measure of fitness. This is an outgrowth of life history theory, utilizes optimal control techniques, and has been applied mainly to plants broken into roots, shoots and reproductive compartments (see Roughgarden, 1986, for a review).

Yet another approach to growth modeling is a systems one, in which a large collection of physiologically detailed process models are coupled. These typically have submodels for light interception and photosynthesis, root activity and nutrient uptake, partitioning of substrates, transpiration, growth and respiration, leaf area expansion, initiation and development of plant organs, and senescence, though not all these may be included in each model. The models then iterate, typically on a daily or hourly time step, keeping track of levels of nutrients and dry weights of various structural compartments. This amounts to solving non-autonomous difference equations and thus is essentially limited to being simulated on a computer. These models have been constructed for a wide variety of crops (Barrett and Peart, 1981; Loomis et al.,1979; Reynolds and Acock, 1985), involve large numbers of parameters that are sometimes difficult to estimate, and are mainly used as research tools to point out which subprocesses are not well understood. As with many large systems models, they are extremely difficult to validate, due to the ability to tune the large number of parameters to the available data. It is only in rare circumstances that data sets independent from those used to estimate the parameters are available for model validation. The models are rarely spatial, merely assuming the variables are uniform over the scale of the plot under consideration. In this sense, the models are limited to monocultures of fairly even age for which spatial heterogeneity in stand structure is not a significant factor for stand growth.

One method to take account of spatial factors is to use individual-based models. These track all individuals in a stand, using some type of growth model for each species in the stand, and take into account the competitive interactions between neighbors through shading and root competition. They have been applied extensively to investigate patterns of succession in a wide variety of forests (Shugart, 1984) by considering species composition in small plots in a stand. Each plot is typically only slightly larger in area than the crown area of a single dominant adult tree. Gaps are created when the dominant dies, and the models track the transients of composition in the plot over a time scale of centuries. From Monte-Carlo runs it is possible to make statistical predictions about the effects of alternative disturbance regimes on forest composition. There is no inherent reason aside from computer limitations why this cannot be applied on larger spatial plots.

These models typically use very simple individual growth models, though more complicated ones have been applied in the case of species for which a good physiological data base is available (Makela and Hari, 1986). Despite their general lack of physiological detail however, these models produce quite realistic predictions for forest dynamics that have been validated in a few cases. Perhaps their weakest component is the handling of competitive effects, for which there is not much general agreement. Recent models of plant populations which take into account explicit neighborhood effects on survivorship, mortality and fecundity may provide some basic theory appropriate in this regard (see Pacala, 1988, for a review). Despite their lack of physiological detail and the fact that these models often include so many parameters that model tuning is a real problem, the approach offers great hope for investigating how altering physiological characteristics of the component species will affect community-level processes (Huston and Smith, 1987).

With regard to resource management, it should be clear from the above that we are still very ignorant about how to scale up from the detailed knowledge available on cellular and organ levels to even a whole plant, let alone to stand and regional scales. But when is it really necessary to do this? Many of the models that are currently used by agronomists and foresters to predict harvests, and schedule fertilization, irrigation and pesticide application are empirical in form. These models work well as long as the data base upon which they are based is adequate. For predictions outside the range of available data however, mechanistically-based models are necessary. One example, discussed below, concerns the long term effects of atmospheric CO_2 increases. For management at the level of individual farmers, an approach which couples a mechanistic model with expert systems methods may well be the best combination of empiricism (from the intuition and experience of the expert opinions solicited) with mechanism (Lemmon, 1986). A major limitation in all these approaches is the unpredictability of the environmental inputs. Stochastic simulators of variables such as rainfall can be included in most approaches, leading to estimates of the variance or even the full probability distribution of yield. Management decisions will then depend upon the manager's assessment of how much risk is acceptable.

It has been argued that in order to make reasonably accurate predictions of the long-term effects of atmospheric CO_2 increases on world productivity, it is necessary to construct systems models which are capable of extrapolative prediction on an ecosystem level, based on mechanistic models for physiological responses to CO_2 (Reynolds and Acock, 1985). While I am sympathetic with the reductionist sentiment which underlies this, I am also quite pessimistic that detailed physiological models are either possible to apply or to validate at regional or world scales. I believe that the best that one could hope for from such models is that they would suggest relatively simple empirical models, that though they lack the details of the mechanistic approach, would still be fairly accurate predictors. This uses a complicated model to determine what parameters really matter, and suggest macrodescriptors that would be fairly robust. On these scales, robust might be defined as within an order of magnitude of the actual.

On world scales, I believe it is much more reasonable to pursue top-down models. These may still include physiologically reasonable formulations, as Landsberg and McMurtrie (1985) argue for in the case of forest management. As an example of how this might be formulated, I'll describe an (admittedly fairly obvious) approach to investigating world productivity changes due to CO_2. My objective is to point out how one might integrate models on different scales, in essence by decoupling them.

Suppose that the world is broken up into several different vegetation types, i = 1,..,n, such as deciduous forest, grasslands of different types, etc. Let

$A_i(u)$ = land area of vegetation type i under world conditions u

$P_i(u)$ = productivity (e.g. biomass production per unit time) per unit area of vegetation type i under world conditions u

$u(t)$ = a time-dependent vector of world conditions (i.e. the distribution of temperature, precipitation, CO_2, etc. over the earth's surface) at time t.

Then u(t) would be generated by a climate model, and if there were several such models then their outputs would each be used to give some estimate of the potential variance in productivity. Then total productivity in year t is

$$P(t) = \sum_i A_i(u(t)) \, P_i(u(t)) \qquad (6)$$

This allows a decoupling of abiotic effects from biotic ones since the $A_i(u)$ might be estimated from Holdridge-type diagrams obtained from a climate model. In this simple case, direct effects of CO_2 on plants are viewed as not being important in determining future world distributions of vegetation types, compared to the effects of climate change.

To estimate $P_i(u)$, consider there to be many species types within the vegetation type i. Then

$$P_i(u(t)) = P_i(u_0) + \delta P_i(u(t)) \tag{7}$$

where $P_i(u_0)$ is the value of productivity in vegetation type i at present and $\delta P_i(u(t))$ is its change from the present to time t. The models may well be much more accurate predictors of changes than of absolute values of productivity and, furthermore, whatever data is collected to validate the models will only be at one or a few CO_2 levels. Then letting

$f_{ij}(u) =$ the fraction of ground area in vegetation type i
occupied by species j under world conditions u

$B_{ij}(u) =$ productivity per unit area of species type j in
vegetation type i under world conditions u

with $\delta B_{ij}(u)$ and $\delta f_{ij}(u)$ representing changes in these when conditions change from present (u_0) to u, we have

$$\delta P_i(u) = \sum_j \{ \delta B_{ij}(u) \ f_{ij}(u) + \delta f_{ij}(u) \ B_{ij}(u_0) \\ + \delta f_{ij}(u) \ \delta B_{ij}(u)\} \tag{8}$$

Here the sum is over all species types in the vegetation type (i.e weeds, pines, grasses, etc.)

In the above, the f_{ij} would presumably come from community-level models taking into account relative competitive ability changes under elevated CO_2, differential abilities for species to adapt to elevated CO_2, etc. The B_{ij} could come from a relatively simple empirical model, possibly derived from complicated physiologically-based models. The above procedure allows one to set up a variety of "null models" since it decouples the community level effects from the direct effects on physiology. For example, one could investigate the assumption that relative species compositions within vegetation types will not change under elevated CO_2 by setting $\delta f_{ij} = 0$. This approach also allows one to estimate how sensitive the larger scale results are to changes in models on smaller scales. In this way it may be useful in providing some confidence interval, or range, for possible effects on world scales from confidence intervals on parameters in the lower-level models. From a public policy perspective, even fairly rough confidence intervals from procedures such as the above would provide a rational basis for analyzing the long-term effects of alternative governmental responses.

5. Spatial Aspects of Plant Epidemiology

The purpose of this section is to point out some relatively unexplored problems in epidemiology that arise from considerations of spatial scale. The vast majority of mathematical work in epidemiology concerns the spread of disease in homogeneously mixing populations. This assumption is reasonable for many animal populations in which the spread of the disease is caused by contact between animals that move about. Even so, it is not realistic in cases for which there is either spatial or some other structure in the population which causes there to be higher contact rates within certain groups than between these groups. There have been a variety of models developed to analyze the effects of such structure in the host population on disease spread (Hethcote, 1978; Post et al., 1983; May and Anderson, 1984). In contrast to the situation in animal epidemiology, there has been relatively little theoretical development in the spatial aspects of plant epidemiology.

Plant epidemiology offers two important differences from the epidemiology of animal diseases, at least as far as modeling is concerned. First, since plants are fixed in space for much of their lifespans (and in essence all of it for crops), disease spread does not occur due to contacts between individuals but rather through the dispersal of the pathogen itself. There is thus an explicit spatial aspect in plant epidemics that has long been noted, but generally ignored from a modeling perspective (Gilligan, 1985). Secondly, plants generally have a continuum of resistance levels to any particular pathogen. The effect of infection by a pathogen on a given plant may range from severe damage or death to no damage at all. In many crop plants, infection leads to reduced growth and yield, but rarely to premature death of the host. This implies that the usual mathematical structure of epidemiological models - classes of infected, susceptible, immune and removed individuals - is inappropriate for most plant situations. The situation is similar to the case of macroparasitic infections in humans (Anderson and May, 1982).

Due to the above, even in monocultures of genetically uniform crop plants, it is inappropriate to assume that disease is spread either uniformly or according to a Poisson distribution over the host population. Despite this, much of the work on the temporal spread of a disease makes this assumption (Rouse, 1985; Hau et al., 1985). What work has been done on the spatial aspects of disease spread tends to be either highly empirical (i.e. statistical models for spore dispersal) or based on physical transport models with very little emphasis on biological effects (McCartney and Fitt, 1985). With few exceptions (i.e. Kampmeijer and Zadoks, 1977) there has been little work which attempts to simulate the dispersal of a crop disease and couple it with the growth of the crop. At the same time, essentially no attempts have been made to analyze the stochastic nature of the spatial spread of crop disease (Gilligan, 1985).

One approach to this is to utilize the individual-based growth models mentioned above. A somewhat preliminary investigation of this method was undertaken by a former student of mine (Bullock, 1986). The objective here was to investigate the effects of alternative spatial patterns of mixtures of resistant and non-resistant plants on the spread of a fungal pathogen. The method used a very simple individual growth model, logistic in form, with no neighborhood competition or physiological effects of environment. Pathogen was introduced according to a random process, with local growth dependent upon leaf area available, and dispersal occuring once pathogen density on a host reached a certain fraction of the host's carrying capacity. Dispersal was determined by an exponential random variable, with the capability to bias the spread due to prevailing winds. Host growth rate was either reduced as a function of pathogen load (for non-resistant plants) or independent of pathogen density (resistant plants). Several alternate spatial patterns were considered including uniform, bordered, striped and checkerboard. Criteria investigated were mean total biomass at end of season as well as probability that biomass at season end was below some threshold (presumably that at which profit is zero). Results indicated that generally the more divided the field, the smaller the amount of damage there was. Also, it was determined that bordering a field with resistant plants, as is sometimes suggested to farmers, had little effect on slowing an epidemic unless pathogen dispersal distances were very small.

It is possible to formulate an analytic approximation to the above situation, if one is willing to make certain assumptions. Consider the situation in one dimension only, and suppose the pathogen spreads according to a diffusion process with local growth dependent upon the local densities of the two host types. Also suppose there is no neighborhood competition in the host, so that plant growth is given by an ordinary differential equation. Then a general form of the problem is

$$\frac{\partial r}{\partial t} = D r_{xx} + f(r, g_1, g_2)$$

$$\frac{dg_1}{dt} = h_1(r, g_1, g_2) \tag{9}$$

$$\frac{dg_2}{dt} = h_2(r, g_1, g_2)$$

with $g_i(x,0) = k_i(x)$, $i=1,2$, $r(x,0) = r_0(x)$, $0 \le k_i \le C$, and

$$\int (k_1 + k_2)\, dx = M.$$

In the above, the $g_i(x,t)$'s are the densities of the two plant types at location x at time t, $r(x,t)$ is the pathogen density there, f is the local growth rate of the pathogen, the h_i's are the growth rates of the plants, the k_i's are the initial planting densities of the two plant types which are bounded by C and total initial planting is M, and $r_0(x)$ is the initial pathogen distribution. In addition one could attach zero boundary conditions for the pathogen density on the plot of length L say. The above becomes a control problem if the object is to choose the initial densities k_i so as to maximize the total biomass at end of season T

$$Y = \int_0^L (g_1(x,T) + g_2(x,T))\, dx. \tag{10}$$

In even the non-control case, this problem is extremely difficult to analyze. In part motivated by this model, R. S. Cantrell and C. Cosner of the University of Miami have investigated a steady-state version of (10) in the simplified case of fixed "good" and "bad" regions for the growth and dispersal of a pathogen. In the control problem, even proving that bang-bang is optimal is very hard. One can pretty much intuit what the answers should be in some special cases depending upon the dispersal rate of the pathogen, the size of the plot, and relative growth rates of the two plant types.

The above is partly meant to show how rapidly mathematical models can become intractable, but also that simplifications of the model can lead to intriguing mathematical problems. There have been a number of other models for disease spread that essentially produce travelling waves. One is a simulation approach similar to the individual model described above (Minogue and Fry, 1983) and another considers the diffusion of pathogen from a focus (van den Bosch et al., 1988).

van den Bosch, F. J. C. Zadoks and J. A. J. Metz. 1988. Focus expansion in plant disease I: the constant rate of focus expansion. Phytopathology 78: 54-58 .

Bullock, M. T. 1986. A Spatial Simulation Model for Disease Spread in a Crop. Master's thesis, Univ. of Tenn., Knoxville.

Charles-Edwards, D. A. 1982. Physiological Determinants of Crop Growth. Academic Press, Sydney.

France, J. and J. H. M. Thornley. 1984. Mathematical Models in Agriculture. Butterworths, London.

Gilligan, C. A. 1985. Introduction to: Mathematical Modelling of Crop Disease, C. A. Gilligan (ed.). Advances in Plant Pathology, Vol. 3, Academic Press, London.

Gross, L. J. 1986. Photosynthetic dynamics and plant adaptation to environmental variability. P. 135-170 in L. J. Gross and R. M. Miura (eds.), Some Mathematical Questions in Biology - Plant Biology. American Math. Soc., Providence.

Hau, B. S. P. Eisensmith and J. Kranz. 1985. Construction of temporal models II: Simulation of aerial epidemics. Chap. 3 in C. A. Gilligan (ed.). Advances in Plant Pathology, Vol. 3, Academic Press, London.

Hethcote, H. W. 1978. An immunization model for a heterogeneous population. Theor. Pop. Biol. 14: 338-349.

Hunt, R. 1982. Plant Growth Curves. Arnold, London.

Huston, M. and T. Smith. 1987. Plant succession: Life history and competition. Amer. Natur. 130:168-198.

Jarvis, P. G. and J. W. Leverenz. 1983. Productivity of temperate, deciduous and evergreen forests. Chap. 8 in Encyclopedia of Plant Physiology, Vol. 12D. Springer-Verlag, Berlin.

Johnson, I. R. 1985. A model of the partitioning of growth between the shoots and roots of vegetative plants. Ann. Bot. 55: 421-431.

Kampmeijer, P. and J. C. Zadoks. 1977. EPIMUL, a Simulator of Foci and Epidemics in Mixtures of Resistant and Susceptible Plants, Mosaics and Multilines. Pudoc, Wageningen.

Kirschbaum, M. U. F., L. J. Gross and R. M. Pearcy. 1988. Observed and modelled stomatal responses to dynamic light environments in the shade plant Alocasia macrorrhiza. Plant, Cell and Envir. 11: 111-121.

Landsberg, J. J. and R. McMurtrie. 1985. Models based on physiology as tools for research and forest management. In J.J. Landsberg and W. Parsons (eds.), Research for Forest Management. CSIRO, Melbourne.

Lemmon, H. 1986. Comax: an expert system for cotton crop management. Science 233: 29-33.

Loomis, R. S., R. Rabbinge and E. Ng. 1979. Explanatory models in crop physiology. Ann. Rev. Plant. Physiol. 30: 339-367.

Makela, A. and P. Hari. 1986. Stand growth model based on carbon uptake and allocation in individual trees. Ecol. Model. 33: 205-229.

Neither of these consider the effects of the pathogen on the plant however.

6. Conclusions

In sum, I have argued that a physiological perspective is often useful, even when the scales of the problem of concern are considerably longer temporally and larger spatially than would normally be addressed by consideration of physiology. My key point might be succinctly stated as "A little reductionism is good for the soul, too much reductionism is bad for the heart". Thus, we gain a mechanistic understanding of the functioning of complex natural systems by taking a physiological perspective. At the same time, there are clear limits to the utility of a reductionist approach, evident from the large number of poorly understood parameters and functional forms which appear in large systems models. I argue for an intermediate approach which uses physiologically-based models to indicate appropriate macrodescriptors for large-scale phenomena. When this is coupled with an analysis of the system's structure according to the rates of the processes appropriate to the questions being addressed (O'Neill et al., 1986), we will have available a truely hierarchical approach to natural systems.

7. Acknowledgements

I thank an anonymous referee for helpful comments. My remarks on the effects of atmospheric CO_2 increases were initiated due to an invitation to attend a U. S. Department of Energy workshop on the topic, and I thank James Reynolds for the invitation to attend. I also thank Carole Levin for urging me to come up with the one sentence summary of my views mentioned in the conclusions above (as well as for a luscious meal during the Conference).

References

Anderson, R. M. and R. M. May. 1982. The population dynamics and control of human helminth infections. Nature 297: 557-563.

Baldocchi, D. D., B. A. Hutchison, D. R. Matt and R. T. McMillan. 1985. Canopy radiative transfer models for spherical and known leaf inclination angle distributions: a test in an oak-hickory forest. J. Appl. Ecol. 22: 539-555.

Barrett, J. R. and R. M. Peart. 1981. Systems simulation in U. S. agriculture. P. 39-59 in Progress in Modeling and Simulation, Academic Press, NY.

May, R. M. and R. M. Anderson. 1984. Spatial heterogeneity and the design of immunization programs. Math. Biosci. _72_: 83-111.

McCartney, H. A. and B. D. L. Fitt. 1985. Construction of dispersal models. Chap. 5 in C. A. Gilligan (ed.). Advances in Plant Pathology, Vol. 3, Academic Press, London.

McCree, K. J. 1970. An equation for the rate of respiration of white clover plants grown under controlled conditions. In: I. Setlik (ed.), Prediction and Measurement of Photosynthetic Productivity, Pudoc, Wageningen.

McMurtrie, R. E. 1985. Upper limits to forest productivity in relation to carbon partitioning. In: The Application of Computer Models in Farm Management, Extension and Research. Austr. Inst. Agric. Sci., Queensland.

Minogue, K. P. and W. E. Fry. 1983. Models for the spread of plant disease: model description. Phytopath. _73_: 1168-1173.

Montieth, J. L. (ed.). 1975. Vegetation and the Atmosphere. Vol. I. Academic Press, London.

Nobel, P. S. 1974. Introduction to Biophysical Plant Physiology. Freeman, San Francisco.

O'Neill, R.V., D. L. DeAngelis, J. B. Waide and T. F. H. Allen. 1986. A Hierarchical Concept of Ecosystems. Princeton University Press, Princeton, NJ.

Pacala, S. W. 1989. Plant population dynamic theory. In J. Roughgarden, S. A. Levin and R. M. May (eds.), Perspectives in Ecological Theory, Princeton Univ. Press, Princeton, NJ.

Post, W. M., D. L. DeAngelis and C. C. Travis. 1983. Endemic diseases in environments with spatially heterogeneous host populations. Math. Biosci. _63_: 289-302.

Reynolds, J. F. and B. Acock. 1985. Predicting the response of plants to increasing carbon dioxide: a critique of plant growth models. Ecol. Model. _29_: 107-129.

Reynolds, J. F. and J. H. M. Thornley. 1982. A shoot:root partitioning model. Ann. Bot. _49_: 585-597.

Ross, J. 1981. The Radiation Regime and Architecture of Plant Stands. Junk, the Hague.

Roughgarden, J. 1986. The theoretical ecology of plants. P. 235-267 in L. J. Gross and R. M. Miura (eds.), Some Mathematical Questions in Biology - Plant Biology. American Math. Soc., Providence.

Rouse, D. I. 1985. Construction of temporal models I: Disease progress of air-borne pathogens. Chap. 2 in C. A. Gilligan (ed.). Advances in Plant Pathology, Vol. 3, Academic Press, London.

Shugart, H. H. 1984. A Theory of Forest Dynamics. Springer-Verlag, NY.

Thornley, J. H. M. 1976. Mathematical Models in Plant Physiology. Academic Press, NY.

ECONOMICS, MATHEMATICAL MODELS AND ENVIRONMENTAL POLICY

Jon M. Conrad
Department of Agricultural Economics
Cornell University
Ithaca, New York 14853

Abstract

This paper briefly reviews several models of externality which provide the theoretical basis of environmental economics. An externality may be defined as a *situation* where the output or action of a firm or individual affects the production possibilities or welfare of another firm or individual who has no direct control over the initial level of the output or activity. Pollution, resulting from the disposal of residual wastes, is a classic example of externality.

Three static models examine the optimality conditions for (1) a two-person externality, (2) a many-person externality (where the externality takes the form of a "pure public bad"), and (3) a two-plant polluter. In the case of a two-person externality negotiation between the affected parties may lead to the optimal level for the externality *regardless* of the initial assignment of property rights. In the many- person case, environmental policies, such as direct controls or economic incentives, may be required to achieve an optimal allocation of resources. Economic incentives may take the form of per unit taxes on emissions or transferable discharge rights. In the third model it is shown how a tax can induce optimal (least cost) treatment from a two-plant polluter.

Two dynamic models examine the cases where (1) a pollution stock may accumulate or degrade according to rates of discharge and biodegradation and (2) a toxic residual must be transported from sites where it is generated to sites where it may be safely stored. The latter problem poses environmental risks from spills in transit or leakage at storage sites.

While radioactive and toxic wastes are likely to continue to be regulated by direct controls some of the more "benign" residuals are suitable for regulation by economics incentives. Effluent taxes in France and the Netherlands, transferable discharge permits on the Fox River in Wisconsin, transferable stove permits in Telluride, Colorado and the EPA's emission-offset policy are indications that economic incentives will play a greater role in the future management of environmental quality.

1. Introduction

The now extensive literature on environmental economics has its theoretical roots in the field of welfare economics and is specifically tied to the concept of externality (Mishan 1971). An externality might be defined as a situation where the output or action of a firm or individual directly affects the production possibilities or welfare of another firm or individual who has no direct control over the level of output or activity. Consider a brewery downstream from a pulp mill. The level of jointly produced pulp waste dumped into the river will directly influence the production process for beer by determining whether the brewery must treat water from the river or use water from an alternative source. As another example, suppose the volume (and type of music) on my stereo adversely affects the welfare of my neighbor. By lacking direct control over the amount of pulp waste or the volume and type of music we mean that the brewery or neighbor has no direct influence over the initial level for the externality. A unilateral decision designed to maximize the profit or welfare of a single individual may impose costs on others (the brewery or neighbor) to such an extent that the situation is nonoptimal; that is, an adjustment in the level of the externality (and possibly side payments) could make one or both parties better off without making anyone worse off.

Externalities which impose costs on others are referred to as negative externalities, while those that convey positive benefits on others are referred to as positive externalities. A well-kept flower garden is thought to be an example of a positive externality. In addition to the satisfaction of the gardener, the garden may be pleasing to passersby, and may increase the value of neighboring property.

It is generally thought that the level or amount of negative externalities will be socially excessive (too much waste, or not enough resources allocated to waste treatment), while the level or amount of positive externalities will be socially deficient (not enough flowers). Economists thought this misallocation of resources might be corrected through a tax on negative externalities and a subsidy on positive externalities. Noneconomists (especially lawyers and politicians) usually prefer direct regulation of negative externalities and perhaps government provision of certain positive externalities. With well defined property rights (or liability rules) some economists felt that neither taxes nor direct regulation were necessary, rather direct negotiation between the affected parties may lead to the optimal level of externality. We will explore the logic behind these and other policies for controlling externalities in the next section. This analysis will be based on some simple static models.

In the third section we will consider some models of dynamic externality were a pollution stock may increase or decrease depending on the rates of residual discharge and biodegradation. The fourth section examines the characteristics of toxic wastes which may require different control policies than those employed for degradable (organic) wastes. The final section attempts to summarize the policy implications that have emerged from the field of environmental economics.

2. Models of Static Externality

When classifying externalities two important considerations are the number of individuals or firms affected (either as polluters or "victims") and the degree of "publicness" or uniformity of external costs or benefits. Our previous examples of the pulp mill/brewery and the raucous stereo might be classified as "small-number" externalities, where the social cost is limited to a few, spatially concentrated, individuals. In contrast, a "purely public" externality occurs when a large number of individuals face the same (uniform) amount of some externality. The small-number externality may be amenable to private negotiation or bargaining of the sort envisioned by Coase (1960). It is possible to talk about an "optimal level" for an externality and examine the likely outcome of negotiations under different property rights or liability rules. An example of a purely public externality might be the current problems of acid rain or ozone depletion which affect a large number of individuals and where the level of externality may be more or less uniform across a large area or population. We will start by considering some static, two-party externality models.

Suppose Individual One would like to engage in an activity whose level is denoted by the continuous variable X and which results in net benefits according to the concave function N(X). This activity, however, imposes costs on Individual Two according to the convex function C(X). Then the welfare of a "community" comprised of Individuals One and Two might be calculated as

$$W = N(X) - C(X) \qquad \qquad (1)$$

and the first order necessary condition for maximization of net social welfare requires $dW/dX = 0$ implying $N'(X) = C'(X)$, and the optimal level of the externality is that which balances net marginal benefits to Individual One with the marginal cost to Individual Two. Figure 1 shows a possible graph of $N'(X)$ and $C'(X)$ and the optimal externality $X = X^*$ where $N'(X)$ and $C'(X)$ intersect.

Now consider the situation when Individual One has the property right to set X at a level which maximizes his or her net benefit. This would require

$N'(X) = 0$ which occurs at $X = \bar{X}$. At $\bar{X}$, however, significant marginal costs are being imposed on Individual Two. While Individual One has the right to set X at a level which maximizes his self-interest he or she might be amenable to a proposal to reduce X if a bribe or side payment were sufficient to compensate for any foregone net benefits. How much would it take and how much would

Individual Two be willing to offer? At $X = \bar{X}$ Individual Two would be willing to pay up to C'(X) for a marginal reduction in X, while Individual One would only require a small amount based on the very low net marginal benefits foregone for initial incremental reductions from $\bar{X}$. A comparison of C'(X) to N'(X) would indicate that Individual Two should be willing to offer up to marginal cost for an incremental reduction in X, while Individual One would require at least compensation of foregone net benefits. While the exact distribution of the vertical difference between C'(X) and N'(X) cannot be deduced, it is positive for reductions in X all the way back to X^*. Thus, there is an incentive for negotiation to lead to a reduction in X from $\bar{X}$ to X^*.

What if Individual Two has the property right to an environment free of cost imposed by X? If Individual Two has the right to initially set the level for X he or she would obviously select X = 0, where marginal cost C'(X) = 0. However at X = 0 Individual One is foregoing significant net benefits and has a strong incentive to approach Individual Two to see if he or she can be bribed into putting up with some costs associated with X > 0. With a similar logic as that governing the negotiated reduction from $\bar{X}$, Individual One would be willing to pay up to N'(X) while Individual Two would require at least a payment to cover the marginal costs, C'(X). The vertical distance between N'(X) and C'(X) (and thus the incentive for negotiation) remains positive from X = 0 up to X = X^*. Thus, there would be a tendency for X to be negotiated upward from 0 toward X^*. What is surprising is that the level of externality after negotiation is the same regardless of the initial assignment of property rights. This is sometimes called the "Coase Theorem" in recognition of Ronald Coase's discussion of this subject in an article entitled "The Problem of Social Cost" (Coase 1960.).

Coase's Theorem hinges on some implicit assumptions, particularly that the transactions costs of negotiating are zero or symmetric when starting from X = $\bar{X}$ or X = 0 (i.e., when property rights are vested with Individual One or Individual Two), and that the net benefit and cost functions are invariant to changes in income. In other words, as side payments are made N'(X) and C'(X) do not shift. This latter assumption is referred to as a "zero income effect."

Coase concludes that there may be no need for government intervention in an externality situation if the social cost were generated and borne by a few individuals and property rights (liability rules) were well defined by common law. In such a case the concerned individuals could be expected to negotiate to a more or less optimal level of externality. These conditions are not likely to be met when an externality results from or adversely affects a large number of individuals. We now turn to this type of public externality. Baumol and Oates (1975) refer to it as an "undepleteable externality".

At a point in time suppose the productive resources for an economy are fixed and $\phi(Q,S) \equiv 0$ is an implicit production possibilities or transformation curve identifying the feasible combinations of a positively-valued commodity, Q, and a negatively-valued residual, S. Thus, the fixed resources must be allocated between production of Q or reduction of S. By convention we assume $\partial\phi(\cdot)/\partial Q > 0$ and $\partial\phi(\cdot)/\partial S < 0$. A graph of $\phi(Q,S) \equiv 0$ is shown in Figure 2. There exists some implicit allocation or resources leading to $S = 0$ and $Q = Q_0$. To increase output above Q_0 requires a reallocation of resources away from residual treatment. While Q can be increased, S increases at an increasing rate. The point (Q_{max}, S_{max}) corresponds to an implicit allocation where all resources are devoted to production of Q.

Suppose there are I individuals with utility functions $U_i = U_i(q_i, S)$ where q_i is the amount of Q going to the ith individual. We assume that $\partial U_i(\cdot)/\partial q_i > 0$ while $\partial U_i(\cdot)/\partial S < 0$. The residual is a "public bad" in the sense that the same level enters everyone's utility function and the "consumption " by one individual does not diminish the amount "consumed" by others. The commodity Q is a private good since an increase in the amount consumed by one individual reduces the amount available for others. Conditions for a Pareto optimum $(Q, S, q_i, i = 1,2 ...,I)$ can be derived by maximizing the utility of one individual subject to (a) indifference curve constraints for all other individuals, (b) a balancing equation for Q and (c) the transformation function $\phi(Q,S) \equiv 0$.

Let $U_i^{\bullet} > 0$ be the utility (indifference) constraint for the ith individual. Then, the Lagrangian may be written as

$$L = \sum_{i=1}^{I} \lambda_i [U_i(\cdot) - U_i^{\bullet}] + \omega \left(Q - \sum_{i=1}^{I} q_i \right) - \mu\phi(\cdot) \qquad (2)$$

where $\lambda_1 = 1$ and $U_1^{\bullet} = 0$. Assuming a solution where $Q > 0$, $S > 0$ and $q_i > 0$ (i.e., an interior solution), then first order conditions require

$$\frac{\partial L}{\partial Q} = \omega - \mu[\partial\phi(\cdot)/\partial Q] = 0 \qquad (3)$$

$$\frac{\partial L}{\partial S} = \sum_{i=1}^{I} \lambda_i [\partial U_i(\cdot)/\partial S] - \mu[\partial\phi(\cdot)/\partial S] = 0 \qquad (4)$$

$$\frac{\partial L}{\partial q_i} = \lambda_i [\partial U_i(\cdot)/q_i] - \omega = 0 \qquad (5)$$

$$\frac{\partial L}{\partial \omega} = Q - \sum_{i=1}^{I} q_i = 0 \qquad (6)$$

and

$$\frac{\partial L}{\partial \mu} = -\phi(Q,S) = 0 \qquad (7)$$

Given the signs for the partials of $\phi(Q,S)$ and $U(q_i,S)$ it can be shown that $\omega > 0$ and $\mu > 0$. Some algebra will reveal that

$$\sum_{i=1}^{I} \frac{[\partial U_i(\cdot)/\partial S]}{[\partial U_i(\cdot)/\partial q_i]} = \frac{[\partial \phi(\cdot)/\partial S]}{[\partial \phi(\cdot)/\partial Q]} \qquad (8)$$

Equation (8) equates the sum of the marginal rates of substitution (of the residual for the commodity) over all individuals to the marginal rate of transformation and is analogous to Samuelson's (1954,1955) condition for optimal provision of a pure public good. In this case, however, we have the condition for the optimal level of an undepleteable externality, or a pure public bad. Equation (4) has an important interpretation. The term

$$-\sum_{i=1}^{I} \lambda_i [\partial U_i(\cdot)/\partial S] > 0$$

is the marginal social damage from an increase in the residual (which negatively affects all individuals). Some additional algebra will reveal that this term must be equated to

$$-\omega \frac{[\partial \phi(\cdot)/\partial S]}{[\partial \phi(\cdot)/\partial Q]} > 0$$

which is the marginal social value of an additional unit of S, which in turn allows the economy to produce more Q (which is what society positively values). Thus, equation (4) requires a balancing of the value of increased output with the marginal social cost resulting from the increase in pollution. The conditions for optimality raise an important practical question: How do you measure the marginal social cost of pollution? The losses are direct utility losses which creates a difficult evaluation problem. Suppose S was an air pollutant with a nonuniform affect over a large urban area. By a careful statistical analysis of property values it might be possible to estimate the disutility of living with alternative levels of S as reflected in reduced property values. Contingent valuation techniques, where an individual is asked to state his willingness to pay for a less polluted environment (or the required compensation for the individual to accept a more polluted environment) might also provide estimates of the marginal social cost of pollution. There are difficulties with both of these approaches and it may be more practical to consider policies which do not require empirical estimates of marginal social cost. The following problem may suggest such a policy.

A corporation has two plants on a large lake. Under normal operating conditions, and without any environmental restrictions, the plant was discharging wastes at rates $\bar{R}_1$ and $\bar{R}_2$ from Plants #1 and #2, respectively. The EPA regards the combined discharge as excessive and requires that the total discharge from both plants not to exceed R. Denoting the untreated discharge from Plants #1 and #2 by R_1 and R_2, respectively, the EPA discharge constraint implies $R_1 + R_2 \leq R$. The amount of waste treated at Plant #1 and #2 is $(\bar{R}_1 - R_1)$ and $(\bar{R}_2 - R_2)$, respectively. Suppose the cost of treatment is a function of the amount treated and that the corporations combined treatment costs may be calculated according to

$$C = C_1(\bar{R}_1 - R_1) + C_2(\bar{R}_2 - R_2)$$

where $C_1(\cdot)$ and $C_2(\cdot)$ are treatment cost functions for Plants #1 and #2, respectively. The firm wishes to minimize the total cost of treatment subject to meeting the EPA discharge constraint. The Lagrangian for this problem may be written

$$L = C_1(\bar{R}_1 - R_1) + C_2(\bar{R}_2 - R_2) - \lambda(R - R_1 - R_2) \qquad (9)$$

Assuming a positive level of treatment at both plants and that the EPA constraint is precisely met (i.e., it holds as an equality), then the first order conditions for minimum treatment costs require

$$\frac{\partial L}{\partial R_1} = -C_1'(\cdot) + \lambda = 0 \qquad (10)$$

$$\frac{\partial L}{\partial R_2} = -C_2'(\cdot) + \lambda = 0 \qquad (11)$$

$$\frac{\partial L}{\partial \lambda} = -(R - R_1 - R_2) = 0 \qquad (12)$$

Equations (10) and (11) imply that $C_1'(\cdot) = C_2'(\cdot)$, that is, the marginal cost of treating the last unit in each plant must be the same. Taken together, equations (10)-(12) constitute a three equation system which may be solved for the cost minimizing rates of discharge R_1, R_2 and the shadow price of the EPA constraint, λ. Note: $\partial L/\partial R = -\lambda < 0$ is the marginal cost of the EPA constraint and as R decreases, treatment costs increase.

How does this problem relate to the difficulty of measuring marginal social damage and the formulation of environmental policy? The EPA specified an allowable discharge, R, hoping that this amount would result in acceptable ambient environmental quality. Alternatively, it could have specified a tax on each unit of R_1 and R_2 and let the corporation decide how much to treat from each plant and thus how much to pay in taxes. If the EPA had solved equations (10) - (12) and set the tax rate $\tau = \lambda$, then the corporation would presumably choose the same values for R_1 and R_2 that it originally chose when faced with the discharge constraint, R. This is true since minimizing

$$TC = C_1(\bar{R}_1 - R_1) + C_2(\bar{R}_2 - R_2) + \tau(R_1 + R_2) \qquad (13)$$

when $\tau = \lambda$ will result in first order conditions requiring $C_1'(\cdot) = C_2'(\cdot) = \tau$, which was the same condition obtained for the original problem. The tax forces the corporation to determine if it can treat the marginal unit of waste for a cost less than the unit tax itself. If it can, it will do so, otherwise it will opt to pay the tax. The tax (called an effluent charge for water pollutants or an emission tax for air pollutants) has the desirable property of achieving a given emission reduction at least cost. Specifically, it causes firms to treat lower cost emissions first. If, after subsequent analysis, the EPA does not view the emission reduction as sufficient to achieve the desired level for ambient environmental quality, it can request a tax increase which will raise the cost of untreated emissions. The tax places the burden of finding the best way to reduce emissions on the corporation and the EPA does not have to spend time discovering the "best practical technology" nor inspecting firms to see that it is has been installed and is being maintained. It will need to monitor emissions to determine the level of untreated residuals in order to calculate the corporation's tax bill, but this should not increase transactions costs above the current costs of monitoring a system of emission standards

To summarize this section so far we have (1) considered a small-number externality (between two individuals) where negotiation between "polluter" and "pollutee" might lead to the same (optimal) level for the externality without government intervention, regardless of the initial assignment of property rights (liability). In the case of a large number externality, uniformly affecting a large number of individuals, we obtained an optimality condition equating the sum of all affected individuals' rate of commodity substitution (RCS) to the rate of transformation of the private good ,Q, for the residual, S. The residual was a "pure public bad" and the optimality condition is analogous to Samuelson's condition for the optimal provision of a pure public good. The marginal social cost of pollution was the negative of the weighted sum of marginal disutility, which while being an interesting concept from a public policy perspective, presented difficult measurement problems. The model of the two-plant corporation, minimizing the cost of treatment subject to a total discharge constraint, revealed that the corporation would allocate treatment so that marginal treatment costs were equal. The shadow price (Lagrange multiplier) on the discharge constraint, if used as a per unit tax on untreated discharges, would induce the same-least cost pattern of treatment. By taxing residual discharges an environmental agency has the ability, in theory, to bring about any desired reduction in emissions at least cost and with lower administration costs than policies that rely on direct regulation.

The three models presented thus far have not contained any detail about the physical characteristics of the residuals being jointly produced nor about the medium in which they might be disposed. A processing or manufacturing plant may often have some control over the chemical structure and form of a residual. It may then be able to dispose of the residual in one or more receiving media. Environmental management becomes a complex problem of simultaneously determining not only the optimal allocation of resources and distribution of output, but the composition and disposition of the many forms of residual waste and the alternative media (air, water, land) into which these residuals may be disposed. To minimize the total cost of disposal one would like to transform and dispose of residuals in such a way that the marginal disposal plus marginal damage cost is the same for all media receiving waste (See Conrad and Clark 1987, Chapter 4, for a more complete static model of multimedia residuals management). It may be optimal to *never* use some disposal options or some media if the marginal disposal cost or the marginal damage cost is always higher than the next best alternative. (For a discussion of this situation with regard to sludge disposal in the New York Bight see Conrad 1985).

In the late 1960s environmental economists became aware of the implications of the First Law of Thermodynamics which dictates a mass/energy balance in a closed system. In most economic systems this means that the mass of raw material inputs ultimately equals the mass of waste (see Figure 3). Kenneth Boulding (1966) likened the earth to a spaceship with not only finite resources but a finite space to dispose of waste (residuals). Maximization of the rate of increase in Gross National Product may not be a desirable objective if the "newly produced goods and services" must ultimately become waste. The notion of a spaceship earth raises a fundamental problem: If we succeed in reducing the amount of Residual #1 being disposed via media #3, might we simply be creating another problem because we now must dispose of more of some other waste via an alternative medium. The management of residuals (which generate negative externalities) may ultimately require a large scale systems approach such as that envisioned by Kneese, Ayres and d'Arge (1970):

> The Administrator of the World Environmental Control Authority sits at his desk. Along one wall of the huge room are real-time displays processed by computer from satellite data, of developing atmospheric and ocean patterns, as well as the flow and quality conditions of the world's great river systems. In an instant, the Administrator can shift from real-time mode to simulation to test the larger effects of changes in emissions of material residuals and heat to water and atmosphere at control points generally corresponding to the locations of the world's great cities and the transport movements among them. In a few seconds the computer displays information in color code for various time periods - hourly, daily, or yearly phases at the Administrator's option....Observing a dangerous reddish glow in the eastern Mediterranean, the Administrator dials sub-control station Athens and orders a step-up of removal by the liquid residuals handling plants there. Over northern Europe, the brown smudge of a projected air quality standards violation appears and sub-control point Essen is ordered to take the Ruhr area off sludge incineration for 24 hours but is advised that temporary storage followed by accelerated incineration - but with muffling - after 24 hours will be admissible. The CO_2 simulator now warns the Administrator that another upweller must be brought on line in the Murray Fracture Zone within two years if the internationally agreed balance of CO_2 and oxygen is to be maintained in the atmosphere.

It is unlikely that sovereign states would be willing to vest an international agency with such powers of intervention and regulation. However, the industrialized countries of western Europe have made some initial attempts to at least better coordinate environmental policies pertaining to transfrontier pollution (OECD 1974, 1976). The U. S. and Canada have a considerable history of diplomacy and cooperation in managing the quality of boundary waters and fishery resources. Recent discussions on acid rain in North America have not led to any substantial agreement on policy, in part because of a reluctance by the current U.S. administration to commit itself to an expensive control program when they regard the consequences and costs of acid deposition as "poorly understood".

While multimedia residuals management makes sense economically, its feasibility from a political point of view must be questioned in light of the difficulty of siting sanitary landfills, nuclear waste repositories and other "nauseous" facilities. In other words, arranging for the transport and disposal of residuals in political jurisdictions different from where they were generated is a difficult and sensitive area of "intergovernmental relations" with very few communities willing to receive wastes from other areas. In a country with many levels of government (e.g., state, county, city or town) the ability to transform residuals for disposal in a medium and at a location with the highest assimilative capacity or with the best storage characteristics may be blocked politically.

3. Dynamic Externality

We now turn to a simple model of dynamic externality where residuals might accumulate as a pollution stock. It is this stock which imposes a social cost and two important questions become (1) Is there an optimal, steady-state level for the pollution stock and the rate of residual discharge? and (2) If the current pollution stock is not optimal what is the optimal rate of residual discharge along an approach path?

Let X_t represent the pollution stock and R_t the rate of residual discharge in period t. We will assume that the pollution stock changes according to

$$X_{t+1} - X_t = R_t - D(X_t) \tag{14}$$

where $D(X_t)$ is a degradation function specifying the rate at which the pollution stock degrades into its (harmless) organic constituents. If $D(X_t) = 0$ then we have a model of pure accumulation.

Suppose Q_t is the output of some good or service sold at a constant per unit price, p. As before let $\phi(Q_t, R_t) \equiv 0$ be an implicit production function expressing the maximum amount of Q_t attainable from the available fixed resources and a given level of R_t. Let $S(X_t)$ denote the social cost from the pollution stock of X_t in period t. Then we are interested in the solution to the problem

$$\text{Maximize} \sum_{t=0}^{\infty} \rho^t \{pQ_t - S(X_t)\}$$

$$\text{Subject to: } X_{t+1} = X_t + R_t - D(X_t)$$

$$\phi(Q_t, R_t) \equiv 0 \text{ and } X_0 \text{ given}$$

The Lagrangian expression for this problem may be written as

$$L = \sum_{t=0}^{\infty} \rho^t \{pQ_t - S(X_t) + \rho\lambda_{t+1}(X_t + R_t - D(X_t) - X_{t+1}) - \mu_t\phi(Q_t, R_t)\} \tag{15}$$

The first order necessary conditions can be shown to imply

$$p - \mu_t \partial\phi(\cdot)/\partial Q_t = 0 \tag{16}$$

$$p\lambda_{t+1} - \mu_t \partial\phi(\cdot)/\partial R_t = 0 \qquad\qquad (17)$$

$$p\lambda_{t+1} - \lambda_t = S'(X_t) + p\lambda_{t+1}D'(X_t) \qquad\qquad (18)$$

With the conventional partials $\partial\phi(\cdot)/\partial Q_t > 0$ and $\partial\phi(\cdot)/\partial R_t < 0$ it will be the case that $\mu_t > 0$ and $\lambda_t < 0$. In steady state $\mu = p/[\partial\phi(\cdot)/\partial Q]$ and $\lambda = (1+\delta)p[\partial\phi(\cdot)/\partial R]/[\partial\phi(\cdot)/\partial Q]$ and we obtain the following system defining the optimal pollution stock, residual discharge and output level

$$-p\frac{[\partial\phi(\cdot)/\partial R]}{[\partial\phi(\cdot)/\partial Q]} = \frac{S'(X)}{[\delta+D'(X)]} \qquad\qquad (19)$$

$$R = D(X) \qquad\qquad (20)$$

$$\phi(Q_t, R_t) = 0 \qquad\qquad (21)$$

The last two equations simply require that the rate of residual discharge equal the rate of biodegradation and the implicit production function holds in steady state.

To see how equations (19)-(21) might define an optimal steady state consider the following numerical example. Suppose

$$\phi(Q,R) \equiv Q - \sqrt{100 + R} = 0$$

$$D(X) = 0.10X$$

$$S(X) = X^2$$

$$\delta = 0.10$$

Then equation (19) implies $X = p(100+R)^{-1/2}/20$, while equation (20) implies $X=10R$. Equating these two expressions and solving for an implicit expression in R yields: $40{,}000R^2(100+R) - p^2 = 0$. Given a value for p, the per unit price of Q, this last equation can be solved numerically for optimal $R > 0$. This was done for $p = 10{,}000$ and (apparently) resulted in a unique optimum with $R^* = 4.88$, $X^* = 48.82$ and $Q^* = 10.24$. (The cubic in R was solved using Newton's method and starting from various positive initial guesses for R the algorithm always converged to $R^* = 4.88$).

We have answered the first of our two questions in that equations (19)-(21) will, with appropriate convexity assumptions, define an optimal level for the pollution stock. Suppose the optimal pollution stock was determined (as in the numerical example) and upon comparison with the initial condition we observe $X_0 > X^*$. What can we say about the optimal discharge policy along an approach path to X^*? In general the optimal discharge policy will be for $R_t < D(X_t)$, and under certain circumstances it will be optimal for $R_t = 0$. What we're looking for is an optimal discharge rate, less than biodegradation, which will induce the pollution stock to optimally decline to X^* (see Figure 4). If the objective function $W_t = pQ_t - S(X_t)$ can be rewritten as an additively separable function in X_t and X_{t+1} and, via reindexing, that additively separable function leads to an objective that is the sum of quasi-concave functions, then a Most Rapid Approach Path (MRAP) is optimal (see Spence and Starrett 1975). In Figure 4 an MRAP from $X_0 > X^*$ has $R_t = 0$ for $0 \le t < \hat{t}$. At $\hat{t}$, $X_t = X^*$ and $R_t = R^*$ for $t > \hat{t}$. If the conditions for MRAP are *not* met the infinite horizon control problem must be solved for the asymptotic approach path along which $0 < R_t < R^* = D(X^*)$, and X_t asymptotically approaches X^* from above while R_t asymptotically approaches R^* from below. These two cases cover the optimal approach from $X_0 > X^*$ provided R_t cannot be negative (i.e., dredging or removal of X_t for disposal via an alternative media is not feasible).

This simple problem captures many aspects of a pollution stock that can accumulate or degrade. As noted earlier if $D(X_t) = 0$ then the pollution stock can only accumulate. If the residual is highly toxic or radioactive then transport and storage so as to preclude leakage into the ambient environment is warranted. (The social cost of disposal via air, water or land are viewed as infinite). We now turn to a discussion of these types of residuals.

4. Management of Toxic Residuals

By a toxic residual we will mean a residual which is not suitable for disposal into (onto) the traditional disposal media: air, water or land. The residual must be transported from the site where it is generated to a storage site. Uncertain social costs arise because the residual may be spilled in transit or it may leak while in storage.

Nuclear wastes would be a good example of the type of residual being considered. Other toxics, such as acids, pesticides or other chlorinated hydrocarbons might be amenable to treatment (e.g., pyrolysis) at a specialized facility with nontoxic constituents discharged into the air or water. We will consider a model of toxics which must be stored.

Let $R_{i,j,t}$ represent the amount of the toxic residual generated at site i which is transported for storage at site j during period t, $i = 1,2,...,I$; $j = 1,2,...,J$; and $t = 0,1,2,...$. Assume that the cost of transport from i to j depends on the volume of waste according to $C_{i,j}(R_{i,j,t})$. Let $X_{j,t}$ represent the amount of the toxic in storage at site j in period t. The change in the stocks of waste in storage will depend on the volume of residuals received from the various generating sites, the amount in storage in the previous period less any unintended leakage. These dynamics are described by the difference equation

$$X_{j,t+1} = (1-\omega_{j,t})X_{j,t} + \sum_{i=1}^{I}(1-\omega_{i,j,t})R_{i,j,t} \qquad (22)$$

where $\omega_{j,t}$ and $\omega_{i,j,t}$ are random variables indicating the fraction of $X_{j,t}$ leaking from storage site j and the fraction of $R_{i,j,t}$ spilled in transit during period t. The distributions for these random variables will depend on the characteristics (including age) of the storage site, the route and method of transport and other exogenous factors (e.g. storms, earthquakes, etc.). The formulation of subjective probability distributions for $\omega_{j,t}$ and $\omega_{i,j,t}$ presents our toxic waste manager with a difficult (impossible?) exercise in risk assessment.

The second difficult element is the assessment of the likely social costs in the event of a spill or a leak. Suppose these costs depend only on the size of the spill or leak in period t so that $S_j(\omega_{j,t}X_{j,t})$ and $S_{i,j}(\omega_{i,j,t}R_{i,j,t})$ represent the social cost of a leak at site j and a spill in transit from site i to site j during period t. In reality these costs may depend on the amount and number of previous leaks or spills and the exact location of the the spill on the route between i and j.

Finally, we will assume that the amount of waste generated at site i is given (exogenously) by $R_{i,t}$ and that the storage capacity at site j is limited to a volume less than or equal to X_j. Then the toxic waste manager may seek to minimize the expected disposal and social costs by solving the problem

$$\text{Minimize } E\left\{\sum_{t=0}^{\infty}\rho^t\left[\sum_{i,j}[C_{i,j}(R_{i,j,t}) + S_{i,j}(\omega_{i,j,t},R_{i,j,t})] + \sum_{j=1}^{J}S_j(\omega_{j,t},X_{j,t})\right]\right\}$$

$$\text{Subject to } X_{j,t+1} = (1-\omega_{j,t})X_{j,t} + \sum_{i=1}^{I}(1-\omega_{i,j,t})R_{i,j,t}$$

$$R_{i,t} - \sum_{j=1}^{J}R_{i,j,t} = 0$$

$$X_{j,0} \text{ given, } R_{i,j,t} \geq 0, 0 \leq X_{j,t} \leq X_j$$

The above problem is a complex stochastic optimization problem and a numerical solution might be obtained using stochastic dynamic programming. In addition to the finite number of generating and disposal sites it may be necessary to assume that the random variables $\omega_{j,t}$ and $\omega_{i,j,t}$ are generated from a set of finite fractions with known, discrete probabilities. (Note, the summation over i,j is over all possible combinations of generation and disposal sites and is also a finite set). The solution will be in a feedback form since the optimal disposal pattern in period t will depend on the pollution stocks at the beginning of period t, which in turn were determined by the random spills and leaks that occurred in period (t-1).

The above model assumes the existence of J disposal sites and thus seeks to minimize the expected sum of transport and social costs. If storage sites do not exist then the problem becomes more a complicated problem of siting and scale of investment in storage facilities. The objective of such a siting-storage problem might be the minimization of the expected present value of the sum of construction (capital),transport and social costs. The fact that alternative storage locations may influence spill or leakage probabilities from the different generating sites makes the problem especially difficult. The current controversy surrounding the location of nuclear waste repositories also points out the difficulty of finding local communities that would even consider the location of such a facility in their district (see Carter 1987). As opposed to transporting wastes from nuclear reactors used to generate electricity, Chapman (1987) estimates that on-site storage coordinated with decommissioning (shutdown and incasement of contaminated materials when the reactor is "retired") may be the least cost means of storing nuclear wastes.

5. Environmental Policy

We noted earlier that the regulation of environmental externalities could basically take two forms: command and control (C&C) or economic incentives (EI) such as effluent taxes, pollution abatement subsidies or marketable pollution permits. Environmental policy in the U. S. has tended to rely on C&C type policies, specifically emission standards and equipment (technology) standards. Subsidies have been provided to municipalities to construct wastewater treatment plants and firms have been given tax deductions or accelerated depreciation for equipment to reduce air pollution. (The plant construction subsidies are not the same as the per unit subsidy for emission reduction that economists tend to think of when considering economic incentives). In the area of oil spills and toxic wastes recent laws have employed the legal principle of *strict liability* for cleanup costs and damages from accidental spills, while a "cradle-to-grave" regulation has been used to control toxic wastes. These policies have produced some significant improvements in the quality of a few lakes, rivers, and "air sheds" but it is generally thought by economists that the accomplishments have been modest and probably achieved at an unnecessarily high cost. Problems which have not been adequately dealt with include transboundary pollution such as acid rain, groundwater contamination, and the disposal or storage of toxic and radioactive wastes. If the current C&C type policies are excessively costly would EI policies offer higher environmental quality at a lower cost to society? For which types of pollutants would EI policies be most appropriate?

We will consider two types of EI policies: a per unit emission tax and a system of transferable pollution permits which allows the holder to discharge or emit a specified amount of some waste into a particular medium (stream, lake or airshed). As was noted in the third model of static externality, a per unit tax on residual discharge will force the cost-minimizing firm to determine if it can treat the marginal unit of residual at a cost less than the per unit tax. A higher tax rate would presumably induce more treatment, lower discharge and higher ambient quality.

Under a system of marketable pollution permits the environmental agency determines a total amount for some residual which might be discharged into a particular medium without resulting in unacceptable ambient quality. This total amount is then divided into smaller units corresponding to the amount or fraction of an amount which might be discharged by a typical firm generating this residual. Permits, entitling the holder to discharge this smaller unit might then be issued gratis to firms who had historically held emission permits (based on installed equipment) or they could be sold at auction. Once distributed the permits could be sold to another firm discharging into the same stream, lake or airshed. If a new firm wished to locate in the area it would have to purchase the necessary pollution permits from existing firms, thus keeping total discharge at the desired level. The opportunity to sell some of their pollution permits to new or existing firms would create an opportunity cost for firms currently holding permits and make them sensitive to the possibility of selling if they could treat (or avoid generation in the first place) at lower marginal cost.

With either of these EI policies the environmental agency needs to know the relationship between total discharge and ambient quality within the receiving medium. With the pollution tax the environmental agency will also need to know (or subsequently learn by trial-and-error) the relationship between the tax rate and the total level of discharge. There are likely to be factors which cause these functional relationships to be stochastic or change over time. A prime consideration for the success of either policy is having a relatively stable relationship between total discharge and ambient environmental quality.

From our current understanding of the various processes whereby residuals are diffused, dispersed and altered within a disposal medium it would appear that the disposal of organic wastes in lakes, streams and estuaries and particulates in a local airshed are the best understood in terms of total discharge (loading) and the resultant ambient quality. While this understanding is far from perfect it is better developed, for example, than our understanding of the movement of toxics in groundwater or the long range transport of oxides of sulfur or nitrogen. Thus, the use of either taxes or transferable pollution permits might be best suited to the more benign organic wastes disposed via water and certain residuals from combustion which are emitted through smoke stacks. In such cases there would exist a relatively long history on the characteristics of the residuals, their transport and behavior within the disposal medium, and the dimensions (metric) of ambient environmental quality.

Toxic and radioactive wastes, because of their more immediate and severe affects on human health, because of the complex way in which they can be transported through soils and groundwater and because of the inability of the individual to readily determine their concentration and the degree of risk to one's health would seem less suitable to control by EI policies and more appropriately controlled by C&C policies.

There have not been many instances of attempts to control pollution using taxes or marketable permits but the instances where EI policies have been tried would seem to support the above observation that they are better suited to nontoxic, degradable residuals. France and the Netherlands have used pollution fees to pay for treatment and to provide an incentive to reduce waterborne residuals. Bower et al. (1981) note that the fees (taxes) charged to date have been too low to create much of an incentive to reduce effluent loadings, but that with the system of fees in place the opportunity to increase fees to create a stronger incentive for reduced loadings does exists.

Experimentation with taxes or transferable permits in the U. S. is limited and restricted primarily to airborne pollutants. In Wisconsin, however, the state Department of Natural Resources approved regulations whereby pulp mills along the lower Fox River were permitted to transfer permits allowing the discharge of a certain amount of waste into the river. The total amount had been reduced from earlier levels in an attempt to raise dissolved oxygen in the river. It was anticipated that firms that could treat wastes at lower cost might sell their permit to higher cost firms (see O'Neil et al. 1983).

In 1979 amendments to the Clean Air Act allowed the EPA to institute the "bubble policy" where a firm in a particular airshed could transfer its pollution permit to another existing firm or to a new firm wishing to locate within the airshed. The number of transfers has not been large and thus there is not an organized market. Rather, transfer has been accomplished by negotiation between firms (see Hartwick and Olewiler 1986, pp. 443-444 for a discussion).

Finally, in Telluride, Colorado, city officials have put a moratorium on the installation of new wood-burning stoves. Individuals wishing to install a new stove must persuade *two* other residents to give up theirs. In late 1986, the market price to get a resident to give up his permit for a stove was about $1,000 (New York Times, 30 November, 1986, p.1).

While a C&C policy for toxic and nuclear wastes seems appropriate it is still possible for the private market to provide transport and storage where competing firms are subject to strict regulation, inspection, and liability in the event of a spill or leak. Under the cradle-to-grave management concept underlying the Comprehensive Environmental Response, Compensation, and Liability Act of 1980 (also known as CERCLA or the Superfund Act) a detailed accounting must be made of the volume of wastes generated and how it is disposed. The act also provides funds for the cleanup of toxic waste sites with subsequent compensation sought (often via litigation) from those responsible for the toxic site.

Scott (1986) believes that a system of marketable permits should be employed to control acid rain in North America. The U. S. and Canada would negotiate total emission rates for a set of regions in each country. Initially, the total emission rate for each region might be similar to the amount currently emitted. Firms would receive marketable permits which could be transferred to existing or new firms within the region. Permits might be sold to firms outside the region but would be subject to an "environmental exchange rate". (A permit to emit X tons of SO_2 in region i might only permit a firm to emit $X/2$ tons of SO_2 if sold and transferred to region j). Governments, individuals or environmental groups would be free to buy permits from firms in regions thought to contribute to acid deposition in their region and simply retire them. Scott feels that the initial total emission rate for each region should be subject to a negotiated schedule of decline which would be reflected in each permit issued to an individual firm. The individual permit would entitle the holder to a *declining* emission rate reflecting the overall regional decline. Firms wishing to maintain their emission rate into the future would have to acquire more permits at probably higher prices. By buying out the emission permits of existing firms interested parties would be compensating firms for an accelerated reduction in emissions.

In summary, environmental policy in the U. S. has relied primarily on direct regulation of firms and individuals generating externalities. At the encouragement of economists there have been a few attempts at using economic incentives to reduce the level of untreated residuals. While the economic incentives created by taxes or marketable permits might be used to control any type of residual emission they are likely to be implemented in efforts to control the more familiar and benign residuals. Toxic and radioactive wastes are likely to continue to be regulated by command and control policies. Because of the limited success and higher cost of command and control policies it is important for environmental administrators to experiment with economic incentives that achieve the desired level of ambient environmental quality but allow firms to search for the least cost way of reducing emissions.

REFERENCES

Baumol, W. J. and W. E. Oates. 1975. *The Theory of Environmental Policy*. Prentice-Hall, Inc., Englewood Cliffs.

Boulding, K. 1966. "The Economics of the Coming Spaceship Earth". in Henry Jarrett, ed., *Environmental Quality in a Growing Economy*, Johns Hopkins Press, Baltimore.

Bower, B. T., R. Barre, J. Kuher, C. Russell, with A. Price. 1981. *Incentives in Water Quality Management: France and the Ruhr Area*. Johns Hopkins Press, Baltimore.

Carter, L. J. 1987. "U.S. Nuclear Waste Program at an Impasse" *Resources* (Summer): 1-4.

Chapman, D. 1987. "Economic Implications of Reactor Decommissioning for Spent Fuel Disposal". Staff Paper No. 87- 4, Department of Agricultural Economics, Cornell University, Ithaca, New York.

Coase, R. H. 1960. "The Problem of Social Cost". *Journal of Law and Economics* 3(Oct.): 1-44.

Conrad, J. M. 1985. "Residuals Management: Disposal of Sewage Sludge in the New York Bight". *Marine Resource Economics* 1(4): 321-344.

Conrad, J. M. and C. W. Clark. 1987. *Natural Resource Economics: Notes and Problems*. Cambridge University Press, New York.

Hartwick, J. M. and N. D. Olewiler. 1986. *The Economics of Natural Resource Use*. Harper & Row, New York.

Kneese, A. V., R. U. Ayres and R. C. d'Arge. 1970. *Economics and the Environment: A Materials Balance Approach*. Johns Hopkins Press, Baltimore.

Mishan, E. J. 1971. "The Postwar Literature on Externalities: An Interpretative Essay". *Journal of Economic Literature* 9(March): 1-28.

O'Neil, W., M. David, C. Moore, and E. Jones. 1983. "Transferable Discharge Permits and Economic Efficiency: The Fox River". Journal of Environmental Economics and Management 10(4): 346-355.

Organization for Economic Co-operation and Development. 1974. *Problems in Transfrontier Pollution*. OECD, Paris.

_____. 1976. *Economics of Transfrontier Pollution*. OECD, Paris.

Samuelson, P. A. 1954. "The Pure Theory of Public Expenditure". *Review of Economics and Statistics* 36(4): 387-389.

_____. 1955. "Diagrammatic Exposition of a Theory of Public Expenditure". *Review of Economics and Statistics* 37(4): 350-356.

Scott, A. D. 1986. "The Canadian-American Problem of Acid Rain". *Natural Resources Journal* 26(2): 337-358.

Spence, A. M. and D. Starrett. 1975. "Most Rapid Approach Paths in Accumulation Problems". *International Economic Review* 16(June): 388-403.

STOCHASTIC NONLINEAR OPTIMAL CONTROL OF POPULATIONS:
Computational Difficulties and Possible Solutions

Christine A. Shoemaker
School of Civil & Environmental Engineering and
Center for Applied Mathematics
Cornell University
Ithaca, NY 14853

Sharon A. Johnson
Department of Management
Worcester Polytechnic Institute
Worcester, MA 01609

Abstract

Computing optimal control policies for the stochastic nonlinear systems associated with population dynamics is computationally and mathematically difficult. We discuss the characteristics of population management problems arising in forestry, fisheries and pest management and how these characteristics affect the selection of optimal control algorithms and the associated computations. We compare the advantages and disadvantages of differential dynamic programming and stochastic dynamic programming. The computational issues associated with linear or nonlinear interpolation for stochastic dynamic programming are discussed. We conclude that for very large-scale nonlinear problems like those arising in forestry, differential dynamic programming would seem to be the most effective method. Most pest control problems require stochastic dynamic programming and nonlinear interpolation. Because of the high degree of parameter uncertainty for fisheries models, these problems require a stochastic procedure like dynamic programming, but in many cases linear interpolation will be adequate.

1. Optimal Control of Pests, Fisheries and Forests

Optimization and simulation methods can assist in the analysis of the management and control of populations, including applications arising in pest control, fisheries management and forestry (Shoemaker, 1981; Curry and Feldman, 1987; Grant, 1987; Clark, 1976). Population control is a dynamic problem, so we will define the following time-varying terms:

x^t state vector
u^t control vector
w^t random variable vector
$R^t(x^t, u^t)$ the net value accrued in period t given the system is in state x^t and the control decision is u^t
n   dimension of the state vector
m   dimension of the control vector
K   number of time periods

The function R is usually an economic function such as the value of the harvest minus the cost of the management decision. The mathematical optimization problem may be stated as:

$$\text{Maximize Expectation} \quad \{ \sum_{t=1}^{K} R^t(x^t, u^t) \} \tag{1.1}$$

$$u^t$$
$$t=1,\ldots,K$$

where x^0 is given and

$$x^{t+1} = T(x^t, u^t, w^t) \tag{1.2}$$

The random vector w^t has a probability distribution, which we will assume is discrete so the probability distribution is

$$P\{w^t = \eta_k\} = p_{kt} \qquad k = 1,\ldots,D \tag{1.3}$$

for specified values of η_k. We would like to develop an adaptive control so that the control at time t is based on an observation of the state x^t at time t or

$$u^{*t} = u^{*t}(x^t)$$

where * denotes the optimizing value of the control.

Optimization methods will maximize (1.1) by using the mathematical structure of the problem to identify the optimal policy. Simulation models may also be used to identify optimal solutions to (1.1) by enumerative search over every possible policy u^t, for t = 1,...,K (e.g. Holling, 1978; Walters, 1969; Reichelderfer and Bender, 1979; Onstad and Shoemaker, 1984). However, if each component of u can assume M values, then the total number of possible values of the control policy is M^{mK}. If Monte Carlo simulation is employed, then the total number of simulations required is $M^{mK}*S$, where S is the number of patterns of random variables w^t, t=1,...,K employed in the Monte Carlo simulation. For a simple case (e.g. M =2, m = 1, and K is small, i.e less than 5) the number of possible values of the control is not large; and hence enumerative searching with a simulation model is a reasonable strategy. However, for cases with a moderate number of time steps, control variables, or discrete values of the control, $M^{mK}*S$ becomes too large for enumerative simulation to be a computationally feasible way to evaluate the policy space. It is these types of problems for which we need to employ optimization analysis.

A major issue in determining the computational feasibility of a specific type of optimization procedure is the dimension of the state and control vectors. For population management problems, the state vector typically consists of the numbers of individuals in each subclass of each species that is considered in the analysis. Hence,

$$x^t= (s^1_{1},s^1_{2},\ldots,s^1_{n1};s^2_{1},s^2_{2},\ldots,s^2_{n2};\ldots;$$

$$s^k_{1},s^k_{2},\ldots,s^1_{nk}) \tag{1.4}$$

where the superscript refers to the species and the subscript refers to the subclass. Examples of subclasses include age or stage classes (such as egg, larval or adult classes), or plant parts (leaves, stems, fruit, roots), or classes describing other heterogeneities such as spatial distribution or genotypes.

For pest management applications the primary species are the pest(s), and the crop. In some cases, a species such as a parasite or predator that attacks the pest will also be included in the analysis. Usually the control variable u^t is the decision whether or not to apply a pesticide, hence the decision variable is a one-dimensional binary variable. It is possible to consider other decisions such as the amount of pesticide applied, but typically the fixed cost of applying a pesticide (costs for application equipment and labor) are so high that usually either no pesticide is applied or the maximum dosage is applied. The function R incorporates the economic value of the damage caused by the pest and the cost of pesticide application. Because most nonlinear programming methods cannot consider binary variables or discontinuous cost functions, previous applications of nonlinear programming to pest management have ignored the fixed cost of pesticide applications (Talpaz et al., 1978; Regev, 1976). Dynamic programming can incorporate fixed costs and binary decision variables.

For forestry problems, the state variables are one or more competing tree species divided by age class and by spatial location, and the decision variable is the locations at which harvesting should occur. Unless clear cutting is the practice, the decision variable may also include information about which age classes of which tree species are cut. In fisheries management, the components of the state vector are one or more fish species that are divided by age class and the decision variable is usually the amount of fishing effort.

Developing the equations which comprise the transition equation T in (1.2) is usually quite difficult because of the data requirements necessary to parameterize the model and to test the validity of the model's description of essential mechanisms of population dynamics. The topics of model development (including design of experiments), verification, and validation are complicated issues that will not be addressed in this paper. We will assume in the following discussion that a description of the dynamics of the system exists as described by $T(x^t, u^t, w^t)$ and that the stochasticity of the dynamics is incorporated in the random vector w^t.

The differences in the type of optimization problem arising in the three areas of pest control, fisheries, and forestry is illustrated in Table 1. The table demonstrates that whereas there is a similar mathematical structure to the control problems arising in these three areas (as given by equation 1.1), there are notable differences in the magnitude of the computational problems that arise when attempting to compute the solutions to these problems.

Pest control problems would appear to be the most difficult from a computational point of view since these are large dimensional, stochastic, nonlinear problems. That the decision variable is binary is a computational advantage for some optimization schemes like dynamic programming, but it is also a disadvantage for other optimization schemes because a large class of nonlinear programming algorithms cannot be used to solve problems with a binary control vector.

Although they have a large number of variables, forestry problems that can be described adequately by deterministic linear transition functions can be solved efficiently by linear programming. Forest (as opposed to agricultural) pest control problems have some of the more difficult aspects of both forestry and agricultural pest control problems. Whereas, agricultural pest control decisions are made over a relatively small area, forest pest control sometimes requires decisions for thousands of acres. Hence, although spatial variation is not a major issue for many agricultural pest control problems, it is very important for migratory pests like the spruce budworm (Stedinger, 1977 and 1984; Fleming et al., 1984). In addition forest pest control problems are nonlinear.

TABLE 1

Differences in Optimization Needs
Among Three Application Areas

	Pest Management	Fisheries	Forestry
Importance of Stochastic Factors	high	medium	low
Dimension of State Variable	medium (10-100)	low (< 5)	high (> 20)
Transition Functions May be Approximated by Linear Function	no	unlikely	possibly
Decision Variable	binary	continuous	continuous or binary
Parameter Uncertainty	low	high	low

Fisheries problems tend to have fewer variables in the state vector, not because the population dynamics are less complex than for forestry or pest control, but rather because it is so difficult to collect data that can be used to develop parameter values for high dimensional models that utilize a large number of parameters. As a result most models used in fisheries have a relatively small number of state variables. Terrestrial systems like agricultural pest control and forestry are easier to study experimentally. Although there is parameter uncertainty for these systems as well, it is not as great as for fisheries.

2. Stochastic Dynamic Programming

A powerful technique for computing the optimal adaptive control of populations subject to nonlinear and stochastic dynamics is Dynamic Programming (e.g. Anderson, 1975;

Shoemaker, 1973 and 1982; Stedinger, 1977; Feldman and Curry, 1982; Walters, 1978). In this section we will describe the formulation of dynamic programming problems for computing the control of populations and will discuss some of the computational obstacles associated with the use of this method for population control. First we define the future value function:

$$F^t(x^t) = \underset{u^t}{\text{Maximum Expectation}} \left[\sum_{s=t}^{K} R^s(x^s, u^s) \right] \qquad (2.1)$$

which leads to the dynamic programming recursive relationship

$$F^t(x^t) = \underset{u^t \ w^t}{\text{Max Exp}} \left[R^t(x^t, u^t) + F^{t+1}(T(x^t, u^t, w^t)) \right] \qquad (2.2)$$

with the end condition

$$F^N(x^N) = \underset{u^t}{\text{Max}} \left[R^N(x^N, u^N) \right] \qquad (2.3)$$

In the case where R^t is deterministic and the random variable w^t has a discrete probability distribution (Equation 1.3), the recursive equation becomes

$$F^t(x^t) = \underset{u^t}{\text{Max}} \left[R^t(x^t, u^t) + \sum_{k=1}^{N_k} P_{kt} \ F^{t+1}(T(x^t, u^t, \eta_k)) \right] (2.4)$$

To understand the computational problems associated with dynamic programming, it is necessary to describe briefly how these problems are solved. The recursive equation given by (2.4) is solved by discretizing the state space and defining the following variable:

x_{ij} - the jth discrete value of the ith component x^t_i of the state vector x^t.

The only values for which the future value function is computed are elements of the set S^t,

$$S^t = \{ (x^t_{1k_1}, x^t_{2k_2}, \ldots, x^t_{nk_n}) ; \ k_1 \in I(N_1), \ldots, k_n \in I(N_n) \} (2.5)$$

where $I(N_i)$ is the set of integers between 1 and N_i. The number of elements of S^t is then

$$\prod_{i=1}^{n} N_i \qquad (2.6)$$

The recursive equation

$$F^t(x^t) = \underset{}{\text{Max}} \ \{ R^t(x^t, u^t) + \sum_{k=1}^{D} P_{kt} \ F^{t+1}(T(x^t, u^t, w_k)) \} \qquad (2.7)$$

is solved only for the discrete values $x^t \in S^t$. Equation (2.7) is solved backward in time by first computing $F^N(x^N)$. The values of F^t are substituted into the recursive equation to find F^{t-1} for t = N, N-1, N-2, ..., 2.

3.   Overcoming the Curse of Dimensionality

If no numerical interpolation or gradient search for the control is implemented, the computation time for solving the dynamic programming problem given by Equation (2.7) is proportional to

$$(\prod_{i=1}^{n} N_i) (\prod_{k=1}^{m} M_k) K \qquad (3.1)$$

where

  N_i - number of discrete values of the ith state variable

  M_k - number of discrete values of the kth control variable

and n,m, and K are defined after (1.1). For example, we recently solved a problem with n = 4, N_i = 15, M_k = 10, m = 3 and K = 50 on an IBM 3090 in about 25 minutes of CPU on an IBM 3090 (Logan and Shoemaker, manuscript). From (3.1) we see that computational time will increase N_i times for each increase in the state variable dimension n. For the example above, if N_i, M_k, m, and K remain constant, then solving a problem with a state dimension of n=10 could be expected to take 4.7 million hours! Hence, we see that dynamic programming applications are limited by the geometric increase in computational effort associated with increases in state variable dimension, a characteristic that Bellman termed the "curse of dimensionality" (Bellman and Dreyfus, 1962).

There are a number of approaches that have been used to combat the computational problems associated with the application of dynamic programming to systems with a large number of variables. The following discusses some approaches that are appropriate for population problems.

Reducing the Value of n With Closed Form Solutions to Submodels

It is clear from Equation (3.1) that reductions in the state variable dimension n will have a geometric impact on computational requirements. It should be noted from Table 1 that a number of fisheries applications have sufficiently small state variable dimension n that additional reductions are not necessary. However, for many problems especially those arising in pest control and forestry, n is sufficiently large to make stochastic dynamic programming computationally infeasible. One way to make n smaller is to develop a simplified version of the system being studied in hopes that it captures the essence of the system dynamics. However, this approach is not always successful since important aspects of ecological interaction may be omitted.

A second approach is to attempt to develop submodels of a large number of state variable components that can be solved analytically in terms of relatively few independent variables. In this way the analytical models can be nested in an optimization model with fewer variables. For example, Shoemaker (1982) used this approach to substitute a three state variable model for a pest-parasite-crop ecosystem that would have required at least 10 state variables to describe if an age class approach had been used. The submodel described the change in age structure within the season as a function of temperature, pesticide application and harvest dates. Shoemaker and Onstad (1983) demonstrated that the three variable model closely mimicked the behavior produced by a much more complicated simulation model upon which the analytical model was based. The closed form solutions of the nonlinear analytical models could then be used to compute the dynamic programming objective function on the basis of the three state variables. As illustrated in the example above, reducing the number of variables from 10 down to 3 reduces the computational requirements by as much as a million-fold, or in this case changes the problem from one that is not computationally feasible to one that can be solved relatively inexpensively.

The analytical approach is appealing to mathematicians and is highly effective computationally. Unfortunately it is not always feasible since our ability to solve submodels analytically depends upon the specific mathematical structure of the system. There are many situations for which the analytical approach will not yield any improvement in the number of independent state variables.

Reducing n by Using Policy Space in $F^t(x^t)$

Another way to try to reduce the state variable dimension n is to replace the state vector describing age classes and species densities by a description of past policies. For example, Shoemaker (1979) describes a procedure for analyzing pest control problems where the vector x^t in $F^t(x^t)$ contains the dates of the last two pesticide applications. Hence,the number of possible values of x^t is relatively small. For a deterministic problem, the actual population density and age structure can be computed from the information about the times of previous applications and input information about migration and oviposition. This approach was later extended to stochastic problems (Shoemaker, 1984). To incorporate the stochastic impact of weather on pesticide residual, it is necessary to estimate the expected population size and age distribution based on previous times of application and weather.

The limitations of the use of policy space in the future value function $F^t(x^t)$ is similar to the limitations associated with the use of analytical functions: the procedure is only applicable for specific mathematical structures. In addition, the use of policy space in $F^t(x^t)$ is usually much more difficult to implement for stochastic problems than for deterministic ones.

Differential Dynamic Programming

A variation on conventional dynamic programming is "differential dynamic programming" which was first introduced by Mayne (1966) and was later developed in a book by Jacobson and Mayne (1970). This technique is a hybrid between dynamic programming and nonlinear programming techniques. It shares with dynamic programming the transformation of a single dynamic optimization problem into many optimization problems over a single time step. This procedure is effective at reducing computational requirements for dealing with a large number of time periods. Both differential dynamic programming (DDP) and conventional dynamic programming have computational requirements that are linearly related to the number of time steps (e.g. Equation (3.1)), which is a much smaller computational requirement than that associated with most nonlinear programming algorithms. DDP has an additional advantage over conventional dynamic programming in that DDP does not require discretization of the state variables. However, like other nonlinear programming techniques, DDP is iterative so that the algorithm repeats a number of times and only stops when it is determined that the current answer is sufficiently close to the optimum. The computational requirements for DDP are proportional to:

$$\text{Max } (n^3, m^3) \text{ K } *(\text{number of iterations}) \qquad (3.2)$$

(Liao and Shoemaker, manuscript), where n,m and K are as defined for equation (1.1-2). For conventional dynamic programming the number of iterations is 1. For DDP the number of iterations required is usually unknown before the computations begin, and hence it is more difficult to predict computational difficulty for DDP. In the authors' experience, the number of iterations has ranged from 2 to 300 for DDP. Hence for large n, comparison of (3.2) and (3.1) indicates that DDP will usually be much more efficient since N_i needs to be at least 3 and is usually 10 or more. DDP has been used to solve problems with 40 state variables and 5 time steps (Yakowitz and Rutherford, 1984), with 238 state variables and 8 time steps (Jones et al., 1987), and with 32 state variables and 20,000 time steps (Liao and Shoemaker, unpublished). The Jones et al. application had linearized transition equations which substantially reduced the number of necessary iterations. The other two references are for nonlinear problems that, like most population control problems, do not lend themselves to linearization. These applications of DDP represent much larger state dimension than any computational applications of conventional dynamic programming, which substantiates the difference in computational effort that would be expected by comparison of equations (3.2) and (3.1). Hence we see that differential dynamic programming is a tool that can be important in solving a range of nonlinear dynamic problems including those arising in population management.

Unfortunately, DDP also has limitations. Most importantly, it is computationally efficient only for deterministic problems. Although Jacobson and Mayne (1970) introduce a stochastic version of DDP, it is not computationally feasible for a large number of time steps because computation grows exponentially with the number of time steps in their scheme. In addition, DDP is designed to consider problems with continuously valued decision variables and differentiable objective and transition functions. As was mentioned earlier, pest control problems either have objective functions with fixed costs (which means they are not differentiable at zero) or have binary decision variables. Hence, the application of DDP is limited for pest control problems. DDP has potential application for fisheries management problems that require a substantial number of state variables (problems with few state variables can be solved by conventional DP while considering stochastic factors) and for forestry problems that require nonlinear transition and objective functions.

Reducing N_i by Interpolation of $F^t(x^t)$

From (2.5) we have defined N_i as the number of discrete values of x_i, the ith component of the state vector and defined the set S^t, which is the set of all values of the state vector for which each component is equal to one of the discrete values. As the DP algorithm works back in time t, the following equation for $F^t(X^t)$ is solved for each $X^t \in S^t$:

$$F^t(x^t) = \underset{u^t}{\text{Max}} \left[R^t(x^t, u^t) + \sum_{k=1}^{D} P_{kt} F^{t+1}(T(x^t, u^t, \eta_k)) \right] \quad (3.3)$$

Note that S^t contains $\prod_{i=1}^{n} N_i$ elements. Unfortunately, in general we have no guarantee that $T(x^t, u^t, \eta_k)$ is an element of S^t. In typical applications, the approach is then to interpolate the value of $F^{t+1}(T(x^t,u^t,\eta_k)$ from nearby values of $x \in S^{t+1}$.

To understand how the interpolation is done, consider a two dimensional problem where $x^{t+1} = (x_1, x_2) = T(x^t, u^t, w_k)$. Then assume that x_{11}, x_{12}, x_{21}, and x_{22} are the nearest co-ordinates to (x_1, x_2) and $P(x_1, x_2)$ is the interpolated value of $F^{t+1}(x^{t+1})$ evaluated at $x^{t+1} = (x_1, x_2)$. Then it follows that

$$P(x_1, x_2) = a_{1g} + a_{2g} (x_1 - x_{11}) + a_{3g} (x_2 - x_{21})$$
$$+ a_{4g} (x_1 - x_{11})*(x_2 - x_{21}), \text{ for } (x_1, x_2) \in \text{ grid } g \quad (3.4)$$

with the requirement that $P(x_{\ell k}, x_{ji}) = F^{t+1}(x_{\ell k}, x_{ji})$ if $(x_{\ell k}, x_{ji})$ is one of the grid points contained in S^t. Hence for each grid g determined by four points, we must solve the following problem for a_{jg}:

$$\begin{bmatrix} 1 & 0 & 0 & 0 \\ 1 & (x_{12}-x_{11}) & 0 & 0 \\ 1 & 0 & (x_{22}-x_{21}) & 0 \\ 1 & (x_{12}-x_{11}) & (x_{22}-x_{21}) & (x_{12}-x_{11})(x_{22}-x_{21}) \end{bmatrix} \begin{bmatrix} a_{1g} \\ a_{2g} \\ a_{3g} \\ a_{4g} \end{bmatrix} = \begin{bmatrix} F^{t+1}(x_{11}, x_{21}) \\ F^{t+1}(x_{12}, x_{21}) \\ F^{t+1}(x_{11}, x_{22}) \\ F^{t+1}(x_{12}, x_{22}) \end{bmatrix}$$

$$(3.5)$$

In this example the dimension of the state vector is n = 2 and the above operation requires 8 floating point operations (flops). In general for multilinear interpolation of an n dimensional function space, the number of flops required to find the coefficients a_{jg} will be $2^{(2n-1)}$. Evaluation of $P(x_1,x_2)$ then requires an additional N-1 flops.

The value of the multilinear approximant $P(x_1,x_2)$ can also be determined without explicitly calculating the coefficients a_{ijk} using tensor product methods (Johnson, 1989, pp. 100-101, Section 2.4.2.1, and Section 3.2.1.2, pp. 163-166). For an n-dimensional function space, the number of flops required for this evaluation is $n + 2(2^n-1)$ flops.

4.   Using Splines for Interpolation of Dynamic Programming

Because of the large number of grid points that are necessary to insure the accuracy of multilinear interpolation, a group at Cornell including the authors and J. Stedinger have been considering the use of higher order piecewise polynomial functions to interpolate the future value function $F^t(x^t)$ arising in stochastic dynamic programming functions. By using multidimensional cubic splines, we are able to approximate $F^t(x^t)$ with fewer grid points and to insure that the approximation is twice differentiable.

The smoothness of the splines can speed up the search for the optimal value of u^t, denoted by u^{*t}, for equation (2.4). Nonlinear programming methods can be used to compute u^{*t} if $F^{t+1}(T(x^t,u^t,w^t))$ is a smooth function, as it is when approximated by splines. When $F^{t+1}(\cdot)$ is approximated by multilinear interpolation, the function will only have first degree continuity, so that the first partial derivatives are discontinuous at the boundaries of each grid element. This smoothness is not an advantage for pest control problems where the control vector is binary since nonlinear programming requires the control variable to have continuous values. The smoothness of $F^t(x^t)$ can be an advantage for problems arising in fisheries and forestry, which do have continuous control variables (see Table 1).

Evaluating a spline requires more computational effort than multilinear interpolation for a given level of discretization. Thus, if both multilinear interpolation and spline approximation are applied to a problem for which each of n state variables are discretized into N nodes, it will take more time to compute the solution with splines. For smooth functions or continuous functions with some curvature, it is expected that the accuracy of the spline approximation will be better, however.

It is more significant to consider the computational effort required to achieve the same level of accuracy in the approximation of the FVF $F^{t+1}(\cdot)$.   Assume that we are considering a problem in which all the elements of the state vector are discretized into the same number of values. For the multilinear interpolation let $N_i = N$ for all i and for the spline interpolation, let $N_i = N'$ for all i. Assume further that N' is selected so that the same level of accuracy is achieved in the approximation of F^{t+1} using the spline with N' discrete levels and multilinear interpolation with N discrete values. We expect that N/N'

> 1 for most functions, but we do not know its exact value. The larger the ratio N/N' is, the more effective the spline is in approximating the function with fewer points.

Since we do not know the exact value of this ratio and it will vary among different types of functions, it is appropriate to ask what the smallest ratio N/N' is such that the computational effort required to use splines does not exceed that required for linear interpolation in a DP algorithm. Table 2 gives this ratio $Y(n)$ as a function of the dimension of the state vector n based on relationships developed in Johnson (1989). Table 2 indicates that the ratio $Y(n)$ diminishes as n increases. This means that the advantages of spline interpolation improve with the dimension of the problem. It is unlikely, for example, that six times as many grid points are required to approximate $F^t(x^t)$ to the same degree of accuracy with multilinear interpolation as for spline interpolation. Hence the results of Table 2 indicate that for a one dimensional problem, it is unlikely that spline approximation will save on computational time. However, it is quite reasonable to assume that one can obtain the same level of accuracy from splines with N'points as from multilinear approximation with 3N'or even 4N'grid points.

The expected advantage of splines can be determined by considering the ratio of estimated computational effort required to solve a DP problem with multilinear interpolation divided by the estimated computational effort with splines. Under certain conditions detailed in Johnson (1989), the computational advantage can be shown to equal:

$$\text{spline computational advantage} = A(n, N/N') = [(N/N')/Y(n)]^n \qquad (4.1)$$

where $Y(n)$ is given in Table 2. It should be noted that the total computational advantage is not incorporated in A, which does not consider a) the effort required for comparisons, b) the decrease in computation time due to the use of nonlinear programming for solving the one stage optimization problem (2.2), or c) some other factors which may change with n, N and N'. The calculation of $Y(n)$ omits $O\{n\}$ terms. The degree to which N/N' is greater than the value in Table 2 will influence the computational advantage. For example assume that N/N' = 3 (a fairly conservative assumption). Then it follows that the predicted computational advantage for n = 4 or 10 is:

$$A(4,3) = (3/2.632)^4 = 1.687 \qquad A(10,3) = 3/2.232)^{10} = 19.24$$
$$\text{if } n = 4 \qquad\qquad \text{if } n = 10 \qquad (4.2)$$

when N/N' = 5, the computational advantage becomes:

$$A(4,5) = (5/2.632)^4 = 13.02 \qquad A(10,5) = (5/2.232)^{10} = 3182$$
$$\text{if } n = 4 \qquad\qquad \text{if } n = 10 \qquad (4.3)$$

TABLE 2

Breakeven values Y(n), which are the minimum ratio of multilinear interpolation points to spline interpolation points necessary to make computational efforts equal. O{n} terms are ignored. Tensor product interpolation and the same grid search algorithm is used for both multilinear and spline procedures in the DP algorithm.

State Variable Dimension n	Y(n)
1	6.00
2	3.464
3	2.884
4	2.632
5	2.491
10	2.232

The computational advantages of splines over multilinear interpolation have been analyzed in much greater depth in Johnson (1989) than is possible to describe in this brief paper. The procedure has not been applied to population problems but shows promise for these problems as well.

In applications to stochastic dynamic programming problems arising in reservoir operation, Johnson (1989) was able to show that the spline approximation (with 4 grid points per dimension) was able to compute the DP solution with better accuracy in only 2% of the time required to solve the problem with multilinear interpolation (and 13 grid points per dimension). Hence for this problem N/N' > 3.2. and the computational advantage is about 50. This actual computational advantage is greater than that predicted by (4.1) and reflects savings in computational time due to the use of nonlinear programming as well as from the factors incorporated in (4.1). The problems arising in reservoir operations reflect systems with linear dynamics and upper bounds on state variable values. As a result the future value function tends to be piecewise differentiable. By contrast, the nonlinear problems arising in population dynamics do not tend to have strict upper bounds on state variable size and restrict amplitude through nonlinear functions describing density dependent factors. It would be expected that the future value function $F^t(x^t)$ would be smoother for population control problems than for the $F^t(x^t)$ arising in reservoir operations, and hence would be even more suited to approximation by smooth functions like splines. This would imply that we could hope that N/N' would be larger for nonlinear population problems although computational studies will be required to verify this conjecture.

5.   Conclusions

The use of optimization methods can be useful to identify policies for population management, but the applications of these methods to realistic problem formulations are limited not only by the data requirements for parameterizing such models, but also the computational difficulties associated with solving stochastic optimization problems for nonlinear dynamical systems. There are different approaches that can be used to address such problems. Each approach has its advantages and disadvantages, and the relative advantage of one method over another depends upon the structure and size of the specific population management problem being considered. Table 1 indicates that the types of problems arising in Pest control, Fisheries, and Forestry display some differences in the problem characteristics that affect the computational feasibility of various approaches. The difficulties of fisheries management analysis center on the large uncertainty associated with parameter values. The size of the state variable is typically small enough for fisheries problems so that stochastic dynamic programming is computationally feasible and can be used if the parameter values can be estimated accurately. For forestry, however, the dimension of the state variable is often so large that dynamic programming is not computationally feasible for many problems. Even the advances in computational efficiency that we would hope to achieve with the spline interpolation are unlikely to enable us to solve conventional dynamic programming problems with greater than 10 or 20 variables given current computing power. Forestry problems typically have more than twenty variables. However, differential dynamic programming would appear to have potential for large dimensional deterministic nonlinear problems arising in forestry. Linear deterministic problems in forestry have already been effectively addressed by linear programming. Pest control problems are nonlinear, stochastic, and typically have a state vector dimension in the middle ranges. As a result pest control problems are also too large in dimension for conventional dynamic programming. In addition they cannot utilize differential dynamic programming because the control variable is binary. For this class of problems, the approach of reducing the state variable dimension with analytical models or a policy space approach coupled with the use of a nonlinear interpolation scheme and conventional dynamic programming appears to show the most potential.

REFERENCES

Anderson, D.R. (1975). "Optimal Exploitation Strategies for an Animal population in a Markovian Environment: A Theory and an Example." Ecology 56: 1281-1298.

Bellman, R.E. and S.E. Dreyfus (1962). Applied Dynamic Programming. Princeton Univ. Press, Princeton, N.J.

Clark, C.W. (1976). Mathematical Bioeconomics: The Optimal Management of Renewable Resources. Wiley-Interscience, New York.

Curry, G.L. and R.M. Feldman (1987). Mathematical Foundations of Population Dynamics. Texas A & M University Press, College Station.

Feldman, R.M. and G.L. Curry (1982). "Operations Research for Agricultural Pest Management." Oper. Res. 30: 601-618.

Fleming, R.A., C.A. Shoemaker, and J.R. Stedinger (1984). "An Assessment of the Impact of Large Scale Spraying Operations on the Regional Dynamics of Spruce Budworm Populations." Can. Ento. 116: 633-644.

Grant, W.E. (1987). Systems Analysis and Simulation in Wildlife and Fisheries Sciences. Wiley-Interscience, New York.

Holling, C.S. (ed.) (1978). Adaptive Environmental Assessment and Management. Wiley-Interscience, New York.

Jacobson, D. and D. Mayne (1970). Differential Dynamic Programming. Elsevier Science Publishers, New York.

Johnson, S.A. (1989). Spline Approximation in Discrete Dynamic Programming With Application to Stochastic Reservoir Systems. Ph.D. Thesis, Cornell Univ., Ithaca, New York.

Jones, L., R. Willis and W. Yeh (1987). "Optimal Control of Nonlinear Groundwater Hydraulics Using Differential Dynamic Programming." Water Res. Res. 23: 2097-2106.

Liao, L-Z and C.A. Shoemaker (manuscript). Convergence in Unconstrained Discrete-Time Differential Dynamic Programming.

Logan, C.J. and C.A. Shoemaker (manuscript). Managing Conjunctive Use Irrigation Systems II: A Case Study in the Indus Plains of Pakistan.

Mayne, D. (1966). "A Second-order Gradient Method for Determining Optimal Trajectories of Nonlinear Discrete-time Systems." Intnl. J. Control 3: 85-95.

Onstad, D.W. and C.A. Shoemaker (1984). "Management of Alfalfa and the Alfalfa Weevil (Hyper postica Gyll.): An Example of Systems Analysis in Forage Production." Agric. Sys. 14: 1-30.

Regev, U., A.. Gutierrez, and G. Feder (1976). "Pests as a Common Property Resource: A Case Study of Alfalfa Weevil and Control." Amer. J. Agr. Econ. 58: 188-196.

Reichelderfer, K.H. and F.E. Bender (1979). "Application of a Simulative Approach to Evaluating Alternative Methods for the Control of Agricultural Pests." Amer. J. Agr. Econ. 61: 258-267.

Shoemaker, C.A. (1973). "Optimization of Agricultural Pest Management III: Results and Extensions of a Model." Math. Bio. 18: 1-22.

Shoemaker, C.A. (1979). "Optimal Timing of Multiple Applications of Pesticides With Residual Toxicity." Biometrics 35: 803-812.

Shoemaker, C.A. (1981). "The Application of Dynamic Programming and Other Optimization Methods to Pest Management." IEEE Trans. Auto. Control 26: 1125-1132.

Shoemaker, C.A. (1982). "Optimal Integrated Control of Univoltine Pest Populations With Age Structure." Oper. Res. 30: 40-61.

Shoemaker, C.A. (1984). "Deterministic and Stochastic Analyses of the Optimal Timing of Multiple Applications of Pesticides with Residual Activity." In Pest and Pathogen Control: Strategy, Tactics, and Policy Models, G.R. Conway (ed.) Wiley-Interscience Publishers, Chichester.

Shoemaker, C.A. and D.W. Onstad (1983). "An Optimization Analysis of the Integration of Biological, Cultural, and Chemical Control of Alfalfa Weevil." Envir. Ento. 12: 286-295.

Stedinger, J.R. (1977). "Spruce Budworm Management Models." Ph.D. Thesis, Harvard University, Cambridge, Massachusetts.

Stedinger, J.R. (1984). "A Spruce Budworm-Forest Model and Its Implications for Suppression Programs." Forest Sci. 30: 597-615.

Talpaz, H., G.L. Curry, P.J. Sharpe, D.W. DeMichele and R.E. Frisbie (1978). "Optimal Pesticide Application for Controlling the Boll Weevil on Cotton." Amer. J. Agr. Econ. 60: 469-475.

Taylor, C.R. and J.C. Headly (1975). "Insecticide Resistance and the Evaluation of Control Strategies for an Insect Population." Can. Ento. 107: 237-242.

Walters, C.J. (1969). "A Generalized Computer Simulation Model for Fish Population Studies." Trans. Am. Fish. Soc. 98: 505-512.

Walters, C.J. (1978). "Some Dynamic Programming Applications in Fisheries Management." In M. Puterman (ed.), Dynamic Programming and Its Applications, Academic Press, New York.

Yakowitz, S. and B. Rutherford (1984). "Computational Aspects of Discrete-time Optimal Control." Appl. Math. Comput. 151: 29-45.

OPTIMAL EVOLUTION OF TREE-AGE DISTRIBUTION FOR A TREE FARM

Henry Wan, Jr.
Economics Department
Cornell University
Ithaca, NY 14853

Abstract

We characterize here the evolution of tree age distribution under optimal forest management. The object is to maximize the discounted sum of the yield in each period, this being an increasing, concave function of the timber harvested. For a family of simple models, it is found that tree age distributions always evolve into either a 'Faustmann uniform distribution' or 'Faustmann cycles' where all trees are cut at the Faustmann age.

Contrary to the conclusions from the optimal growth literature, cyclical orbits can be optimal under any small but positive discount rate. Furthermore, such cyclicality is structurally stable, due to the reacheability constraints as implied by the underlying vintage asset structure.

1. An Overview

Traditionally economists illustrate capital theory with the optimal timing of cutting a tree in isolation [e.g., Jevons (1957), Wicksell (1934)]. The work of Faustmann (1849), on the other hand, considers the site value when reforestation calls for a shorter optimal tree life. In all such analyses, the yield from harvest is assumed to be proportional to the timber obtained. When one considers forests rather than single trees, it is natural to ask what happens, if diminishing returns apply to the timber harvested. By simulation, Kemp and Moore (1979) find for a continuous-time case, that optimal tree age distribution converges to a uniform Faustmann distribution. Mitra and Wan (1985, 1986) have deduced for the discrete time world, that the rate of discount matters greatly. With no discounting, a linear yield function calls for Faustmann cycles, and a strictly concave yield leads to a uniform Faustmann distribution. With discounting, all that is known then is that examples exist where both the uniform Faustmann distribution and the Faustmann cycles are consistent with a strictly concave yield function. Related results have been reviewed both by Clark (1976) and by Dasgupta (1982).

In this study, we will relate the evolutionary pattern of tree ages to both the growth profile and the discount rate, for a class of simple examples. To make our analysis transparent, we assume that the tree life may be adequately represented with two equal periods and the periodic yield is logarithmic in the quantity of timber harvested.

Analytically, the novelty is that unlike the optimal growth models [e.g., Cass and Shell (1976)], there is no positive threshold rate of discount here, below which the convergence to a uniform Faustmann distribution is guaranteed.

At the 'physical' level, what makes the difference is the underlying vintage capital structure. Trees are replaced by harvest and reforestation; the growth process is unalterable by human action. Thus, starting from an arbitrary age distribution, there are always distributions which can never be reached next period. For example, any distribution where all trees are of age n > 0, can never be succeeded by the same distribution next period. One cannot expect all trees are n periods old, next period, if none is n-1 periods old at the present. This gives rise to the reacheability constraints, in terms of state-to-feasible controls correspondence. Such restrictions are all important in the network flow type of dynamic programming, but absent from the optimal growth models a la Ramsey. The effect of such constraints in forest management is seen below.

To begin with, the uniform Faustmann distribution has two desirable properties. Each tree is cut at an age maximizing its growth potential. Each period, an equal amount is harvested, so as to minimize the effect of diminishing returns. Moreover, it is always reacheable from any initial age distributions in a finite number of periods, e.g., through an immediate clearcutting, followed by a sequence of harvests of equal acreages. However, it is possible that from anywhere along an optimal path, there is no justifiable sequence of steps leading to the uniform Faustmann distribution. Some sequence of steps is physically unreacheable; others call for the harvest of trees prior to the Faustmann age. Such current sacrifice in potential tree growth may outweigh the present value of the stream of benefits from reaching the uniform Faustmann distribution. The Optimal Faustmann cycle arises from the repetitive interplay of both the physical and the economic considerations outlined above.

To be sure, there are optimal cycles in the literature which do not involve the reacheability constraints [see, e.g., Benhabib and Nishimura (1985)]. Furthermore, in our examples, the feature which seems to intrigue economists most is the absence of a positive threshold discount rate. This is also shared by an early example of Weitzman [see Samuelson (1973)]. However, Benhabib and Nishimura have shown that such cycles disappear under smallest perturbations: they are not generic. By analogy, the periodicity of such orbits is similar to the presence of a conjugate pair of pure imaginary roots for the Euler equation if the latter happes to be linear. Any minor perturbation of the model structure will destroy such conservative orbit.

What is unique for our optimal Faustmann cycle is the role played by the reacheability constraints. They generate Euler conditions in terms of strict inequalities, and not equalities. Except for the negligible class of cases, an arbitrary but small perturbation in the growth curve, the periodic yield function, or the discount rate will not invalidate the crucial inequalities. It simply displaces the optimal orbit from one neutrally stable Faustmann cycle to another. This explains the morphological stability of the optimal Faustmann cycles.

Viewed in this manner, the optimal Faustmann cycles are ersatz optimal stationary solutions: further adjustments toward the uniform distribution are thwarted by the reachable constraints. Alternatively, the optimal uniform distribution is a limit form of the Faustmann cycles, where no reacheability condition intervenes. The periodicity of such cycles is always an integer, equal to the Faustmann tree life.

On a broader plane, forest management must respond to three types of forces: the biological, which underlies the growth curve, the economic, which is manifested in the periodic yield function and the discount rate, and, finally, the logical, which is represented by the reacheability constraints. The interaction of the growth curve and the discount rate decides the Faustmann age, at which all trees should preferably be cut. The strictly concave yield function implies that the timber harvested should preferably be kept at a constant level, and not an alternating feast and famine pattern, to ameliorate the impact of diminishing returns. The tug of war between these two considerations tends to favor a gradual evolution toward a uniform Faustmann distribution. However, sometimes such a tendency may be blocked, by the reacheability constraints: trees of a desirable age do not spring into life overnight. In other times, adjustment through premature cutting is too costly. The recurrence of such a scenario leads to the Faustmann cycle. The totality of these orbits may be defined as the 'attractor' of our dynamic adjustment process. From within this attractor, inertia impedes any further gravitation toward the uniform Faustmann distribution. From without, any optimal path must evolve toward this attractor.

Our optimal decision rule lends to some heuristic interpretation: the tree age distribution should always be adjusted toward the uniform Faustmann distribution, at some due deliberate speed, provided this does not incur excessive current opportunity cost.

We shall specify our simple model in Section 2, and report our principal findings in Section 3. The methods of derivation is commented upon in Section 4 and concluding remarks are offered in Section 5.

2. The Model

We consider now a tree farm with a unit area. A portion x_{nt} is planted with trees of age n at the end of period t, or, of age n - 1 at the beginning of period t. State variable $x_t = (x_{nt})$, the tree-age distribution at t, is an element of X, the state space. X takes the form of an (N-1) - simplex, where N is the maximal tree life.

The timber quality is assumed to be homogeneous for all trees, and the timber content of a tree is nondecreasing in its age. Thus, the maximum harvest is obtained from an entire farm of trees all at the age of N periods. This is taken as unity. The harvest is of size a_n if all trees are at age n instead and are cut down at once. Hence, (a_n) is a nondecreasing sequence of fractions.

For simplicity, reforestation is assumed to be automatic and costless.

Let k_{nt} be the portion of x_{nt} cut at the end of period t, thus,

$$0 \leq k_{nt} \leq 1, \qquad \text{for } 1 \leq n < N,$$
$$k_{Nt} = 1. \tag{1}$$

The dynamics of the system is then:

$$x_{1,t+1} = k_{1t} + k_{2t}x_{2t} +\ldots\ldots\ldots+k_{Nt}x_{Nt}$$
$$x_{n+1,t+1} = (1-k_{nt})x_{nt}, \qquad \text{for } 1 \leq n < N. \tag{2}$$

Also, the harvest in timber content at period t is,

$$c_t: = a_1 k_{1t} x_{1t} + \ldots\ldots + a_N k_{Nt} x_{Nt} \tag{3}$$
$$= a_1 (x_{1t} - x_{2,t+1}) + a_2 (x_{2t} - x_{3,t+1}) + \ldots a_N x_{Nt}.$$

The yield function is $u(.)$ and the yield in period t is,

$$u(c_t)$$

while the discounting factor is a proper fraction b. Therefore, the payoff is:

$$\sum_{t=0}^{\infty} b^t u(c_t) \tag{4}$$

Given the initial distribution, x_0, the forest management problem requires the maximization of (4), subject to (1), (2) and (3), by selecting

$$k_t = (k_{nt}) \text{ in } [0,1]^{N-1} x\{1\}$$

as the control vector for period t.   Once $k = (k_t)$ is chosen, the history $x = (x_t)$ is completely decided.   For subsequent analysis, we shall directly regard x_{t+1} as the control variable and eliminate k_t from (2) and (3).   One must replace (1) with the reacheability constraints below:

$$0 \le x_{1t} \le 1, \tag{5}$$
$$0 \le x_{nt} \le x_{n-1,t-1}.$$

Likewise, we may rewrite the yield function in 'reduced form':

$$\underline{u}(x_t, x_{t+1}) := u(c_t),$$

where c_t depends linearly on x_t and x_{t+1}, as in (3).

At the heart of our analysis is the long run behavior of the tree age distribution, and the central concept is the Faustmann age, N^*, decided as follows,

$$a_{N^*} b^{N^*}/(1-b^{N^*}) \ge a_n b^n/(1-b^n) \text{ for all } n, \ 1 \le n \le N. \tag{6}$$

A uniform Faustmann distribution may be written as,

$$x^* = (1/N^*, \ldots 1/N^*, 0, \ldots, 0). \tag{7}$$

In contrast, a Faustmann cycle is completely determined by the ordered pair of a vector,

$$x^{**}_0 = (x^{**}_{10}, x^{**}_{20}, \ldots x^{**}_{N*0}, 0, \ldots, 0), \qquad (8)$$

and a harvest rule:

$$k_{nt} = 1 \text{ if } n = N*, \text{ and } 0 \text{ otherwise, for all } n, \qquad (9)$$
$$1 \leq n < N.$$

We shall see that for our examples, the long term behavior belongs to either one or the other of (7) or (8) - (9).

Up to this point, the model follows exactly the formulation in Mitra and Wan (1985). To demonstrate the robustness of optimal Faustmann cycles in the simplest possible circumstances, we now introduce two special assumptions:

(a) $N = 2$,

and

(b) $u(c): = \log c$.

To ease notations, we redefine x_t as the scalar x_{2t}, and a as the scaler a_1. We next introduce the expression,

$$h: = (1/a) - 1.$$

Thus, we have,

$$
\begin{aligned}
c_t &= x_t + a[(1-x_t) - x_{t+1}] \\
&= a + (1-a)x_t - ax_{t+1} \qquad (10) \\
&= a(1 + hx_1 - x_{t+1}).
\end{aligned}
$$

h may be interpreted as the own rate of interest for keeping an 'old' tree. To see this, note that the timber content for a tree farm with all trees one year old is a. The incremental gain for these trees to grow into two year olds is (1-a). h of course is defined as, (1-a)/a.

Now we can write the reduced form yield function as,

$$\underline{u}(x_t, x_{t+1}): = \log a + \log (1+hx_t - x_{t+1})$$

The state-to-feasible control set may be written as,

$C(x_t): = [0, 1-x_t]$.

G, the graph of C, is displayed in Figure 1, where the two lines $x_{t+1} = 0$ and $x_{t+1} = 1 - x_t$ form the two boundaries of G.

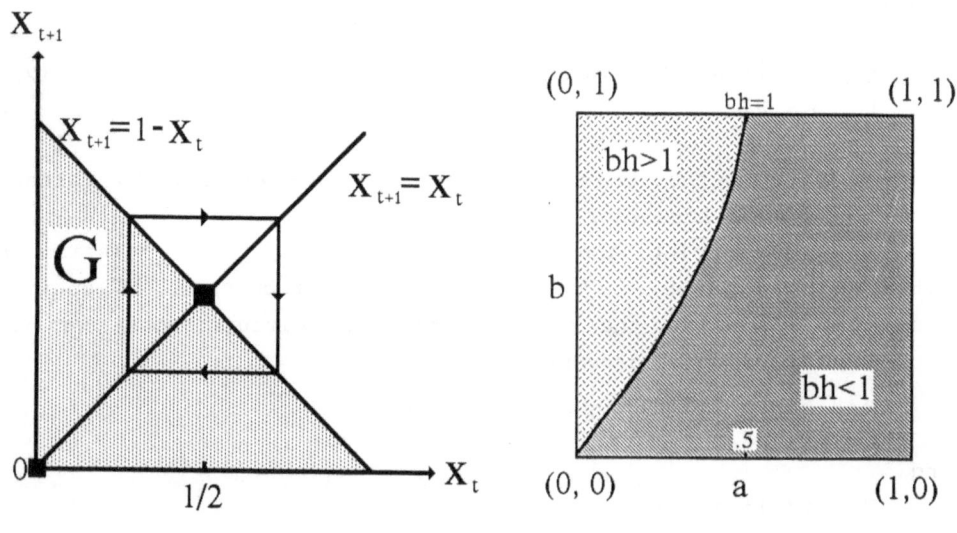

Figure 1   Figure 2

We can consider the entire class of models indexed by the ordered pairs (a,b) in the open unit square. This may be trichotomized by the sign of the criterion: $a/(1-b) - b/(1-b^2)$, or its equivalent, the sign of 1-bh. A positive (resp. negative) sign implies that the Faustmann age is one period (resp., two periods). A zero sign means that both one and two periods qualify as the Faustmann age. We shall refer to the zero, positive and negative signs of the criterion as Categories 1,2, and 3, respectively. These are dipicted in Figure 2 above.

Back to Figure 1, we also include in the diagram the 'updating line', $x_{t+1} = x_t$. The two heavy dots on that line, representing,

$x_t = 0 = x_{t+1}$,

and

$x_t = 1/2 = x_{t+1}$

turn out to be positions for optimal stationary programs, in fact, uniform Faustmann distributions, under Categories 1 and 2, and Categories 1 and 3, respectively. These satisfy the condition for optimal stationary programs,

$$P(x_t) = x_t,$$

A squared-shaped directed path illustrates a 2-period cycle. This can be shown as optimal under Category 3, and has the characterization,

$$P(x_t) = 1 - x_t.$$

These will also be discussed below. The use of the updating line here to trace out the successive values of x_t is like the use of the supply curve in the standard cobweb diagram. The graph of the policy function, which acts like the demand curve in the standard cobweb diagram is not drawn here, to avoid congestion. See however, Figures 3,4 and 5 below.

3. The Results

Ours is a very well behaved dynamic programming problem. As a consequence, it has a number of very desirable properties:
1. From any initial distribution, there exists a unique optimal path, x*.
2. There exists a value function, $V(\cdot)$, which is continuous over the state space, X.
3. There is a unique optimal policy, $P(\cdot)$, which is continuous over the state space, X.
Next, we may set up a Lagrangian function for our problem,

$$L = \Sigma_t b^t \{\log(a + ax_t - ax_{t+1}) + q_t x_{t+1} + Q_t(1 - x_t - x_{t+1})\} \qquad (11)$$
$$:= \log a/(1-b)$$
$$+ \Sigma_t b^t \{Q_t + [u(x_t, x_{t+1}) - Q_t x_t + (q_t - Q_t)x_{t+1}]\},$$

and note that:
4. There is a set of necessary conditions for optimality as follows,

$$\partial L/\partial x_t = b^t\{[[(h/c_t) - (1/bc_{t-1})] - Q_t + (q_{t-1}/b)$$
$$- (Q_{t-1}/b)\} = 0, \tag{12}$$
$$\text{for all } t \geq 1$$
$$q_t \geq 0, \ x_{t+1} \geq 0, \ q_t \ x_{t+1} = 0$$
$$Q_t \geq 0, \ 1 - x_t - x_{t+1} \geq 0, \ Q_t(1 - x_t - x_{t+1}) = 0.$$

Recall now a stationary program is (x_t) where x_t is constant. By (10), c_t must be constant, also. By (11), it can then be shown that,

5. For Category 2, the only stationary program which may be optimal is, $x_t = 0$, all t, and for Category 3, the only optimal stationary program is $x_t = 1/2$, all t.

We now draw upon the method of supporting prices, to determine the optimal policy for each category. First recall G is the graph of the state-to-feasible control correspondence, with the expression:

$$G := \{x_t, x_{t+1}\}: \ 0 \leq x_t \leq 1, \ 0 \leq x_{t+1} \leq 1 - x_t\}.$$

To prove that (x_t^*) is optimal, it is sufficient to find supporting prices, (p_t), such that for all t,

$$F(x_t, x_{t+1}; p_t, p_{t+1}) := \underline{u}(x_t, x_{t+1}) - p_t x_t + b p_{t+1} x_{t+1} \tag{13}$$

reaches it maximum over G at (x_t^*, x_{t+1}^*).

We now follow a two-step procedure, first by deciding the long term behavior for the three Categories, and then applying backward induction to trace out the optimal paths leading from any initial position in X to the long term orbits.

By long term behavior, we mean here the orbits within a set, X_p, in the state space with the following properties: (a) any orbit starting within that set will always be inside that set, and (b) for each state in the set, there is some orbit which will pass through this state infinitely many times. We shall refer to this set as the persistent set, in analogy with the nomenclature of the theory of the Markov process. One may refer to this set as the attractor, even though there is yet no common usage for the latter term. The complement of the persistent set will be called the transient set, X_T. Optimal orbits outside the persistent set will be referred to as the transient behavior.

In this manner, we have found the following results:

6. (On the persistent set, X_p)

Category 1

Category 2

Category 3

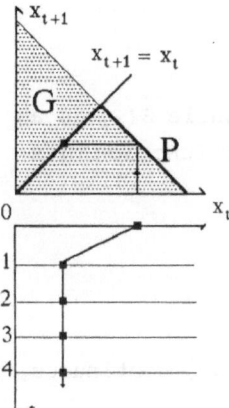

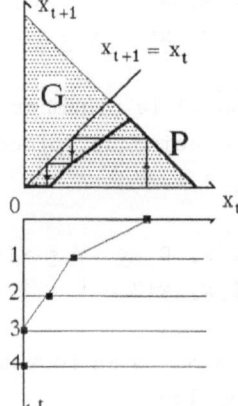

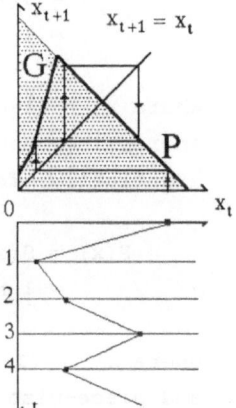

Figure 3   Figure 4   Figure 5

For Catetory 1, the persistent set is a continuum of optimal stationary programs, where

$x_t = x_0$ for any x_0, such that $0 \le x_0 \le 1/2$;

for Category 2, the persistent set is the singleton, $\{0\}$; and

for Category 3, the persistent set is the interval, $[(1-z)/2, (1+z)/2]$, where

$z := (1-b)(hb-1)/(1+b)(hb+1)$.

Next, to study the transient behavior of the system, we shall determine the optimal policy function under the three separate Categories, as follows:

7. (On the optimal policy)

One can determine the optimal policy $P(\cdot)$ by using the Bellman equation for Category 1 and the supporting prices for Categories 2 and 3. In summary:

For Category 1,

$P(x) = x$,   for all x, $0 \le x \le 1/2$,

     $= 1-x$,   for all x, $1/2 \le x \le 1$;

for Category 2,

$$P(x) = 0, \quad \text{for all } x, \ 0 \le x \le x'$$
$$= S(x), \quad \text{for all } x, \ x' \le x \le x''$$
$$= 1-x, \quad \text{for all } x, \ x'' \le x \le 1,$$

where $x' = (1-hb)/h^2b$, and x'' is a number in $(1/2,1)$ while $S(\cdot)$ is an increasing, nonnegative, continuous, piece-wise linear function of x; and finally, for Category 3,

$$P(x) = S(x), \quad \text{for all } x, \ 0 \le x \le x''',$$
$$= 1-x, \quad \text{for all } x, \ x''' \le x \le 1,$$

where $x''' = (1-z)/2$ while $S(x)$ is an increasing, positive, continuous, and piece-wise linear function.

We can now display the graphs of $P(\cdot)$ above in Figure 3,4 and 5 respectively, with the graph of P in heavy line.

An age distribution x within X_T, with $P(x) = 1-x$ may be regarded as superannuated: at such x, trees at and only at their maximal life will be harvested. For all Categories, only an initial distribution may be a superannuated tree distribution under the optimal policy. If a tree age distribution is within X_T but not superannuated, it will evolve monotonically toward the persistent set of distributions, X_P, and enter it in finite many periods, with progressively more younger trees, period after period, under Category 2 and more older trees under Category 3. In either case, some trees will be cut before reaching their maximal life span. Summing up, for all Catgegories, initially, one may either have a distribution already in the persistent set, or start with a monotonic adjustment process which eventually reaches the persistent set in a finite number of periods, or begin with a superannuated forest where all old trees but none other are cut in one period, prior to the monotonic adjustment phase which starts next period.

Moreover, a heavily discounted program, with a low value of b, implies monotonic convergence to a unique uniform Faustmann distribution. For a lightly discounted program, with a large, positive value for the proper fraction b, then almost everywhere, one embarks on an optimal path which leads to a cyclical orbit, i.e., an optimal Faustmann cycle. The optimal stationary program, or the uniform Faustmann distribution, is simply a limit form of the cycles, and can be reached only by coincidence. Cycles vary in amplitude, and the maximum amplitude occurs if the discount factor and the own rate of interest of trees are such that $bh^2 = 1$. This amplitude converges toward zero when b approaches either 1, the no discount case, or the

knife edge value of 1/2, where any weighted average of the two uniform
Faustmann distributions is an optimal stationary program. Recall
the expression of z, when a and b are such that $b^2h = 1$ and b
approaches zero, one can have the cyclical amplitude arbitrarily
close to its supremum, 1/2. However, under no circumstances, can
the amplitude of 1/2 be actually realized. Otherwise, one will have
harvest sizes alternating between zero and one, incurring infinitely
negative payoffs when the harvest is zero, under the logarithmic
yield function. This is clearly nonoptimal, since it is dominated
by the always feasible solution of a clear cut every period. Figure
6 provides a bifurcation map which displays how the persistent set
varies with respect to b, for any given value of a.

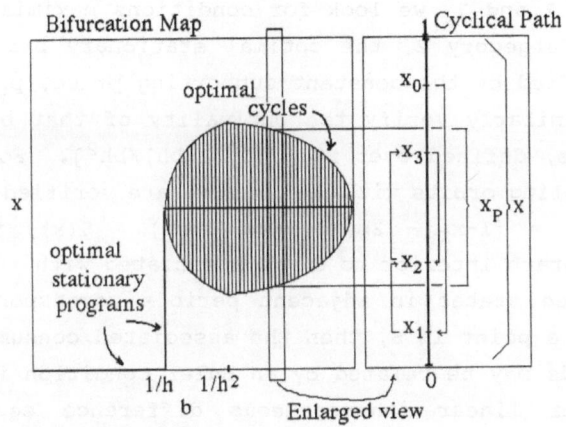

Figure 6

4. Method of Analysis

For the sake of completeness, we shall now outline how the above
results may be derived. For details one may refer to Wan (1987).

Since (i) both the state space, X, and the value of the state-to-
feasible control correspondence, C(x), at each x in X are compact
and convex, (ii) C(·) is continuous over X, and so is the yield

function, $\underline{u}(\cdot)$, over all pairs (x_t, x_{t+1}) in G with the only exception at $(0,1)$, and (iii) this position $(0,1)$ is always dominated in yield by the equally feasible choice, $(0,0)$, then, by a Banach space fixed point theorem, we can prove (a) the existence of a continuous value function, (b) the existence of a solution to this model of optimal forest management, for any initial distribution, and (c) the existence of an optimal policy rule, $P(x)$, which is upper hemicontinuous. By the strict concavity of the yield function, $u(\cdot)$, one can deduce that (d) the optimal policy, $P(\cdot)$, is unique and hence continuous. This completes our commentary on the derivation of Results 1,2, and 3.

Result 4 follows from the standard complementary slackness conditions in the Kuhn-Tucker theory of nonlinear programming. Routine reasoning then leads to Result 5. For Results 6 and 7, the policy P for Category 1 has a closed form. It implies a candidate value function which may be verified as the solution of a Bellman equation. For Categories 2 and 3, we look for conditions maximizing F in (13) over G. For Catgegory 2, the optimal stationary program, $x_t = 0$, all t, is verified by the constant supporting price, $p_t = h$, for all t. One can similarly verify the optimality of that branch of $P(x)$ with zero-value, defined over $x_H := (0, (1-bh)/bh^2]$. For Category 3, the optimal cycling orbits with $x_{t+1} = 1-x_t$ are verified by supporting prices with $p_t = (1-x_t - 2b)/2(1-x_t)(1-b^2)$. $S(x)$, the branch of $P(x)$ with its graph interior to G, is associated with interior maxima of F. If three states in adjacent periods correspond to the two coordinates of a point in S, then the associated consumptions in two adjacent periods may be related by an Euler condition in the form of a second order linear inhomogeneous difference equation: $x_{t+2} - (bh+h)x_{t+1} + bh^2 x_t = 1 - bh$. S corresponds to a continuum of solution sequences for this Euler equation, each sequence with a distinctive boundary condition, anchored upon the graph of P over X_H for Category 2 and the graph of P over the persistent set X_P for Category 3. The graph of S may be shown to be continuous and piecewise linear. The closure of the graph must be bounded away from the 'updating line', $x_{t+1} = x_t$. Otherwise, by the continuity of $P(x)$ in Result 3, there must be an optimal stationary program in $(0, 1/2)$ and Result 5 is violated. This can be shown to imply the monotonicity of $S(\cdot)$. This monotonicity of $S(\cdot)$, in turn, helps the verification of the branch of $P:P(x) = 1-x$, by showing that certain constructed prices qualify as supporting prices.

In principle, an interested reader can follow this sketch to derive all the results.

5. Conclusion

We have provided an entire class of examples to illustrate the presence of optimal cycles at any positive discount rate. We believe that such a result is robust. This is because in these models, the vintage capital structure implies reacheability constraints. As stated before, such constraints may keep some of the Euler conditions in strict inequality form. This, in turn, implies qualitative stability in that the inequalities continue to hold, _mutatis mutandis_, for _all_ small structural displacement. Hopefully, such speculations can be analytically studied in the future. A possible approach is outlined below.

Let us continue in our two-period world, where each model is completely determined by the ordered triplet, (a, b, $\underline{u}(\cdot)$). Both the heterogeneity of timber quality and the recognizance of benefits accruing from standing timber (as it is done in an on going study by Mitra, Ray and Roy) can be subsumed by a different specification of the $\underline{u}(\cdot)$ function. Now, for our class of examples, an optimal policy may be regarded as a point in some policy space consisting of continuous functions. Variations of $\underline{u}$, as well as a and b will displace the optimal policy within that policy space. Any member of this policy space must have its graph contained within G. Moreover, for our examples, any optimal policy has a graph containing at most three branches, the first (which only exists for Category 2), coincides with the horizontal axis, the second, the graph of $S(\cdot)$, is in the interior of G, and the last is along the line of 1-x. The graph of $S(\cdot)$ coincides with the updating line for the knife-edge case of Category 1, implying a continuum of neutrally stable optimal stationary programs. Generically, the graph of $S(\cdot)$ is bounded away from the updating line. If it lies below as in Category 2, it means global convergence to the unique optimal stationary program in finitely many periods. If it lies above, as in Category 3, it means convergence to some optimal cycling orbit, in a finite number of periods: a limiting case happens to be an optimal stationary program. What is now known is that if we assume that

$$\underline{u}(x_t, x_{t+1}) = u(a + (1-a)x_t - ax_{t+1}),$$

with u' > 0 > u", then actually most of the qualitative results can be proved without assuming the logarithmic form of the yield function.

In particular, optimal Faustmann cycles exist at all positive discount rates, and our bifurcation map holds. See Wan (1987).

The applicability of our analysis is not limited to forest management only. Any problem where assets have a vintage structure, be it housing, other durable assets, or human capital, has the similar reacheability constraints. The only exceptions turn out to be the simple intertemporal models focused upon by economists over the past six decades, e.g., the Ramsey model where all capital is working capital, like cash or such inventories which can never become shopworn. Methodologically speaking, the concentration on the Ramsey model has provided much insight, some of which facilitate our exploration here. On the other hand, after 60 years, we probably have learned enough by now to turn attention toward other problems which all share the vintage capital structure in our model.

In conclusion, I acknowledge my debts to M.C. Kemp who renewed my interest in forestry economics, and to M. Mitra whose earlier joint research with me has benefitted me greatly, and our discussion has encouraged me to complete the research here in the best way I can. Remaining mistakes remain to be my responsibilities.

REFERENCES

Benhabib, J. and Nishimura, K., Competitive equilibrium cycles, Journal of Economic Theory 35: 284-306 (1985).

Cass, D. and Shell, K. (eds.), The Hamiltonian approach to dynamic economics, New York: Academic Press (1976).

Clark, C., Mathematical Bioeconomics, New York: Wiley (1976).

Dasgupta, P.A., The Control of Resources, Oxford: Blackwell (1982).

Faustmann, M., Berechnung des werthes, welchen walboden sowie nach haubare halzbestande fur die weltwirtschaft besitzen, Allgemeine Forest und Jagd Zeitung 25 (1849): with English translation in Gane, M. (ed.), Martin Faustmann and the Evolution of Discounted Cash Flow, Institute Paper No. 42, Commonwealth Forestry Institute, University of Oxford, 27-55 (1968).

Jevons, W.S., The Theory of Political Economy, New York, Kelley and Millman (1957).

Kemp, M.C. and Moore, E.G., Biological capital theory: a question and a conjecture, Economic Letters 4, 141-144 (1979).

Mitra, T. and Wan, H., Jr., Some theoretic results on the theory of forestry, Review of Economics 52, 263-282 (1985).

Mitra, T. and Wan, H. Jr., On the Faustmann solution to the forest management problem,Journal of Economic Theory 40, 229-249 (1986).

Samuelson, P.A., Optimality of profit, including prices under ideal planning, Proceedings of National Academy of Science, U.S.A. 70, 2109-2111 (1973).

Wan, H., Jr., Optimal Paths without GAS at any Discount Rates: New Examples from Forest Management, CAE working paper, Economics Department, Cornell University, Ithaca, NY, USA (1987).

Wicksell, K., Lectures on Political Economy, vol. 1, 2nd edition, New York, Kelley and Millman (1934).

Postscript

It is my pleasure to present this paper at this symposium. This study illustrates some of the issues economists face in their research. Hopefully, this study also ties well to the other papers in the symposium, specifically, to the paper of Clark which provides an overview about the interface between mathematics, ecology and economics, the paper of Shoemaker, which discusses in great detail the method of dynamic programming, a method also employed here, and to the paper of Conrad, which reviews the perscriptive aspects of economics, while this paper focuses upon the descriptive side of the same coin. In economics, to describe is often for the sake of to prescribe. This postscript has the additional purpose of relating what we economists do to what is done by other researchers who are not primarily economics-oriented, in applications, in concepts and in methodology.

Present day eco-system is shaped partly by Nature, and - like it or not - partly also by human action. Human beings are purposeful and adaptive. They need not have the capacity for elaborate computations. However, those who survive and thrive-those who count-modify their behavior by experience, for goals often identified as economic gains. Their action may be regarded as 'rational', with at least as much justification as the invocation of the Principle of Least Action in physics or the postulation by biologists that species adopt breeding strategies to maximize the chance of species survival. Hence, Hamiltonian dynamics becomes extensively used in economics.

Like in mathematical biology, scale is important. When decision-makers are large relative to their environment, their actions are modeled as dynamic games; when they are very small, they compete impersonally, taking market prices as given. In the kind of medium case which we are concerned with, they operate independently, yet they are concerned that market will be spoiled if too much is sold at once.

Economists are pre-occupied with the robustness of their results. This is because we face a dearth of empirical data with general representativeness. In contrast with the universal constants measured by successive generations of experimental physical scientists, with ever increasing refinement, what economists can measure are often not relevant at other points of space or time. The behavior patterns of present American consumers share few, if any, invariant and quantifiable parameters with those observed by Adam Smith in pre-revolutionary France. Thus, economists must be content with qualitative findings, like the Engel's Law. These hopefully will remain valid over broad classes of possible worlds, even though which of these we actually dwell in, we may never know. Naturally, morphological stability is the property we hold dear for our theoretical constructs. Our study is a specimen of the questions economics is about.

The qualitative findings we are concerned with are usually patterns. For the forest management problem which we consider, there is the configuration of age distribution at any period, and there is the evolution of such distributions over time. The evolutionary patterns we expect to observe are presumably optimal from the viewpoint of the forest management. Since these are the effective decision makers, the consequences of their optimization exercises must be taken seriously, just as biologists scrutinize seriously the food gathering problems of various animal species.

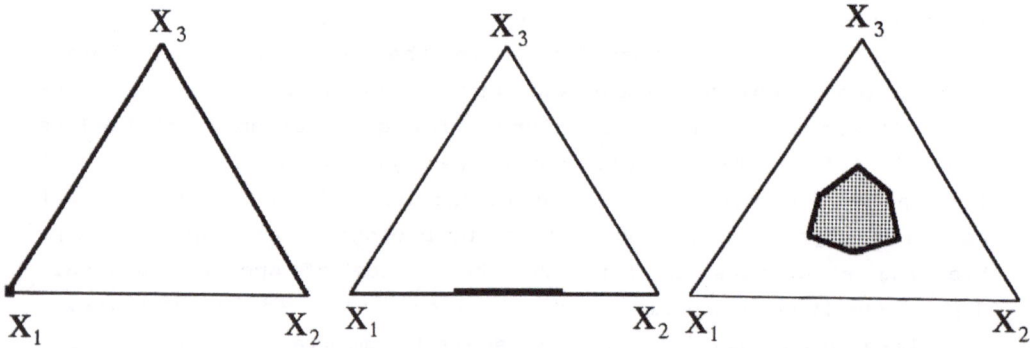

Figure 7

The overall conclusion we obtain is that, even in a model with a multi-period (rather than two-period) maximum tree life, there tends to be attractors of zero, one, two or higher dimensions, as depicted above in Figure 7, which contain three alternative, generic cases in a three period model. Which of the three prevails depends upon the structural parameters of the model. Any tree-age distribution is presented as a point $x = (x_1, x_2, x_3)$ in the state space which is a regular triangle or 2-simplex. Except for the case of the zero-dimension attractor, i.e., a unique optimal steady state of clear cut, each period, perpetual cycling within each attractor is to be expected. Forest managers are not likely to go after the optimal Faustmann uniform age distribution at any cost. In fact, they do not get exactly there at all in most event. Faustmann cycles take the form of two-point oscillations or three-point counterclockwise trangular paths. The hexagonal boundaries of the attractor set in this case marks the limit of tolerable differences in harvest levels, in view of the adjustment cost. Overall, unless the discount rate is very high, and all paths converge to the trivial uniform Faustmann solution of a clean cut every period, it is an 'almost sure' outcome that the system ends up in cycles rather than otherwise. Furthermore, which of the infinitely many neutrally stable cycles one ends up with depends upon the initial state: historical accidents and external shocks all matter very much.

Ours is a stylized model. Practical decision-makers in forest management must consider additional considerations, when circumstances warrant it. One such example is the treatment of forest fire and risk management in M.J. Reed, Protecting a Forest Against Fire: Optimal Protection Patterns and Harvest Policies, Natural Resource Modeling 2:33-54.

Part III.   Infectious Diseases

MATHEMATICAL MODELS OF INFECTIOUS DISEASES IN MULTIPLE POPULATIONS

Wei-min Liu

Department of Mathematical Sciences

Indiana University-Purdue ·University at Indianapolis

Indianapolis, IN 46205

Abstract

Some infectious diseases can exist in several populations or species and transmit from one to another. After a short review of several mathematical models for this phenomenon, it is pointed out that in the model of Holt and Pickering (1985), where the disease-caused mortality is taken into account, all equilibria can be unstable simultaneously and a periodic solution can appear through Hopf bifurcation.

Some infectious diseases can or have to transmit from one species to another (Dobson and May, 1986). For example, malaria infection is caused by single-celled blood parasites, protozoa, in the genus *Plasmodia*. The disease also appears in birds, reptiles and monkeys. The life cycle of the human malaria parasite involves a phase of asexual growth and development in man and a phase of sexual proliferation in the mosquito. Plague, schistosomiasis, and typhus are some other examples which involve more than one host species. Section 1 is a brief review of the malaria model of Ross (1911), which is perhaps the first mathematical model for an infectious disease in more than one population. In this model and some other host-vector relationship models, only the interpopulation transmission is taken into account. Section 2 points out that a heterogeneous population model of Hethcote (1978) can be considered as a model of infectious disease in n populations, whose population sizes are assumed to be constants. Section 3 introduces a model of Holt and Pickering (1985), where the disease-induced mortality is included and the population sizes are no longer assumed to be constants. This section also shows that all equilibria in this model can be unstable simultaneously and a stable periodic solution can occur through Hopf bifurcation.

1. Ross's malaria model

A pioneering mathematical work on the dynamics of infectious disease is the malaria model of Ross (1911) (also see Bailey, 1975), which involves a human population and a mosquito population. Suppose that each of these two species has two subpopulations: susceptible and infected. Let N_1 and N_2 be respectively the sizes of human and mosquito populations; let i_1 and i_2 be their infected subpopulation sizes. The Ross model can be expressed as

$$\frac{di_1}{dt} = \frac{bf_2i_2(N_1 - i_1)}{N_1} - (\gamma_1 + \mu_1)i_1,$$

$$\frac{di_2}{dt} = \frac{bf_1i_1(N_2 - i_2)}{N_1} - (\gamma_2 + \mu_2)i_2, \tag{1}$$

where b is the man biting rate of mosquito, f_1 and f_2 are proportions of infected individuals who are infectious, γ_1 and γ_2 are recovery rates, μ_1 and μ_2 are death rates, which are assumed to be equal to the birth rates. When the population sizes N_1 and N_2 are constants, denoting the infected fractions i_1/N_1 and i_2/N_2 by I_1 and I_2, we can write the system (1) in the form of

$$\frac{dI_1}{dt} = abf_2 I_2(1 - I_1) - (\gamma_1 + \mu_1)I_1,$$

$$\frac{dI_2}{dt} = bf_1 I_1(1 - I_2) - (\gamma_2 + \mu_2)I_2,$$

$$(2)$$

where $a = N_2/N_1$ is the overall mosquito density per person. Lotka (1923) studied these equations in detail. Ross further assumes that in human population μ_1 is negligible compared with γ_1 and that in mosquito population γ_2 is negligible compared with μ_2. However these assumptions do not change the qualitative analysis of the system, and we do not use them. The equilibria of system (2) can be obtained explicitly. There is always a trivial equilibrium $(I_1, I_2) = (0,0)$, which is stable when $R \leq 1$, and is unstable when $R > 1$, where the disease reproductive rate R is

$$R = \frac{ab^2 f_1 f_2}{(\gamma_1 + \mu_1)(\gamma_2 + \mu_2)}. \tag{3}$$

When $R > 1$, there is a nontrivial stable equilibrium given by

$$\bar{I}_1 = \frac{ab^2 f_1 f_2 - (\gamma_1 + \mu_1)(\gamma_2 + \mu_2)}{bf_1(abf_2 + \gamma_1 + \mu_1)},$$

$$\bar{I}_2 = \frac{ab^2 f_1 f_2 - (\gamma_1 + \mu_1)(\gamma_2 + \mu_2)}{abf_2(bf_1 + \gamma_2 + \mu_2)}. \tag{4}$$

Ross's model shows that an effective measure to eradicate malaria is to reduce the mosquito density a.

2. Models with constant population sizes

Some models for a heterogeneous population can also be considered as models for multiple populations. Hethcote (1978) proposed an immunization model for a heterogeneous population:

$$\frac{dS_i}{dt} = -\sum_{j=1}^{n} \frac{\lambda_{ij} N_j}{N_i} I_j S_i + (1 - \phi_i)\delta_i - \delta_i S_i - \theta_i S_i + \epsilon_i R_i,$$

$$\frac{dI_i}{dt} = \sum_{j=1}^{n} \frac{\lambda_{ij} N_j}{N_i} I_j S_i - \gamma_i I_i - \delta_i I_i, \tag{5}$$

where $i = 1, 2, \ldots, n$, S_i, I_i, and R_i are respectively the fractions of susceptible, infective, and removed classes in group i, $S_i + I_i + R_i = 1$, N_i is the number of individuals in group i and is assumed to be constant, λ_{ij} is the contact rate of an infective in group j with individuals in group i, γ_i is the recovery rate, δ_i is the birth rate which equals the death rate in group i, θ_i is the rate that immunization removes individuals from the susceptible class and ϕ_i (< 1) is the rate that immunization removes newborns from the susceptible class.

If we assume the immunization constants $\theta_i = \phi_i = 0$ and denote $\lambda_{ij} N_j/N_i$ by β_{ij}, we obtain a model for infection dynamics in multiple populations, each of which is a SIRS system with constant population size:

$$\frac{dI_i}{dt} = \sum_{j=1}^{n} \beta_{ij} S_i I_j - (\gamma_i + \delta_i) I_i,$$

$$\frac{dR_i}{dt} = \gamma_i I_i - (\delta_i + \epsilon_i) R_i. \tag{6}$$

Note that we use the equations of dI_i/dt and dR_i/dt instead of dS_i/dt and dI_i/dt, because the equations of dR_i/dt are linear. The Jacobian matrix at equilibrium

$$(\bar{I}_1, \ldots, \bar{I}_n, \bar{R}_1, \ldots, \bar{R}_n)$$

can be written as

$$J = \begin{pmatrix} J_{11} & J_{12} \\ J_{21} & J_{22} \end{pmatrix}, \tag{7}$$

where

$$J_{11} = diag(\bar{S}_1, \ldots, \bar{S}_n)B + J_{12} - diag(\gamma_1 + \delta_1, \ldots, \gamma_n + \delta_n), \tag{8}$$

$$J_{12} = -diag((\gamma_1 + \delta_1)\bar{I}_1/\bar{S}_1, \ldots, (\gamma_n + \delta_n)\bar{I}_n/\bar{S}_n), \tag{9}$$

$$J_{21} = diag(\gamma_1, \ldots, \gamma_n), \tag{10}$$

$$J_{22} = -diag(\delta_1 + \epsilon_1, \ldots, \delta_n + \epsilon_n), \tag{11}$$

$$B = (\beta_{ij}), \tag{12}$$

and $diag(\gamma_1, \ldots, \gamma_n)$ is the diagonal matrix with diagonal elements $\gamma_1, \ldots, \gamma_n$, etc. For the trivial equilibrium, $\bar{I}_i = \bar{R}_i = 0$ $(i = 1, \ldots, n)$ and J_{12} is a zero matrix. Because J_{22} is a diagonal matrix with negative diagonal elements, the stability of the trivial equilibrium is determined by the eigenvalues, $z_1, \ldots, z_n$, of J_{11}. When $\max_i Re(z_i) < 0$, the trivial equilibrium is locally asymptotically stable; when $\max_i Re(z_i) > 0$, it is unstable. Hethcote further mentions that in the latter case there is a unique nontrivial equilibrium. Its locally asymptotical stability is proved by Hethcote and Thieme (1985).

Liu and Levin (1989) find out that for $n = 2$ if the interpopulation transmission rates, β_{12} and β_{21}, are 0 below c_i, 1 above d_i, and cubic functions in the intervals $[c_i, d_i]$ with continuous first order derivatives at c_i and d_i, i.e.,

$$\beta_{ij} = b_{ij} h_j(I_j), \tag{13}$$

$$h_j(I_j) = \begin{cases} 0, & \text{when } I_j \leq c_j; \\ (I_j - c_j)^2(3d_j - c_j - 2I_j)/(d_j - c_j)^3, & \text{when } c_j < I_j < d_j < 1; \\ 1, & \text{when } I_j \geq d_j; \end{cases} \tag{14}$$

for some parameter values, the nontrivial equilibrium is locally asymptotically stable, but sustained oscillation is observed outside its attractive domain.

3. Periodic solutions in the model of Holt and Pickering

Holt and Pickering (1985) proposed a model with varying population sizes, where the disease caused mortalities in two populations are taken into account. Their model is

$$\dot{S}_1 = r_1 S_1 - \beta_{11} S_1 I_1 - \beta_{12} S_1 I_2 + e_1 I_1,$$

$$\dot{I}_1 = \beta_{11} S_1 I_1 + \beta_{12} S_1 I_2 - d_1 I_1,$$

$$\dot{S}_2 = r_2 S_2 - \beta_{22} S_2 I_2 - \beta_{21} S_2 I_1 + e_2 I_2,$$

$$\dot{I}_2 = \beta_{22} S_2 I_2 + \beta_{21} S_2 I_1 - d_2 I_2, \tag{15}$$

with parameters

$$r_i = a_i - b_i,$$
$$d_i = \alpha_i + b_i + \gamma_i,$$
$$e_i = a_i(1 - f_i) + \gamma_i,$$

where a_i is the per capita birth rate in uninfected hosts, $a_i(1 - f_i)$ is the per capita birth rate in infected individuals ($0 \leq f_i \leq 1$), b_i is the inherent death rate of the susceptibles, α_i is the parasite-induced per capita death rate, γ_i measures the per capita rate at which infected individuals recover and reenter the susceptible portion of the population, $\beta_{ij} S_i I_j$ is the transmission rate from population j to population i. It is assumed that $d_i > e_i$, otherwise the population size will grow exponentially. They showed that there is a trivial equilibrium $(\tilde{S}_1, \tilde{I}_1, \tilde{S}_2, \tilde{I}_2) = (0,0,0,0)$ and two boundary equilibria $(\hat{S}_1, \hat{I}_1, 0, 0)$ and $(0, 0, \hat{S}_2, \hat{I}_2)$ where

$$\hat{S}_i = \frac{d_i}{\beta_{ii}}, \qquad \hat{I}_i = h_i \hat{S}_i, \tag{16}$$

and

$$h_i = \frac{r_i}{d_i - e_i}.$$

The nontrivial equilibrium $(S_1^*, I_1^*, S_2^*, I_2^*)$ exists when

$$\frac{\beta_{11}}{\beta_{21}} > \frac{h_1 d_1}{h_2 d_2} > \frac{\beta_{12}}{\beta_{22}}, \tag{17}$$

or

$$\frac{\beta_{11}}{\beta_{21}} < \frac{h_1 d_1}{h_2 d_2} < \frac{\beta_{12}}{\beta_{22}}, \tag{18}$$

and is given by

$$I_i^* = \frac{\beta_{jj} d_i h_i - \beta_{ij} d_j h_j}{\beta_{11} \beta_{22} - \beta_{12} \beta_{21}}, \tag{19}$$

$$S_i^* = \frac{I_i^*}{h_i}, \tag{20}$$

where $i = 1, 2$ and $j = 3 - i$. Note that condition (17) is equivalent to

$$I_1^* > 0, \qquad I_2^* > 0, \qquad \beta_{11} \beta_{22} - \beta_{12} \beta_{21} > 0, \tag{21}$$

and condition (18) is equivalent to

$$I_1^* > 0, \qquad I_2^* > 0, \qquad \beta_{11} \beta_{22} - \beta_{12} \beta_{21} < 0, \tag{22}$$

Holt and Pickering proved that when (17) holds, the two boundary equilibria are unstable; while when (18) holds, they are locally asymptotically stable. Holt and Pickering do not exclude the possibility of existence of periodic solutions, but they conjecture that the nontrivial equilibrium is locally asymptotically stable when (17) holds and is unstable when (18) holds. In the following, it is proven that the second part of the conjecture is true, i.e., the nontrivial equilibrium is unstable when (18), or equivalently, (22) is true. However, the first part of the conjecture is not always true. Based on the general analysis, we have found an example which satisfies (17) and (21) but its nontrivial equilibrium is unstable. In this example, all equilibria are unstable, and numerical solution and Poincaré map strongly suggest that there is an asymptotically stable periodic solution (Fig. 1). The parameter values in this example may be not realistic, but it reveals that the model of Holt and Pickering can show more complicated behavior than stable and unstable equilibria.

The Jacobian matrix of (15) is

$$
J = \begin{pmatrix}
r_1 - \beta_{11}I_1 - \beta_{12}I_2 & -\beta_{11}S_1 + e_1 & 0 & -\beta_{12}S_1 \\
\beta_{11}I_1 + \beta_{12}I_2 & \beta_{11}S_1 - d_1 & 0 & \beta_{12}S_1 \\
0 & -\beta_{21}S_2 & r_2 - \beta_{22}I_2 - \beta_{21}I_1 & -\beta_{22}S_2 + e_2 \\
0 & \beta_{21}S_2 & \beta_{22}I_2 + \beta_{21}I_1 & \beta_{22}S_2 - d_2
\end{pmatrix}.
$$

At the nontrivial equilibrium, it becomes

$$
J = \begin{pmatrix}
-e_1 h_1 & -\beta_{11}S_1^* + e_1 & 0 & -\beta_{12}S_1^* \\
d_1 h_1 & -\beta_{12}I_2^*/h_1 & 0 & \beta_{12}S_1^* \\
0 & -\beta_{21}S_2^* & -e_2 h_2 & -\beta_{22}S_2^* + e_2 \\
0 & \beta_{21}S_2^* & d_2 h_2 & -\beta_{21}I_1^*/h_2
\end{pmatrix}. \tag{23}
$$

If the characteristic polynomial of J is

$$
z^4 + c_1 z^3 + c_2 z^2 + c_3 z + c_4, \tag{24}
$$

then the Routh-Hurwitz conditions for all eigenvalues of J to have negative real parts are

$$
c_1 > 0, \quad c_2 > 0, \quad c_3 > 0, \quad c_4 > 0, \quad c_1 c_2 - c_3 > 0, \quad c_1 c_2 c_3 - c_3^2 - c_1^2 c_4 > 0. \tag{25}
$$

The independent ones in (25) are

$$
c_1 > 0, \quad c_3 > 0, \quad c_4 > 0, \quad c_1 c_2 c_3 - c_3^2 - c_1^2 c_4 > 0.
$$

Obviously,

$$
c_1 = e_1 h_1 + e_2 h_2 + \beta_{12}I_2^*/h_1 + \beta_{21}I_1^*/h_2 > 0. \tag{26}
$$

Careful calculations show that

$$
c_2 = \beta_{11}(d_1 - e_1)I_1^* + \beta_{22}(d_2 - e_2)I_2^* + e_1 e_2 h_1 h_2 + e_1 I_1^* \beta_{21} h_1/h_2 + e_2 I_2^* \beta_{12} h_2/h_1 > 0, \tag{27}
$$

because of the assumptions $d_1 > e_1$ and $d_2 > e_2$. Moreover,

$$
c_3 = (d_1 - e_1)I_1^*(e_2 h_2 \beta_{11} + \beta_{21} d_1 \frac{h_1}{h_2}) + (d_2 - e_2)I_2^*(e_1 h_1 \beta_{22} + \beta_{12} d_2 \frac{h_2}{h_1}) > 0. \tag{28}
$$

However,

$$
c_4 = (d_1 - e_1)(d_2 - e_2)I_1^* I_2^*(\beta_{11}\beta_{22} - \beta_{12}\beta_{21}), \tag{29}
$$

which is positive when (21) holds, and is negative when (22) holds. Therefore, under condition (22), the nontrivial equilibrium is unstable.

For the parameter values $\beta_{11} = 0.000021$, $\beta_{12} = 0.000019$, $\beta_{21} = \beta_{22} = 0.0001$, $d_1 = 0.02$, $d_2 = 0.01$, $h_1 = 0.1$, $h_2 = 1$, $e_1 = e_2 = 0.0005$, the condition (21) is valid, but $c_1 c_2 - c_3 < 0$. Therefore, the nontrivial equilibrium $(500, 50, 50, 50)$ is unstable. The eigenvalues of its Jacobian matrix are -0.0000839, -0.0151, $0.0000508 \pm 0.00856i$. For initial conditions near this nontrivial equilibrium (Fig. 1a), near the boundary equilibrium $(952, 95.2, 0, 0)$ (Figs. 1b), and near the boundary equilibrium $(0, 0, 100, 100)$ (Figs. 1c, d), the numerical solutions strongly suggest that there is an asymptotically stable periodic solution. The Poincaré map on the 3-dimensional hyperplane $I_2 = 50$ also suggests a stable fixed point corresponding to the stable periodic solution in the 4-dimensional phase space.

If we let the parameter e_1 (or e_2) approach d_1 (or d_2) from below (d_1 and d_2 are invariant), then h_1 (or h_2) tends to infinity. If we assume h_1 (or h_2) is a first order infinity, then the maximal order of the positive terms in $c_1c_2 - c_3$ is at least 2, while the negative terms are at most of order 1. Thus, when e_1 (or e_2) is large enough, $c_1c_2 - c_3$ is positive. The same argument also applies to $c_1c_2c_3 - c_3^2 - c_1^2c_4$. Therefore, once we find a set of parameter values for which the nontrivial equilibrium is unstable while (21) is true, we can make it stable by increasing e_1 (or e_2). Note that in this process, the constant term, c_4, of the characteristic polynomial (24) remains positive, and there does not exist zero eigenvalue. Thus, there must be a pair of imaginary eigenvalues of J passing through the imaginary axis, and this is just the condition for Hopf bifurcation.

Acknowledgments

This work started when the author was supported by the 1987 summer fellowship from the Graduate School of Cornell University and Simon A. Levin from the NSF grant DMS 8406472 and McIntire Stennis grant 183568. The graphs were plotted with several program subroutines written by Moshe Braner.

REFERENCES

Bailey, N. T. J.: The mathematical theory of infectious diseases and its applications, 2nd edition, London: Griffin, 1975.

Dobson, A. P. and R. M. May: Patterns of invasions by pathogens and parasites, In Ecology of Biological Invasions of North America and Hawaii (H. A. Mooney and J. A. Drake eds.), 59-76, New York: Springer-Verlag, 1986.

Hethcote, H. W.: An immunization model for a heterogeneous population, Theor. Pop. Biol. **14**, 338-349 (1978).

Hethcote, H. W. and H. R. Thieme: Stability of the endemic equilibrium in epidemic models with subpopulations, Math. Biosci. **75**, 205-227 (1985).

Holt, R. D. and J. Pickering: Infectious disease and species coexistence: a model of Lotka-Volterra form, Amer. Nat. **126**, 196-211 (1985).

Liu, W. and S. A. Levin: Influenza and some related mathematical models, In Applied Mathematical Ecology (S. A. Levin, T. G. Hallam and L. J. Gross eds.), Biomathematics **18**, Heidelberg: Springer-Verlag, In press.

Lotka, A. J.: Contributions to the analysis of malaria epidemiology, Amer. J. Hyg. **3** (Suppl. 1), 1-121 (1923).

Ross, R.: The prevention of malaria, 2nd edition, London: Murray, 1911.

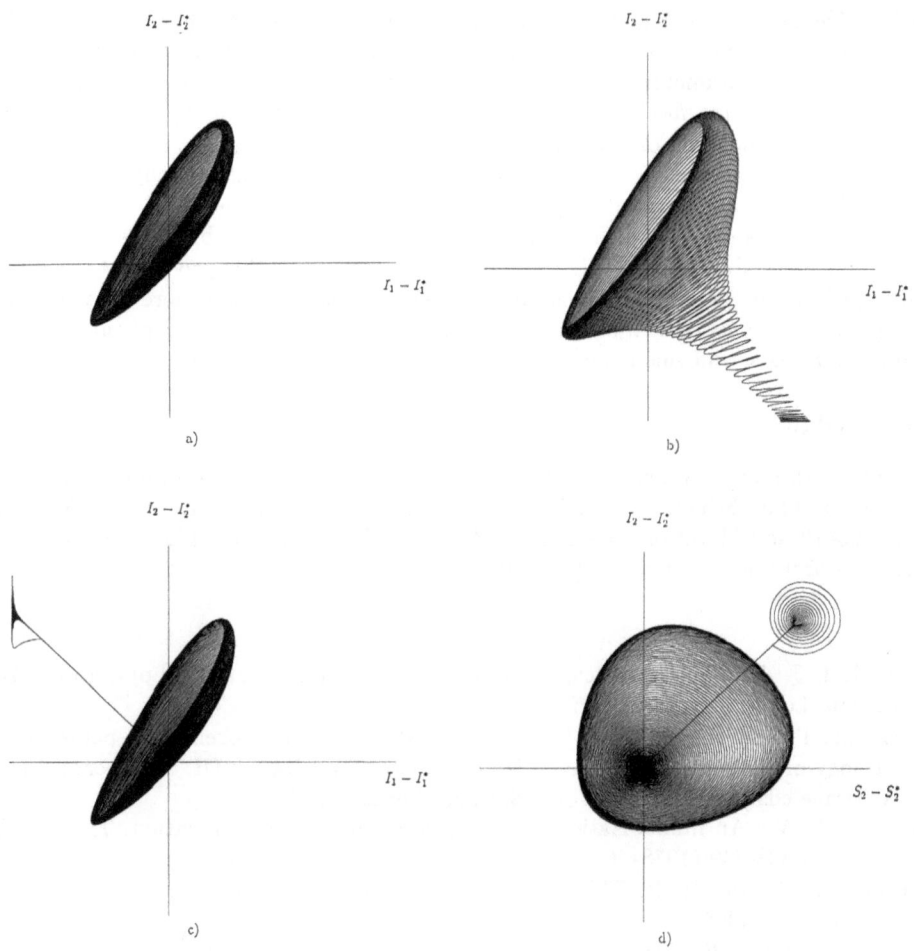

Fig. 1 Projection of the solution curves, a) starting from $(501, 50.1, 50.1, 50.1)$ to the (I_1, I_2) plane, b) starting from $(950, 100, 0.1, 0.1)$ to the (I_1, I_2) plane, c) starting from $(10, 10, 90, 90)$ to the (I_1, I_2) plane, d) starting from $(10, 10, 90, 90)$ to the (S_2, I_2) plane.

EPIDEMIC MODELS IN POPULATIONS OF VARYING SIZE

Fred Brauer
Department of Mathematics
University of Wisconsin
Madison, WI 53706

Abstract

We consider simple compartmental models for infectious diseases with exposed and infective periods of fixed length. In particular, we examine the effect of incorporating nonlinear population dynamics into such models on threshold phenomena and the stability of endemic equilibria.

1. Introduction

The mathematical description of the spread of infectious diseases is often given by compartmental models. It is customary to let $S(t)$ denote the number of members of a population susceptible to the disease, $E(t)$ the number of members exposed but not yet infective, $I(t)$ the number of infective members, and $R(t)$ the number of members who have been removed from the possibility of infection through isolation, immunization, recovery with full immunity or death caused by the disease. The model is then described in terms of the rates of transitions between classes. An SIR model is one in which infectives pass directly to the removed class, as in measles (through immunity) or rabies (through death), while an SIS model is one in which infectives return to the susceptible class on recovery, as in gonorrhea.

One of the oldest and most ubiquitous principles in mathematical epidemiology is the existence of a threshold which distinguishes between possible outcomes. This principle is usually attributed to Kermack and McKendrick (1927), although a threshold phenomenon appears in the earlier work of Ross (1911) on the transmission of malaria.

The nature of a threshold depends on the type of compartmental model. The original threshold principle of Kermack and McKendrick was for an SIR model in a closed population, with no births or deaths. There is a dimensionless quantity, often called the contact number, which determines the course of the disease if a small number of infectives is introduced into the population. If the contact number is less than 1, the function $I(t)$ decreases monotonically to zero (the disease dies out), while if the contact number is greater than 1, the function $I(t)$ first increases and then decreases to zero (an epidemic occurs). If new susceptibles are introduced into a population, either through births or through recovery in an SIS model, the threshold quantity may distinguish between the disease's dying out and an endemic state in which $I(t)$ remains positive at all times. It has been observed that measles is an endemic disease in many communities. By including births of new susceptibles (with an equivalent number of deaths from the removed class to keep the total population size constant) in the Kermack–McKendrick model, this phenomenon can be explained. However, the periodic fluctuations which have been observed in the number of infectives can not be explained in this way.

For a disease which may be endemic, the assumption of a constant total population size is unrealistic because an endemic equilibrium means a state sustained long enough for total population size to vary significantly. We shall investigate compartmental models for infectious diseases which include density–dependent population dynamics to determine whether such extensions can improve understanding. It is necessary to consider two different SIR models (or SEIR models because the inclusion of an exposed class does not add further complication) corresponding to a recovered class as in measles which plays a role in the birth and death processes and a removed (through death) class as in rabies. When births and deaths are included in an SIR model, it is necessary to modify the principle that "so far as transmission of disease is concerned, recovery is a comparatively unimportant event that happens to some cases who have been isolated" [Bailey (1975)].

We are restricting our attention to simple compartmental models, making no attempt to consider models with age structure, different risk groups, or vertical transmission, to mention only a few of the factors which may be important in the study of a specific disease. Our purpose is to consider whether the addition of a birth and death process alters a model's behavior substantially, not to give an overview of epidemic models.

2. SIR models with recovery

The original SIR model of Kermack and McKendrick (1927) is

$$S' = -\beta IS$$
$$I' = \beta IS - \gamma I \qquad\qquad [1]$$
$$R' = \gamma I .$$

It is assumed that the rate of infection is proportional to the number of susceptibles and to the number of infectives; this mass action spread assumption was originally formulated by Hamer (1906) and has been used almost universally since then, although recently some more general models have been studied [Liu, Levin & Iwasa (1986)]. It is also assumed that the recovery rate is γ, meaning that a fraction γ of the number of infectives recover in unit time. This should not be interpreted literally: The differential equation $I' = \beta IS - \gamma I$ in (1) is equivalent to the integral equation

$$I(t) = \int_0^t \beta S(x)I(x)e^{-\gamma(t-x)}dx \qquad\qquad [2]$$

which may be interpreted as saying that the probability of remaining infective at time t after having become infective at time x is $e^{-\gamma(t-x)}$, so that the average length of the infective period is $1/\gamma$. More generally, we could replace (2) by

$$I(t) = \int_0^t \beta S(x)I(x)P(t-x)dx \; , \qquad\qquad\qquad [3]$$

where $P(t-x)$ is the probability of remaining infective at time t after having become infective at time x. In this paper we shall use $P(s) = 1 \; (0 \leq s \leq \tau)$, $P(s) = 0 \; (s > \tau)$, corresponding to an infective period of fixed length τ instead of an exponentially distributed infective period. These two forms for the infective period appear to lead to similar behavior in models of the type we are considering, and there is reason to believe that these models are quite insensitive to changes in the distribution of infective periods. Good descriptions of the properties of models with exponentially distributed infective periods may be found in [Anderson & May (1979), Hethcote, Stech & van den Driessche (1981b), May & Anderson (1979)].

If new susceptible members are added to the population, even if the total population size is held constant, there is a significant change in the behavior of the system. The first formulation of such a model was made by Soper (1929) in an attempt to explain the possibility of an oscillatory endemic equilibrium which had been observed in some studies of measles. Soper's model is

$$S' = -\beta IS + \mu$$
$$I' = \beta IS - \gamma I \qquad\qquad\qquad [4]$$
$$R' = \gamma I - \mu \; ,$$

assuming a death rate μ from the removed class and a birth rate of new susceptibles which is μ greater than the death rate of susceptibles. The assumption that the birth rate of susceptibles is linked in this way to the death rate of removed members is unreasonable biologically. Soper's model is also flawed mathematically; because R can become negative it is not properly posed. The predictions of Soper's model are not consistent with observation because the model implies that there must be an endemic equilibrium which is stable with damped oscillations while observation indicates that there may or may not be an endemic equilibrium, and there may be sustained oscillations about an endemic equilibrium.

The difficulties with Soper's model were partially resolved by Hethcote (1974), who assumed deaths in each class and obtained a threshold phenomenon distinguishing between dying out of the disease and a stable endemic equilibrium; more complicated assumptions such as seasonal variations in the contact rate must be made to produce sustained oscillations.

Another line of work based on Soper's idea is the epidemic model of Wilson & Burke (1942)

$$S'(t) = A - \beta S(t)[A\tau - S(t-\tau-\sigma) - S(t-\sigma)] \qquad\qquad\qquad [5]$$

which assumed an infective period of fixed length τ following an exposed or latent period of fixed length σ, and a constant rate A of recruitment of new susceptible members. The Wilson–Burke equation is derived from the system

$$S'(t) = A - \beta S(t)I(t)$$

$$E(t) = \int_{t-\sigma}^{t} \beta S(x)I(x)dx$$

$$I(t) = \int_{t-\tau-\sigma}^{t-\sigma} \beta S(x)I(x)dx$$

[6]

$$R'(t) = \beta S(t-\tau-\sigma)I(t-\tau-\sigma) - A$$

or the differentiated form of (6)

$$S'(t) = A - \beta S(t)I(t)$$
$$E'(t) = \beta S(t)I(t) - \beta S(t-\sigma)I(t-\sigma)$$
$$I'(t) = \beta S(t-\sigma)I(t-\sigma) - \beta S(t-\tau-\sigma)I(t-\tau-\sigma)$$
$$R'(t) = \beta S(t-\tau-\sigma)I(t-\tau-\sigma) - A .$$

Observe that $E(t)$ and $R(t)$ are determined when $S(t)$ and $I(t)$ are known and that

$$I(t) = \int_{t-\tau-\sigma}^{t-\sigma} \beta S(x)I(x)dx$$

[7]

$$= \int_{t-\tau-\sigma}^{t-\sigma} [A-S'(x)]dx = A\tau - S(t-\sigma) + S(t-\tau-\sigma) .$$

Substitution of (7) into $S'(t) = A-\beta S(t)I(t)$ gives (5).

The Wilson–Burke equation (5) has an endemic equilibrium $S = \frac{1}{\beta\tau}$ if $A > 0$, shown to be asymptotically stable by Wilkins (1945). The case $A = 0$, discussed by Wilson & Worcester (1944) is actually more complicated mathematically; the problem (5) with $A = 0$ is not properly posed because every constant function is a solution. By restricting (5) to act only on functions satisfying the initial data this problem could be avoided [Busenberg & Cooke (1980)]. However, the model is sensitive to changes in initial data. It is not difficult to show that if monotone decreasing data is given on an initial interval $-(\sigma+\tau) \leq t \leq 0$, then $S(t)$ is monotone decreasing on $0 \leq t < \infty$ and tends to a limit which depends on the initial data. It is also known [Cooke & Yorke (1973)] that if there is no exposed period ($\sigma = 0$) and no source of new susceptibles ($A = 0$) but an arbitrary distribution of infective periods, then every solution tends to a limit as $t \to \infty$.

3. SIR models for fatal diseases

The Wilson–Burke model shares the flaws of the Soper model, although this is less apparent because the model is described by a single equation for $S(t)$. The basic problem is the linkage which must be assumed between the birth rate of new susceptibles and the death rate in the removed class. However, for a disease like rabies or rinderpest which kills almost all its victims, so that the removed class becomes the class of members killed by the disease it is reasonable to assume that births and deaths other than those caused by the disease depend on the number of susceptible members. Thus it is appropriate to consider a model

$$S'(t) = g\{S(t)\} - \beta S(t)I(t)$$

$$E(t) = \int_{t-\sigma}^{t} \beta S(x)I(x)dx$$

$$I(t) = \int_{t-\tau-\sigma}^{t-\sigma} \beta S(x)I(x)dx$$

$$R'(t) = \beta S(t-\tau-\sigma)I(t-\tau-\sigma)$$

[8]

with g(S) representing the excess of births over deaths in unit time when the susceptible population is S. Grossman (1980) considers similar equations with deaths allowed in the classes E and I.

Like the Wilson–Burke model (5), the system (8) can be reduced to a single equation by observing that E(t) and R(t) are determined when S(t) and I(t) are known and that

$$I(t) = \int_{t-\tau-\sigma}^{t-\sigma} \beta S(u)I(u)du = \int_{t-\tau-\sigma}^{t-\sigma} [g\{S(u)\}-S'(u)]du$$

$$= \int_{t-\tau-\sigma}^{t-\sigma} g\{S(u)\}du - S(t-\sigma) + S(t-\tau-\sigma) .$$

The resulting equation for S(t) is

$$S'(t) = g\{S(t)\} - \beta S(t)[S(t-\tau-\sigma)-S(t-\sigma) + \int_{t-\tau-\sigma}^{t-\sigma} g\{S(u)\}du] .$$

[9]

In order to study the possible equilibria of the equation (9), we assume that the population has a carrying capacity K in the absence of the disease, that is

$$g(0) = 0, \ g(K) = 0, \ g(S) > 0 \ \text{for} \ 0 < S < K .$$

[10]

It will also be necessary to require

$$g(S) > Sg'(S), \ 0 < S < K .$$

[11]

If g"(S) < 0 for 0 < S < K, the condition (10) is satisfied; it is also satisfied for other functions g such as $g(S) = rSe^{1-\frac{S}{L}} - dS$ with $K = L(1-\log d/r)$, suggested by Blythe, Nisbet & Gurney (1982) for modelling populations with intra–specific competition of scramble type.

The equilibria of the equation (9) are the solutions S_∞ of

$$g(S_\infty) - \beta\tau S_\infty g(S_\infty) = 0 .$$

Thus either $g(S_\infty) = 0$, which implies either $S_\infty = 0$ (extinction) or $S_\infty = K$ (disappearance of the disease), or $S_\infty = \frac{1}{\beta\tau}$ (endemic equilibrium). To analyze the stability of each equilibrium, we follow the standard procedure of linearizing the equation about the equilibrium and then constructing the characteristic equation — the condition on λ which makes $S(t) = ce^{\lambda t}$ a solution of the linearized equation. An equilibrium is stable if all roots of this characteristic equation have negative real part (except possibly for zero roots which have been introduced by using a differentiated form of the model for which every constant is a solution). The linearization of (9) about an equilibrium S_∞ is

$$u'(t) = [g'(S_\infty)-\beta\tau g(S_\infty)]u(t) + \beta S_\infty u(t-\sigma)$$

[12]

$$-\beta S_\infty u(t-\tau-\sigma) - \beta S_\infty g'(S_\infty) \int_{t-\tau-\sigma}^{t-\sigma} u(x)dx$$

and the corresponding characteristic equation is

$$\lambda + a = be^{-\lambda\sigma}(1-e^{-\lambda\tau}) + c \int_{-\tau-\sigma}^{-\sigma} e^{\lambda x}dx$$

[13]

$$= (b\lambda+c)e^{-\lambda\sigma}\left[\frac{1-e^{-\lambda\tau}}{\lambda}\right]$$

with

$$a = \beta\tau g(S_\infty) - g'(S_\infty), \quad b = \beta S_\infty , \quad c = -\beta S_\infty g'(S_\infty) .$$

[14]

The linearization at the equilibrium $S_\infty = 0$ is $u'(t) = g'(0)u(t)$, and the hypothesis (10) implies that this equilibrium is unstable. In analyzing the stability of the other two equilibria, we will require the following special case of a result of Hethcote, Stech, and van den Driessche (1981a) for the characteristic equation (13).

Lemma 1: All roots of the equation (13) with $\tau \geq 0$, $\sigma \geq 0$, $a > |c|\tau > 0$, $|b|\tau \leq 1$ satisfy $R\lambda < 0$. If $\tau \geq 0$, $\sigma \geq 0$, $b\tau = 1$, $a > 0$, there are values of c (necessarily satisfying $-c\tau > a > 0$) such that (13) has roots with $R\lambda \geq 0$. If $a-c\tau < 0$, there are roots of (13) with $R\lambda \geq 0$.

For the equilibrium $S_\infty = K$, we have $a = -g'(K) > 0$, $b = \beta K$, $c = -\beta K g'(K) > 0$ since $g(K) = 0$, $g'(K) < 0$. By Lemma 1, this equilibrium is stable if $\beta \tau K < 1$ and unstable if $\beta \tau K > 1$. Behavior near the equilibrium $S_\infty = \frac{1}{\beta \tau}$ is more complicated. At this equilibrium,

$a = \beta \tau g(\frac{1}{\beta \tau}) - g'(\frac{1}{\beta \tau})$, positive by the hypothesis (11), $b = 1/\tau$, and $c = -g'(\frac{1}{\beta \tau})/\tau$. Thus $a-c\tau > 0$ if $\beta \tau K > 1$ so that $g(\frac{1}{\beta \tau}) > 0$, and it follows that if $g'(\frac{1}{\beta \tau}) < 0$, so that $c > 0$, the equilibrium is stable. If $g'(\frac{1}{\beta \tau}) > 0$, so that $c < 0$, we still have stability if $a+c\tau > 0$, or $\beta \tau g(\frac{1}{\beta \tau}) - 2g'(\frac{1}{\beta \tau}) > 0$. In the logistic case, $g(S) = rS(1-\frac{S}{K})$, this is equivalent to $\beta \tau K < 3$. In general, there will always be a transition to instability of this equilibrium, corresponding to sustained oscillations about the equilibrium if $\beta \tau K$ is sufficiently large. We conclude that there is a threshold phenomenon: If $\beta \tau K < 1$ the only stable equilibrium is $S_\infty = K$, and the disease disappears, but if $\beta \tau K > 1$ there is an endemic equilibrium $S_\infty = \frac{1}{\beta \tau}$ which is stable if $\beta \tau K$ is sufficiently close to 1 and unstable, with oscillations about the equilibrium if $\beta \tau K$ is sufficiently large. This result is the same qualitatively as results obtained for the model

$$S' = rS - \gamma SN - \beta SI$$
$$E' = \beta SI - (\sigma+b+\gamma N)E \qquad [15]$$
$$I' = \sigma E - (\alpha+b+\gamma N)I$$

proposed for studying fox rabies in Europe [Anderson, Jackson, May & Smith (1981)]. Here, N is the total population size, $N = S+E+I$, a is the birth rate of new susceptibles, b is the natural death rate in each class, $1/\sigma$ is the average time in the exposed class, and $1/\alpha$ is the average time in the infective class. The fox population is assumed to grow logistically in the absence of disease with a carrying capacity $K = \sigma/\gamma$.

Field studies for fox rabies in Europe indicate that there is a critical carrying capacity K_T of approximately 1 fox/km^2 below which rabies dies out and above which rabies becomes endemic. Our model (9) gives $K_T = \frac{1}{\beta \tau}$, and since τ is approximately 5 days, or $\frac{1}{73}$ year, we may use this data to estimate the transmission coefficient β as 73 km^2/year. For the model (15), $K_T = \frac{(\sigma+a)(\alpha+a)}{\beta \sigma}$, which gives $\beta = 79.7$, using the observed values $a = 1$, $\sigma = 13$, $\alpha = 73$. The difference between the values of β given by the two models is well within the uncertainty of experimental data. The two models are similar in their incorporation of population dynamics although they differ in the specifics as well as in the distribution of exposed and infective periods. They agree with each other and with observed reality in predicting that rabies will die out in regions of low fox density and will be endemic otherwise, with stable oscillations about an endemic equilibrium in regions of high fox density. The model (9) may be studied analytically more easily than the model (15) because it is described by a single equation.

From the original model (8), it is easy to calculate the equilibrium values of I and E at an endemic equilibrium along with $S_\infty = \frac{1}{\beta\tau}$, namely

$$I_\infty = \tau g(\tfrac{1}{\beta\tau}), \ E_\infty = \sigma g(\tfrac{1}{\beta\tau}) \ .$$

Then the total equilibrium population in the logistic case is

$$S_\infty = E_\infty + I_\infty = \frac{1}{\beta\tau}[\frac{r(\sigma+\tau)}{\beta\tau K}\frac{(\beta\tau K-1)}{} + 1]$$

and the reduction in the total equilibrium population size from the carrying capacity is

$$K - (S_\infty + E_\infty + I_\infty) = \frac{\beta\tau K-1}{\beta\tau}[1 - \frac{r(\sigma+\tau)}{\beta\tau k}] \ .$$

When K is large, this approaches K. Even though the equilibrium $S_\infty = 0$ is unstable, there may be oscillations about an endemic equilibrium in regions of high fox density which bring the fox population low–enough that a small perturbation may cause extinction of the fox population in the region.

The models (9) and (15) are completely inappropriate for a disease such as measles in which recovered members of the population may be expected to play a role in the birth and death process (even though it was developed from a model which was originally designed for measles and even though it exhibits behavior similar to what is observed for measles). To model such a disease, we may continue to assume that all births are in the susceptible class, but we must apportion the deaths among the various classes. However, it may be verified that such a model will not exhibit behavior different from what is predicted by a model with new births in the susceptible class and constant total population size, [Hethcote (1974)] but we shall not attempt to describe more elaborate models here. Our main conclusion is that for SEIR models of rabies type, the introduction of population dynamics is essential, but for SEIR models of measles type the introduction of more realistic population dynamics does not yield more realistic results. In order to construct a model which predicts sustained oscillations we must make other hypothesis on the disease, such as seasonal periodicity in the transmission parameter β [see, for example, London & Yorke (1973), Dietz (1976), Grossman (1980)].

4. SIS models

For an infectious disease which confers no immunity on recovery, the possibility of an endemic equilibrium arises even for simple models without births or deaths [Hethcote, Stech, & van den Driessche (1981a)]. Gonorrhea is a disease of this type which has been modelled mathematically; influenza could also be cited as an example. Let us examine the effect of including density–dependent births and deaths in such a model. We assume, as before, that there is an exposed period of fixed length $\sigma \geq 0$ followed by an infective period of fixed length $\tau > 0$. Now, however, we assume that infective members recover and return immediately to the susceptible class. We assume that all newly born members of the population are susceptible, and that the exposed and infective periods are sufficiently short that all deaths occur while members are in the susceptible class, and that the birth and death rates depend on total population size.

The total population size $N = S + E + I$ satisfies an ordinary differential equation $N' = g(N)$, and we assume that the population has a carrying capacity K in the absence of the disease, so that (10) is satisfied. The disease is modelled by the system

$$S'(t) = g\{N(t)\} - \beta S(t)I(t) + \beta S(t-\tau-\sigma)I(t-\tau-\sigma)$$

$$E(t) = \int_{t-\sigma}^{t} \beta S(x)I(x)dx \qquad\qquad [16]$$

$$I(t) = \int_{t-\tau-\sigma}^{t-\sigma} \beta S(x)I(x)dx \ .$$

It is convenient to replace the equation for E in (16) by an equation for N, and to write the model as

$$S'(t) = g\{N(t)\} - \beta S(t)I(t) + \beta S(t-\tau-\sigma)I(t-\tau-\sigma)$$

$$I(t) = \int_{t-\tau-\sigma}^{t-\sigma} \beta S(x)I(x)dx \qquad\qquad [17]$$

$$N'(t) = g\{N(t)\} \ .$$

The equilibria $(S_\infty, I_\infty, N_\infty)$ of (17) are given by $N_\infty = K, I_\infty = \beta\tau S_\infty I_\infty$. The value E_∞ of E corresponding to an equilibrium is given by $E_\infty = \frac{\sigma}{\tau}I_\infty$. The relation $I_\infty = \beta\tau S_\infty I_\infty$ implies that either $S_\infty = \frac{1}{\beta\tau}$ or $I_\infty = 0$. Thus there are two possibilites: $S_\infty = K, I_\infty = 0, N_\infty = K$ (with $E_\infty = 0$), and $S_\infty = \frac{1}{\beta\tau}, I_\infty = \frac{\beta\tau K - 1}{\beta(\sigma+\tau)}, N_\infty = K$ (with $E_\infty = \frac{\sigma(\beta\tau K - 1)}{\beta\tau(S+\tau)}$). The possibility of an endemic equilibrium with $I_\infty > 0$ arises only if $\beta\tau K > 1$. The linearization of (17) about an equilibrium (S_∞, I_∞, K) is

$$u'(t) = -\beta I_\infty u(t) - \beta S_\infty v(t) + g'(K)w(t) + \beta I_\infty u(t-\tau-\sigma) + \beta S_\infty v(t-\tau-\sigma)$$

$$v(t) = \int_{t-\tau-\sigma}^{t-\sigma} [\beta S_\infty v(x) + \beta I_\infty u(x)]dx \qquad [18]$$

$$w'(t) = g'(K)w(t) .$$

The corresponding characteristic equation is

$$[g'(K)-\lambda]\det \begin{vmatrix} -\beta I_\infty\{1-e^{-\lambda(\tau+\sigma)}\} - \lambda & -\beta S_\infty\{1-e^{-\lambda(\tau+\sigma)}\} \\ \beta I_\infty \int_{-\tau-\sigma}^{-\sigma} e^{\lambda x}dx & \beta S_\infty \int_{-\tau-\sigma}^{-\sigma} e^{\lambda x}dx-1 \end{vmatrix} = 0 .$$

The factor $g'(K) - \lambda$ gives a negative root because of (10), and may be removed. Thus we obtain a characteristic equation

$$\det \begin{vmatrix} -\beta I_\infty\{1-e^{-\lambda(\tau+\sigma)}\} - \lambda & -\beta S_\infty\{1-e^{-\lambda(\tau+\sigma)}\} \\ \beta I_\infty \int_{-\tau-\sigma}^{-\sigma} e^{\lambda x}dx & \beta S_\infty \int_{-\tau-\sigma}^{-\sigma} e^{\lambda x}dx - 1 \end{vmatrix} = 0 . \qquad [19]$$

As this equation does not involve the function g in any way, the stability properties of the model (16) must be identical with those of the model (16) but with $g(N) \equiv 0$. In other words, the assumption of density–dependent births and deaths does not alter the behavior of the model. The possibility of an endemic equilibrium depends on a flow of new susceptibles, which may come through recovery from the disease rather than through new births.

The characteristic equation (19) has the form

$$1 = -a\frac{1-e^{-\lambda(\tau+\sigma)}}{\lambda} + be^{-\lambda\sigma}\frac{1-e^{-\lambda\tau}}{\lambda} \qquad [20]$$

with $a = \beta I_\infty \geq 0$, $b = \beta S_\infty > 0$ (after the removal of a common factor λ caused by the use of the differentiated form of the equation for S). In analyzing the stability of each of the equilibria of (17) we will require the following result about the characteristic equation (20), whose proof may be found in the appendix.

Lemma 2: All roots of the equation (20) satisfy $R\lambda < 0$ if $\tau \geq 0$, $\sigma \geq 0$, $a \geq 0$, $b \geq 0$, and

$$b(b-a)\tau^2 < 1 . \qquad [21]$$

For the equilibrium $S_\infty = K$, $I_\infty = 0$, the condition (21) becomes $\beta\tau K < 1$, and this equilibrium is stable if $\beta\tau K < 1$. For the equilibrium $S_\infty = \frac{1}{\beta\tau}$, $I_\infty = \frac{\beta\tau K-1}{\beta(\sigma+\tau)}$, we have

$$b(b-a)\tau^2 = \beta^2\tau^2 S_\infty(S_\infty-I_\infty) = \frac{(\sigma+\tau)-\tau(\beta\tau K-1)}{\sigma+\tau}$$

and this is less than 1 if and only if $\beta\tau K > 1$. Thus the endemic equilibrium is stable if $\beta\tau K > 1$ and the equilibrium corresponding to vanishing of the disease is stable if $\beta\tau K < 1$. Again, we have a threshold phenomenon, with the threshold quantity depending on the transmission rate, the length of the infective period, and the population size. Unlike the rabies–type SEIR model, there is no loss of stability when the threshold quantity increases, and thus no possibility of sustained oscillations.

5. Conclusions

Mathematical models for the spread of infectious diseases may be quite simple, looking for qualitative conclusions about a class of diseases, or complex, looking for quantitative conclusions about a specific disease. While complex models are essential for the planning of detailed control schemes, the importance of studying simple models should not be underestimated. Here, we have taken a minimalist view, asking the question whether the incorporation of population dynamics into simple compartmental models improves qualitative understanding.

In general terms, the answer to this question depends on the nature of the disease being studied. For a universally fatal disease, the incorporation of density–dependent birth and death rates allows the possibilities of a stable endemic equilibrium and of sustained oscillations about an endemic equilibrium for large contact numbers of the infection. These possibilities are observed in real life, and the incorporation of relatively simple population dynamics adds a great deal to the correlation between model and observation. On the other hand, for a non–fatal disease which confers no immunity the incorporation of population dynamics adds absolutely nothing. This suggests that in modelling any infectious disease, a continuing flow of new susceptibles either through birth or recovery without immunity is an essential feature.

For a disease from which victims recover with immunity, the incorporation of population dynamics adds the possibility of a threshold between disappearance of the disease and a stable endemic equilibrium, but not oscillations about an endemic equilibrium. In order to model such a disease more realistically it is necessary to make more specific assumptions on the transmission of the disease. It would be of interest to explore the extent to which more realistic population dynamics combined with such transmission assumptions affect the predictions of the model.

It is a broad principle in population dynamics that long delays, such as in recruitment, may product oscillatory behaviour [Blythe, Nisbet & Gurney (1982)]. There is no evidence that long delays, such as an exposed period, tend to produce oscillations in simple epidemiological models but an incubation period in a vertically transmitted disease [Busenberg, Cooke & Pozio (1983)] may lead to oscillations. The mathematical question of what form a delay equation must have to produce oscillations and the corresponding biological question of what types of delays tend to produce oscillations are not completely understood. From a mathematical point of view we have been studying models with two delays, even though τ is not a delay from a biological point of view. Two–delay models have not been studied extensively.

We have not explored the effect of the introduction of population dynamics on the outcome of control strategies for various types of diseases. While detailed strategies for the control of specific diseases should be developed from more detailed models, it is possible that general principles on the effectiveness of various approaches may be modified by the inclusion of birth and death processes in a model.

REFERENCES

Anderson, R.M., H.C. Jackson, R.M. May, and A. M. Smith. 1981. Population dynamics of fox rabies in Europe. *Nature* **289**, pp. 765–771.

Anderson, R.M., and R.M. May. 1979. Population biology of infectious diseases: I. *Nature* **280**, pp. 361–367.

Bailey, N.T.J. 1975. *The Mathematical Theory of Infectious Diseases*, 2nd ed., Hafner.

Blythe, S.P., R.M. Nisbet, and W.S.C. Gurney. 1982. Instability and complex dynamic behavior in population models with long time delays. *Theor. Pop. Biol.* **22**, pp. 147–176.

Busenberg, S. and K.L. Cooke. 1980. The effect of integral conditions in certain equations modelling epidemics and population growth. *J. Math. Biology* **10**, pp. 13–32.

Busenberg, S., K.L. Cooke, and M.A. Pozio. 1983. Analysis of a model of a vertically transmitted disease. *J. Math. Biology* **17**, pp. 305–329.

Cooke, K.L. and J.A. Yorke. 1973. Some equations modelling growth process and gonorrhea epidemics. *Math. Biosciences* **16**, pp. 75–101.

Dietz, K. 1976. The incidence of infectious diseases under the influence of seasonal fluctuations, in *"Mathematical Models in Medicine"*, Lecture Notes in Biomathematics No. 11, Springer–Verlag, pp. 1–15.

Grossman, Z. 1980. Oscillatory phenomena in a model of infectious diseases. *Theor. Pop. Biol.* **18**, pp. 204–243.

Hamer, W.H. 1906. Epidemic disease in England. *Lancet* **1**, pp. 733–739.

Hethcote, H. 1974. Asymptotic behavior and stability in epidemic models, in *"Mathematical Problems in Biology"*, Lecture Notes in Biomathematics No. 2, Springer–Verlag, pp. 83–92.

Hethcote, H., H.W. Stech, and P. van den Driessche. 1981a. Stability analysis for models of diseases without immunity. *J. Math. Biol.* 13, pp. 185–198.

_____. 1981b. Periodicity and stability in epidemic models: a survey, in *"Differential Equations and Applications in Ecology, Epidemics, and Population Models"*. Academic Press, pp. 65–82.

Kermack, W.O. and A.G. McKendrick. 1927. *A contribution to the mathematical theory of epidemics.* Proc. Royal Soc. of London, Series A, 115, pp. 700–721.

Liu, Wei–min, S.A. Levin and Y. Iwasa. 1986. Influence of nonlinear incidence rates upon the behavior of SIRS epidemiological models. *J. Math. Biol.*, 23, pp. 187–204.

London, W.P. and J.A. Yorke. 1973. Recurrent outbreaks of measles, chickenpox and mumps. I. Seasonal variation in contact rates. *Am. J. Epidemiology*, 98, pp. 453–468.

May, R.M. and R.M. Anderson. 1979. Population biology of infectious diseases, II. *Nature* 280, pp. 455–461.

Ross, R. 1911. *The Prevention of Malaria.* 2nd ed., Murray.

Soper, H.E. 1929. Interpretation of periodicity in disease–prevalence. *J. Royal Statistical Soc.* 92, pp. 34–73.

Wilkins, Jr., J.E. 1945. The differential–difference equation for epidemics. *Bull. Math. Biophys.* 7, pp. 149–150.

Wilson, E.B. and M.H. Burke. 1942. The epidemic curve. *Proc. Nat. Acad. Sci.* 28, pp. 361–367.

_____ and J. Worcester. 1944. A second approximation to Soper's epidemic curve. *Proc. Nat. Acad. Sci.* 30, pp. 37–44.

Appendix

In Section 4, we stated a result (Lemma 2) for a characteristic equation of a type originally considered in [Hethcote, Stech, & van den Driessche (1981a)]. Here, we shall prove this lemma and describe the extensions which can be obtained by similar methods.

Lemma 2: If $\tau \geq 0$, $\sigma \geq 0$, $a \geq 0$, $b \geq 0$, and

$$b(b-a)\tau^2 < 1 ,$$ [A1]

then all roots of the equation

$$1 = -a\frac{1-e^{-\lambda(\tau+\sigma)}}{\lambda} + be^{-\lambda\sigma}(\frac{1-e^{-\lambda\tau}}{\lambda})$$ [A2]

satisfy $R\lambda < 0$.

Proof: If $a \geq 0$, $b \geq 0$, then for $R\lambda \geq 0$,

$$\left|-a(\frac{1-e^{-\lambda(\tau+\sigma)}}{\lambda}) + be^{-\lambda\sigma}(\frac{1-e^{-\lambda\tau}}{\lambda})\right| \leq a(\tau+\sigma) + b\tau .$$

Thus if a and b are sufficiently small, for example if $a(\tau+\sigma) \leq \frac{1}{2}$ and $b\tau \leq \frac{1}{2}$,

$$\left|-a(\frac{1-e^{-\lambda(\tau+\sigma)}}{\lambda}) + be^{-\lambda\sigma}(\frac{1-e^{-\lambda\tau}}{\lambda})\right| < 1$$

for $R\lambda \geq 0$ and all roots of (A2) satisfy $R\lambda < 0$. All roots of (A2) continue to satisfy $R\lambda < 0$ as a and b are increased so long as neither $\lambda = 0$ nor $\lambda = iy$ with $y > 0$ is a root of (A2). The condition that $\lambda = 0$ be a root of (A2) is $1 + a(\tau+\sigma) = b\tau$. If $\lambda = iy$ is a root of (A2), by writing (A2) in the form

$$\lambda + a(1-e^{-\lambda(\tau+\sigma)}) = be^{-\lambda\sigma}(1-e^{-\lambda\tau})$$

we see that

$$iy + a\{1-\cos y(\tau+\sigma)\} + ia\sin y(\tau+\sigma)\}$$

$$= b(\cos y\sigma - i\sin y\sigma)\{(1-\cos y\tau)+i\sin y\tau\} .$$ [A3]

We obtain

$$a = b\cos y\sigma - (b-a)\cos y(\tau+\sigma)$$

$$y = -b\sin y\sigma + (b-a)\sin y(\tau+\sigma) .$$

This implies

$$a^2 + y^2 = (b-a)^2 + b^2 - 2b(b-a)\cos y\tau \,,$$

$$y^2 = 2b(b-a)(1-\cos y\tau) = 4b(b-a)\sin^2 \tfrac{y\tau}{2} \,. \tag{A4}$$

If $b < a$, A4 can not be satisfied and if $b \geq a$, then $4b(b-a)\sin^2 \tfrac{y\tau}{2} < b(b-a)\tau^2 y^2$ because $\sin^2 \tfrac{y\tau}{2} < (\tfrac{y\tau}{2})^2$ for $y > 0$; thus the equation (A4) can not be satisfied if (A1) holds.

We think of σ and τ as fixed and consider the (a,b) plane. The roots of (A2) satisfy $R\lambda < 0$ for all (a,b) in the region containing the origin and bounded by the line $b\tau = 1 + a(\tau+\sigma)$ and the hyperbola $b(b-a)\tau^2 = 1$ because if (a,b) does not cross these curves no root can cross the imaginary axis. For $a \geq 0, b \geq 0$ the hyperbola lies below the line and thus all roots of (A2) satisfy $R\lambda < 0$ if $a \geq 0, b \geq 0, \tau \geq 0, \sigma \geq 0$ and (A1) is satisfied.

The argument used to establish Lemma 2 does not depend on the conditions $a \geq 0, b \geq 0$; these are imposed merely to permit us to restrict our attention to the first quadrant of the (a,b) plane. The full region of the (a,b) plane for which all roots of (A2) satisfy $R\lambda < 0$ is the region containing the origin and bounded by the line $b\tau = 1 + a(\tau+\sigma)$ and the hyperbola $b(b-a)\tau^2 = 1$. The portion of this region in the first quadrant is the set of points below the hyperbola. The portion in the second quadrant is the triangle bounded by the axes and the line $b\tau = 1 + a(\tau+\sigma)$, joining the points $(-\tfrac{1}{\tau+\sigma}, 0)$ and $(0, \tfrac{1}{\tau})$. In the third quadrant the desired region is bounded above by the line $b\tau = 1 + a(\tau+\sigma)$ and below by the hyperbola $b(b-a)\tau^2 = 1$, to the right of their intersection at $(-\tfrac{\tau+2\sigma}{\sigma(\tau+\sigma)}, -\tfrac{\tau+\sigma}{\tau\sigma})$, and in the fourth quadrant the desired region is the set of points above the hyperbola $b(b-a)\tau^2 = 1$.

STABILITY AND THRESHOLDS IN SOME
AGE-STRUCTURED EPIDEMICS

Stavros Busenberg
Department of Mathematics
Harvey Mudd College
Claremont, CA 91711
U.S.A.

Kenneth Cooke
Department of Mathematics
Pomona College
Claremont, CA 91711
U.S.A.

Mimmo Iannelli
Dipartimento di Matematica
Università di Trento
38050 Povo (Trento)
Italy

Abstract

Age-structured models for diseases that can be transmitted both horizontally and vertically are derived and analyzed. The relation between these models and the catalytic curve models of epidemics is explicitly given. The possibility of using the catalytic curve to deduce information about the age-dependent contact rate and to identify the presence of vertical transmission is demonstrated. For certain age-dependent forms of the force of infection terms, explicit endemic threshold criteria are derived, and the stability of the steady states is determined.

1. Introduction and description of results.

In this paper we report on continuing studies of age-structured models of epidemics which are formulated in the McKendrick [7] type of equations. The models are of the $s \to i \to r$ and $s \to i \to s$ types, that is, where the disease either imparts total immunity or else does not provide any immunity. In both cases we will also consider the possibility that the disease can be vertically transmitted. We shall summarize and extend analytical results that we have recently obtained, [2], and describe additional computer simulations.

We use the following equations describing the dynamics of the transmission of the disease:

$$\frac{\partial s}{\partial t} + \frac{\partial s}{\partial a} + \mu s = -\lambda s + \gamma i$$

$$\frac{\partial i}{\partial t} + \frac{\partial i}{\partial a} + \mu i = \lambda s - (\gamma + \delta)i \qquad (1.1)$$

$$\frac{\partial r}{\partial t} + \frac{\partial r}{\partial a} + \mu r = \delta i$$

$$i(0,t) = b_1(t), \quad s(0,t) = b_2(t), \quad r(0,t) = 0$$

$$i(a,0) = i_0(a), \quad s(a,0) = s_0(a), \quad r(a,0) = r_0(a).$$

The basic variables in our model are the age specific density of susceptible, infective and immune individuals, $s(a,t)$, $i(a,t)$ and $r(a,t)$, respectively, of age a at time t. Thus

---

The work of the first author was partially supported by NSF Grant No. DMS-8703631 and that of the second author by NSF Grant No. DMS-8603450.

the total number of susceptibles $S(t)$ at time t is given by $S(t) = \int_0^\infty s(a,t)da$. The parameters $\beta(a)$, $\gamma(a) + \delta(a)$, and $\mu(a)$ denote the age-specific birth, recovery and death rates, and $\lambda(a,t)$ the age-specific force of infection, that is, the probability for a susceptible of age a to become infective in a unit time interval. The birth functions $b_1(t)$ and $b_2(t)$ are taken as linear and have the forms

$$b_1(t) = \int_0^\infty q\beta(a)i(a,t)da, \quad b_2(t) = \int_0^\infty \beta(a)[s(a,t) + (1-q)i(a,t) + r(a,t)]da$$

where $0 \le q \le 1$ is the probability that the disease be vertically transmitted. When there is no vertical transmission $q = 0$, and hence, $i(0,t) = 0$, that is, all newborns are susceptible. In this model we have not included the possibility of maternally transmitted immunity, hence, $r(0,t) = 0$. When the disease does not impart immunity, then $\delta(a) \equiv 0$, and the model becomes of the $s \to i \to s$ type; while if there is no recovery, then $\gamma = 0$, and the model becomes of the $s \to i \to r$ type.

In this model the disease does not significantly affect the death rate, so μ is taken to be the same for all three subpopulations. Hence, adding the equations in (1.1) we obtain the following familiar problems for the total population density $p(a,t) = s(a,t) + i(a,t) + r(a,t)$,

$$\frac{\partial p}{\partial t} + \frac{\partial p}{\partial a} + \mu p = 0$$

$$p(0,t) = \int_0^\infty \beta(a)p(a,t)da \qquad (1.2)$$

$$p(a,0) = p_0(a).$$

This is the standard McKendrick-von Foerster equation for which we make the following typical hypotheses:

(H_1) β, μ and p_0 are non-negative, piecewise continuous functions on $[0,\infty)$.

(H_2) $\beta(\alpha) > 0$ for $a \in (A_0, A)$ and $\beta(a) = 0$ for $a \notin (A_0, A)$.

(H_3) $\int_0^\infty e^{-\int_0^a \mu(\sigma)d\sigma}\, da < \infty$.

Note that H_3 implies that $\int_0^\infty \mu(\sigma)d\sigma = \infty$.

Under these hypotheses, (1.2) has a steady state solution (see [1], for example)

$$p(a,t) = p_\infty(a) = b_0 e^{-\int_0^a \mu(\sigma)d\sigma}, \quad t \in [0,\infty), \qquad (1.3)$$

if and only if, the net population reproduction rate R is equal to one:

$$R = \int_0^\infty \beta(a)e^{-\int_0^a \mu(\sigma)d\sigma}\, da = 1. \qquad (1.4)$$

We consider the two following forms of the force of infection:

(i) *Intracohort*: $\lambda(a,t) = \kappa(a)i(a,t)$

(ii) *Intercohort*: $\lambda(a,t) = \kappa(a)I(t) = \kappa(a)\int_0^\infty i(a,t)da$

For the parameters γ and κ, we make the following hypothesis:

(H_4) γ and κ are non-negative piecewise continuous functions on $[0,\infty)$ and κ is bounded.

In the next section we give our results on the relation between the $s \to i \to r$ age-dependent model and the catalytic curve models discussed in Muench [8]. This analysis extends out previous work [2] to a slightly more general case. In Section 3 we state the precise results obtained in [2] for the intracohort force of infection for the $s \to i \to s$ model, and indicate the methods used in the proofs. In Section 4 we do the same for the intercohort force of infection. Section 5 describes some computer simulations of cases for which the stability analysis has been too complicated to complete.

In the remainder of this section, we briefly indicate the threshold and stability conditions and we give detailed interpretations, in epidemiological language, of the threshold conditions. In the intracohort case, an endemic state exists and is globally asymptotically stable if, and only if, $T > 1$ where

$$T = q \int_0^A \beta(a) e^{\int_0^a [-\mu(\sigma)-\gamma(\sigma)+\kappa(\sigma)p_\infty(\sigma)]d\sigma} da. \tag{1.5}$$

Note that T depends linearly on the vertical transmission coefficient q and in an exponential way on the cure rate γ and the horizontal transmission rate κ. So, control strategies that increase the cure rate or decrease the horizontal contact rate are more effective than strategies that change the vertical transmission rate. There also can be no endemic steady state if either there is no vertical transmission ($q = 0$, hence, $T = 0$) or else no horizontal transmission ($\kappa = 0$, hence, $T < 1$).

It is possible to give a meaningful intuitive interpretation of the quantity T in (1.5). Suppose that a uniform density $i(a,t) = 1$ of infectives is introduced into a population of susceptibles in steady state. Then the rate of infection is $\lambda s = \kappa(a)p_\infty(a)$ and by integrating the steady state form of (1.1) we get, since $i(0) = 1$,

$$i(a) = exp \int_0^a [\kappa(\sigma)p_\infty(\sigma) - \mu(\sigma) - \gamma(\sigma)]d\sigma.$$

We may interpret $\exp[\int_0^a \kappa(\sigma)p_\infty(\sigma)d\sigma]$ as the density of infectives of age a that are produced by horizontal contacts, and $\exp[-\int_0^a (\mu(\sigma)+\gamma(\sigma))d\sigma]$ as the fraction that remain in the infective class up to age a. Each of these infectives gives birth to $q\beta(a)$ infected newborn. Hence, T can be interpreted as the density of infected in the next generation due to an intitial density one in an otherwise totally susceptible population. We may thus regard T as the basic reproductive number of the infection. One may ask what would happen if infectives were introduced into a susceptible population with non-uniform density. In that case, because of the vertical transmission, the infection will spread through all ages, and at time $t > A$ will have a positive lower bound that is uniform in a. The preceding argument then applies once more and shows that if $T > 1$, the infection will persist.

For the intercohort case with no vertical transmission ($q = 0$), the threshold parameter is given by

$$T = \int_0^\infty \int_0^\infty \kappa(a)p_\infty(a) e^{-\int_a^{a+\tau} (\mu(\sigma)+\gamma(\sigma))d\sigma} d\tau \, da. \tag{1.6}$$

It is shown in [2] that if $T \leq 1$, there is a unique equilibrium, with no disease present, which is locally asymptotically stable. If $T > 1$, it is unstable, and a positive endemic equilibrium exists. Stability of the endemic equilibrium has been shown only in special cases and the general question of stability has not been resolved. Here the dependence of T on the horizontal transmission rate κ is linear while the dependence on the cure rate γ is again exponential. So, in this instance, control strategies that vary the cure rate are potentially more efficient than those that affect the horizontal transmission rate. This contrasts with the results for the intracohort case, as does the fact that an endemic equilibrium is possible even when $q = 0$ and vertical transmission is absent. Consequently, the form of the horizontal interaction rate λ is an important factor in the design of control strategies for the epidemic models that we are considering. In both the intracohort and the intercohort cases the age distribution of the population and of the various model parameters clearly affects the threshold parameter T.

A biological interpretation of the parameter T in (1.6) can be attained in two ways. First, suppose that one average infective ($I = 1$) is introduced into a wholly susceptible population at steady state. Noticing that the ages of these infectives are irrelevant in horizontal transmission, we see that $\kappa(a)p_\infty(a)$ will be the expected number per unit age and time of infections in individuals of age a. Then $\kappa(a)p_\infty(a)\exp\left(-\int_a^{a+\tau}[\mu(\sigma) + \gamma(\sigma)]d\sigma\right)$ is the expected number of those who are still in the infectious class at age $a + \tau$. Integrating over τ gives the age distribution of those infected at a who are still infectious at various later times, and integrating over a gives the number of secondary cases produced from the introduction of $I = 1$ average infectives.

In the special case in which there is no cure or removal, and κ is constant, so that the model is of $s \to i$ type, Kim and Aron [6] showed that the threshold T in (1.6) is equal to $T = \kappa N E$, where N is the total population size and E is the average expectation of remaining lifetime in the stationary population. They further showed that E equals the average age of the stationary population. We now use similar arguments to obtain an analogous result for our general case. Let $-m'(a) = (\mu(a) + \gamma(a))m(a)$, the probability density function for removal from the infectiv e class due to death or cure. Then $m(a) = \exp\left(-\int_0^a(\mu(\sigma) + \gamma(\sigma))d\sigma\right)$ is the "infective survival function", where survival now means retention in the infective class up to age a. The probability of removal from class i at age $a + x$ conditioned on being in class i at age a, is $-m'(a + x)/m(a)$. Therefore, the expectation of remaining time infectious, given infectiousness at age a, is

$$e(a) = -\int_0^\infty \frac{xm'(a + x)}{m(a)}dx = -\frac{1}{m(a)}[xm(a + x)|_0^\infty - \int_0^\infty m(a + x)dx]$$

Under the assumption that $xm(x)$ tends to zero as $x \to \infty$, we get

$$e(a) = \frac{\int_a^\infty m(x)dx}{m(a)}.$$

Now we have

$$T = \int_0^\infty \int_0^\infty \kappa(a) p_\infty(a) \frac{m(a+\tau)}{m(a)} d\tau\, da = \int_0^\infty \int_a^\infty \kappa(a) p_\infty(a) \frac{m(x)}{m(a)} dx\, da$$

$$= \int_0^\infty \kappa(a) p_\infty(a) e(a)\, da.$$

The interpretation of this formula is straightforward: as above, $\kappa(a) p_\infty(a)$ is the expected density of infections of age a per unit time in a wholly susceptible population when $I = 1$, and $e(a)$ is the expected time infectious at age a. The product gives the expected density of infections of age a during the infectious period, and the integral over a gives the total number of secondary cases.

We finally note that threshold results for some age-dependent epidemic models in various cases have been obtained by Greenhalgh [5], Dietz and Schenzle [4], Thieme [10], and others.

2. Age structure and the Catalytic Curve.

Muench [8] called attention to the usefulness of the so-called catalytic curve in the study of epidemics. In a steady state population, this is the curve that shows the fraction of individuals who show evidence of having had the infection, as a function of the age of the individuals. Muench formulated simple ordinary differential equations, analogous to those used to model chemical reaction rates, to simulate the observed data for several diseases. These equations generally contained one or two parameters, which he was able to estimate by curve-fitting methods.

Our purpose here is to derive an expression for the catalytic curve, starting from general age-dependent model equations, and to show that the catalytic curve contains enough information to determine whether or not a disease is vertically transmitted. By doing this, we demonstrate that the catalytic curve follows directly from the assumptions underlying the age-dependent model. We assume, as did Muench, that the population has reached a steady state, that mortality due to disease is negligible, and that there is no migration in or out of the population.

We shall consider two basic models, the first being the $s \to i \to r$ model described by Eq. (1.1). The second model has the diagram $s \quad \to \quad i \quad \begin{smallmatrix} \nearrow \ r \\ \\ \searrow \ \tilde{s} \end{smallmatrix}$ where $\tilde{s}$ denotes "tagged" susceptibles who are counted either through antibody tests or in some other manner as having had the disease. The results in these two cases have basic similarites but also some differences. Consequently, we will discuss them separately starting with the $s \to i \to r$ model. In this case the steady state form of (1.1) gives the following system.

$$\frac{ds}{da} + \mu(a)s = -\lambda s + \gamma(a)i$$

$$\frac{di}{da} + \mu(a)i = \lambda s - \delta(a)i - \gamma(a)i$$

$$\frac{dr}{da} + \mu(a)r = \delta(a)i \tag{2.1}$$

$$s(0) = \int_0^\infty \beta(a)[s(a) + (1-q)i(a) + r(a)]da = \int_0^\infty \beta(a)[p(a) - qi(a)]da$$

$$i(0) = q \int_0^\infty \beta(a)i(a)da, \quad r(0) = 0.$$

We have assumed in these equations that all offspring of susceptible or immune parents are susceptible. According to the steady state assumption, the reproductive rate R satisfies (1.4) and $p(a)$ is given by (1.3):

$$p(a) = b_0 e^{-\int_0^a \mu(\sigma)d\sigma}.$$

We now integrate the equation for s, and obtain

$$s(a) = e^{-\int_0^a [\mu(\sigma) + \lambda(\sigma)]d\sigma}[s(0) + \int_0^a \gamma(\tau)i(\tau)e^{\int_0^\tau [\mu(\sigma) + \lambda(\sigma)]d\sigma} d\tau]. \tag{2.2}$$

We are interested in the fraction of individuals of age a who have been infected and have recovered into the immune class or who are still infectious. Calling this $C(a)$, we have

$$C(a) = \frac{i(a) + r(a)}{p(a)} = \frac{p(a) - s(a)}{p(a)} = 1 - \frac{s(a)}{p(a)}.$$

Using the known form of $p(a)$, we may write

$$C(a) = 1 - \frac{W(a)}{b_0} e^{-\int_0^a \lambda(\sigma)d\sigma}, \tag{2.3}$$

where

$$W(a) = s(0) + \int_0^a \gamma(\tau)i(\tau)e^{\int_0^\tau [\mu(\sigma) + \lambda(\sigma)]d\sigma} d\tau. \tag{2.4}$$

Equation (2.3) is valid for any form of the force of infection. For example, for the intracohort form (i) we have

$$C(a) = 1 - \frac{W(a)}{b_0} e^{-\int_0^a \kappa(\sigma)i(\sigma)d\sigma}.$$

For the intercohort form (ii), we have

$$C(a) = 1 - \frac{W(a)}{b_0} e^{-I\int_0^a \kappa(\sigma)d\sigma},$$

where

$$I = \int_0^\infty i(a)da.$$

In order to obtain more explicit forms, it would be necessary to solve the differential equation for $i(a)$ explicitly. If $\gamma(a) = 0$, $C(a)$ is an increasing function with finite limit, as in the catalytic curve of Muench, but this is not necessarily so in the more general case we are considering here. In general $W(0) = s(0) = \int_0^\infty \beta(a)p(a)da - q\int_0^\infty \beta(a)i(a)da = b_0 - q\int_0^\infty \beta(a)i(a)da$, hence,

$$C(0) = 1 - \frac{W(0)}{b_0} = \frac{q}{b_0}\int_0^\infty \beta(a)i(a)da = \frac{i(0)}{b_0}. \tag{2.5}$$

¿From (2.5) we see that, if there is no vertical transmission, $C(0)$ is 0, but if $q > 0$ then $C(0) > 0$. Therefore, the model agrees with the intuition that the presence of vertical transmission may be inferred if a measurable fraction of newborn individuals exhibit antibodies to the infectious agent.

On the other hand, knowing the catalytic curve does not provide sufficient information to identify the form of the force of infection. Of course, from (2.3) we obtain

$$\int_0^a \lambda(\sigma)d\sigma = -\log\frac{b_0}{W(a)}[1 - C(a)], \quad \lambda(a) = \frac{C'(a)}{1 - C(a)} + \frac{W'(a)}{W(a)}. \tag{2.6}$$

However, we cannot determine $\lambda(a)$ from this equation unless $i(a)$ is known or $\gamma(a) = 0$, so that W is constant and we can compute $\lambda(a)$ from (2.6) once $C(a)$ is known. But, from this information alone it is not possible to tell whether λ has the intracohort, the intercohort, or some other form.

Suppose that the rate of reporting of new infectives is known as a function of age. Then $s(a)\lambda(a)$ is known. If the recovery rates $\gamma(a)$ and $\delta(a)$ are known, the equations in (2.1) can be integrated to yield explicit expression for i and r in terms of known quantities. Then $s(a)$ can be found, because $s + i + r = p$ is known. Thus, the percentages of susceptibles, infectives, and immune will be known as functions of age.

We now consider the $s \rightarrow i \begin{smallmatrix} \nearrow r \\ \downarrow \\ \searrow \tilde{s} \end{smallmatrix}$ model which has the following steady state form

$$\frac{ds}{da} + \mu(a)s = -\lambda s$$

$$\frac{di}{da} + \mu(a)i = \lambda s + \tilde{\lambda}\tilde{s} - \delta(a)i - \gamma(a)i$$

$$\frac{dr}{da} + \mu(a)r = \delta(a)i - \eta(a)r$$

$$\frac{d\tilde{s}}{da} + \mu(a)\tilde{s} = -\tilde{\lambda}\tilde{s}(a) + \gamma(a)i + \eta(a)r \tag{2.7}$$

$$s(0) = \int_0^\infty \beta(a)[s(a) + \tilde{s}(a) + (1-q)i(a) + r(a)]da = \int_0^\infty \beta(a)[p(a) - qi(a)]da$$

$$i(0) = q\int_0^\infty \beta(a)i(a)da, \quad r(0) = 0, \quad \tilde{s}(0) = 0.$$

The equation for the total population $p(a)$ is again given by (1.3), while the equation for s now becomes

$$s(a) = s(0)e^{-\int_0^a [\mu(\sigma)+\lambda(\sigma)]d\sigma}, \qquad (2.8)$$

The fraction $C(a)$ of all individuals who have had the disease is given by

$$C(a) = \frac{i(a) + r(a) + \tilde{s}(a)}{p(a)} = \frac{p(a) - s(a)}{p(a)} = 1 - \frac{s(a)}{p(a)}.$$

Using the expressions (1.3) and (2.8) for $p(a)$ and $s(a)$ we obtain

$$C(a) = 1 - \frac{s(0)}{b_0}e^{-\int_0^a \lambda(\sigma)d\sigma}, \qquad (2.9)$$

hence,

$$\lambda(a) = \frac{C'(a)}{1 - C(a)}, \qquad (2.10)$$

and knowledge of the catalytic curve yields the force of infection λ as a function of age. This contrasts with the result for the $s \to i \to r$ model discussed above. However, knowing λ as a function of a does not determine its dependence on i. Finally, it is worth noting that this result is independent of whether or not the recovered or immune individuals can relapse ($\eta \neq 0$) or not ($\eta = 0$).

3. Analysis of the Intra-Cohort Infection Model. Here we sketch the treatment of the intercohort $s \to i \to s$ infection model introduced in the previous sections. We assume that the net reproduction rate $R = 1$, and that the total population $p = i + s$ has reached its equilibrium value p_∞ given by (1.3). Setting $s(a,t) = p_\infty(a) - i(a,t)$ we obtain the following problem with the parameters $\beta(a)$, $\mu(a)$, $\kappa(a)$, $\gamma(a)$ satisfying the assumptions (H_1)-(H_4):

$$\begin{cases} \text{a)} & \frac{\partial i}{\partial t} + \frac{\partial i}{\partial a} + \mu(a)i = \kappa(a)[p_\infty(a) - i(a,t)]i(a,t) - \gamma(a)i(a,t), \\ \text{b)} & i(0,t) = q\int_0^\infty \beta(a)i(a,t)da, \\ \text{c)} & i(a,0) = i_0(a), \end{cases} \qquad (3.1)$$

Equation (3.1a) can be explicitly solved along the characteristic lines $t - a =$ constant getting the following formula

$$i(a,t) = \begin{cases} \dfrac{i_0(a-t)e^{\int_0^t a(a-t+\sigma)d\sigma}}{1+i_0(a-t)\int_0^t e^{\int_0^\tau a(a-t+\sigma)d\sigma}\kappa(a-t+\tau)d\tau} & \text{if } a > t, \\[4mm] \dfrac{i(0,t-a)e^{\int_0^a a(\sigma)d\sigma}}{1+i(0,t-a)\int_0^a e^{\int_0^\tau a(\sigma)d\sigma}\kappa(\tau)d\tau} & \text{if } a < t, \end{cases} \qquad (3.2)$$

where

$$\alpha(\sigma) = -\mu(\sigma) - \gamma(\sigma) + \kappa(\sigma)p_\infty(\sigma). \qquad (3.3)$$

By the renewal condition (3.1 b), formula (3.2) does not determine $i(a,t)$ explicitly except in the special case $q = 0$. In fact, in this case we have

$$i(a,t) = \begin{cases} \dfrac{i_0(a-t)e^{\int_0^t a(a-t+\sigma)d\sigma}}{1+i_0(a-t)\int_0^t e^{\int_0^\tau a(a-t+\sigma)d\sigma}\kappa(a-t+\tau)d\tau} & \text{if } a > t \\ 0 & \text{if } a < t. \end{cases} \tag{3.4}$$

So the evolution of the disease is explicitly obtained. From (3.4) we can also deduce that, along the characteristic line $a - t = a_0$, we have $i(a_0 + t, t) \to 0$ as $t \to \infty$. That is, the infective cohort which has age a_0 at time $t = 0$ eventually vanishes. We conclude that, when there is no vertical transmission of the disease, the epidemic dies off due to the aging process.

For the case $q > 0$ we define the infectives' birth rate:

$$u(t) = i(0,t). \tag{3.5}$$

In fact, substituting (3.2) into (3.1b) we get an integral equation of the form

$$u(t) = F(t) + \int_0^t G(a, u(t - a))da, \tag{3.6}$$

where

$$F(t) = \int_t^\infty \frac{q\beta(a)E(a)i_0(a - t)}{E(a - t) + i_0(a - t)\int_{a-t}^a E(\tau)\kappa(\tau)d\tau} \quad t \geq 0, \tag{3.7}$$

$$G(a, z) = \frac{q\beta(a)E(a)z}{1 + z\int_0^a E(\tau)\kappa(\tau)d\tau} \quad a \geq 0, \quad z \geq 0. \tag{3.8}$$

Note that, because of the assumptions (H_1)-(H_4), we have

$$F(t) = 0 \quad \text{for} \quad t \geq A, \tag{3.9}$$

$$G(a, z) > 0 \quad \text{for} \quad a \in (A_0, A); \quad G(a, z) = 0 \quad \text{for} \quad a \notin (A_0, A). \tag{3.10}$$

Furthermore, setting:

$$G(a, z) = D(a, z)z, \tag{3.11}$$

we note that for $a \in (A_0, A)$ we have

$$z \to G(a, z) \quad \text{is an increasing function,} \tag{3.12}$$

$$z \to D(a, z) \quad \text{is a decreasing function.} \tag{3.13}$$

Equation (3.6) has a unique continuous non negative solution whose behavior we are going to investigate. Because of (3.9) and (3.10), Equation (3.6) has the following limiting equation:

$$v(t) = \int_0^A G(a, v(t - a))da \quad t \in (-\infty, +\infty). \tag{3.14}$$

The asymptotic behavior of (3.6) is related to the constant solutions $v(t) = V$ of (3.14) which must satisfy

$$V = V \int_0^A D(a, V) da. \tag{3.15}$$

So $V = 0$, or

$$1 = \int_0^A D(a, V) da. \tag{3.16}$$

Now, the function $\Delta(V) = \int_0^A D(a, V) da$ is decreasing with limit zero as $V \to \infty$; thus (3.16) has one and only one solution, if and only if, the threshold condition:

$$T = \int_0^A D(a, 0) > 1, \tag{3.17}$$

is satisfied.

The threshold parameter

$$T = \int_0^A q\beta(a)E(a) da = \frac{q}{b_0} \int_0^A \beta(a) p_\infty(a) e^{\int_0^a [-\gamma(\sigma) + \kappa(\sigma) p_\infty(\sigma)] d\sigma} da, \tag{3.18}$$

can be interpreted as a net-infection-reproduction rate. We summarize:

Proposition 3.1. *Let the net-infection-reproduction rate T be defined in (3.18). Then, if $T \le 1$, Eq. (3.15) has only the trivial constant solution $V \equiv 0$. If $T > 1$ then it has also a nontrivial solution $V_\infty > 0$ which is determined by (3.16).*

Now we can investigate the ultimate behavior of $u(t)$ which depends on whether $T \le 1$ or $T > 1$. We have the following result

Theorem 3.2. *If $T \le 1$, then $\lim_{t \to \infty} u(t) = 0$, and if $T > 1$ $\lim_{t \to \infty} u(t) = V_\infty$.*
The proof of the first part of this result is obtained by defining

$$M_n = \max\{u(t) | t \in [nA, (n+1)A]\}$$

It can then be shown that

$$M_n < M_{n-1}. \tag{3.19}$$

Thus M_n is decreasing and it can be shown that the limit is zero.

The proof of the second part of this result relies on the same method as above but is a little more complicated. Here, together with M_n, we have to consider the behavior of $m_n = \min\{u(t) | t \in [nA, (n+1)A]\}$. We refer the reader to [2] for the details.

We emphasize that this is a global result in the sense that if $T > 1$ then for all parameter values and intial functions $i_0(a)$, the birth function $u(t) = i(0, t)$ tends to V_∞. Once this is known, equation (3.2) gives the asymptotic behavior of $i(a, t)$.

4. Analysis of the inter-cohort model.

In this section we sketch the analysis of the $s \to i \to s$ model with the force of infection taking the form $\kappa(a)I(t)$ and with the assumptions that there is no vertical transmission ($q = 0$) and that the total population is at its equilibrium distribution. The main results of this section establish local stability or instability of equilibria and show that under certain conditions no periodic solutions can arise from a Hopf bifurcation from an endemic steady state. Our analysis is based on the reduction of the problem to an integral equation whose derivation is described below.

The basic assumptions in this case are that $\lambda(a,t) = \kappa(a)I(t)$, $q = 0$, $r = 0$, $R = 1$ and $p(a,t) = p_\infty(a)$. Setting $s(a,t) = p_\infty(a) - i(a,t)$ in the second equation of (1.1) we obtain the following problem:

$$
\begin{cases}
\frac{\partial i}{\partial t} + \frac{\partial i}{\partial a} + \mu(a)i = \kappa(a)[p_\infty(a) - i(a,t)]I(t) - \gamma(a)i(a,t), \\
i(0,t) = 0, \\
i(a,0) = i_0(a).
\end{cases}
\tag{4.1}
$$

The parameters in this problem satisfy the conditions imposed in Section 1, and in particular, conditons $(H_1)-(H_4)$. Note that $i(0,t) = 0$ because there is no vertical transmission. Now, assuming $I(t) = \int_0^\infty i(a,t)da$ as known, (4.1) can be integrated along the characteristics $a - t =$ constant to yield

$$
i(a,t) =
\begin{cases}
i_0(a - t)e^{-\int_0^t \alpha(a-t+\sigma,I(\sigma))d\sigma} + \int_0^t I(\tau)e^{-\int_\tau^t \alpha(a-t+\sigma,I(\sigma))d\sigma} \\
\quad \times \kappa(a - t + \tau)p_\infty(a - t + \tau)d\tau & \text{for } a > t \\[2ex]
\int_0^a I(t - a + \tau)e^{-\int_\tau^a \alpha(\sigma,I(t-a+\sigma))d\sigma}\kappa(\tau)p_\infty(\tau)d\tau, & \text{for } a < t
\end{cases}
\tag{4.2}
$$

with

$$
\alpha(s,I(t)) = \gamma(s) + \mu(s) + \kappa(s)I(t).
\tag{4.3}
$$

Clearly, (4.2) is not an explicit expression for i since it involves $I(t)$ on the right-hand side. However, integrating (4.2) from $a = 0$ to ∞ we obtain the following equation for $I(t)$:

$$
I(t) = F(t,I(\cdot)) + \int_0^t H(t,s,I(\cdot))I(t - s)ds,
\tag{4.4}
$$

where

$$
F(t,I(\cdot)) = \int_0^\infty i_0(a)e^{-\int_0^t \alpha(a+\sigma,I(\sigma))d\sigma}da,
\tag{4.5}
$$

and

$$
H(t,\tau,I(\cdot)) = \int_0^\infty \kappa(a)p_\infty(a)e^{-\int_a^{a+\tau}(\mu(\sigma)+\gamma(\sigma))d\sigma}e^{-\int_0^\tau \kappa(a+\tau-\sigma)I(t-\sigma)d\sigma}da.
\tag{4.6}
$$

The derivation of (4.4) involves a few non-obvious changes in the variables of integration; see [2]. We note that equation (4.4) is a non-linear Volterra integral equation of a form that has been studied extensively and on which we will base our analysis of the epidemic problem. What is important here is the special form of the functionals F and G that are dictated by the structure of our model.

We set $v(t) = \lim_{h \to \infty} I(t + h)$, a limit which exists for any constant or eventually constant solution. The limiting equation is

$$v(t) = \int_0^\infty H(t, \tau, v(\cdot))v(t - \tau)d\tau, \tag{4.7}$$

with H given by (4.6) and which we rewrite as

$$H(t, \tau, v(\cdot)) = \int_0^\infty A(a + \tau, a)e^{-\int_0^\tau \kappa(a + \tau - \sigma)v(t - \sigma)d\sigma}da. \tag{4.8}$$

We look for the steady-state solutions of (4.7) which are constants V satisfying

$$V = V \int_0^\infty H(t, \tau, V)d\tau = V \int_0^\infty \int_0^\infty A(a + \tau, a)e^{-V \int_a^{a+\tau} \kappa(\sigma)d\sigma}da \, d\tau. \tag{4.9}$$

So, either $V = 0$, or

$$1 = \int_0^\infty \int_0^\infty A(a + \tau, a)e^{-V \int_a^{a+\tau} \kappa(\sigma)d\sigma}da \, d\tau. \tag{4.10}$$

The right-hand side of (4.10) is a monotone decreasing function of V whose maximum over the non-negative reals (we are interested in non-negative solutions only) is at $V = 0$. Hence, we obtain the following threshold criterion:

Theorem 4.1. *Let*

$$T = \int_0^\infty \int_0^\infty \kappa(a)p_\infty(a)e^{-\int_a^{a+\tau}(\mu(\sigma) + \gamma(\sigma))d\sigma}d\tau \, da. \tag{4.11}$$

Then if $T \leq 1$, (4.1) has the unique trivial equilibrium $i(a, t) \equiv 0$. If $T > 1$, then (4.1) has both the trivial equilibrium and a positive endemic equilibrium

$$i^*(a) = I^* \int_0^a \kappa(\tau)p_\infty(\tau)e^{-\int_\tau^a(\mu(\sigma) + \gamma(\sigma))d\sigma}e^{-I^* \int_\tau^a \kappa(\sigma)d\sigma}d\tau. \tag{4.12}$$

where $I^* > 0$ is the unique constant satisfying

$$1 = \int_0^\infty da \int_0^\infty \kappa(a)p_\infty(a)e^{-\int_a^{a+\tau}(\mu(\sigma) + \gamma(\sigma))d\sigma}e^{-I^* \int_a^{a+\tau} \kappa(\sigma)d\sigma}d\tau.$$

Proof. The equilibrium values of I^* satisfy (4.10) which has a unique positive solution, if and only if, $T > 1$. The expression (4.12) for the equilibrium distribution $i(a)$ is obtained by substituting the constant I^* in the expression (4.2) and noting that the equilibrium distribution is independent of t, hence, it must hold for $t \to \infty$.

The local stability of the steady states 0 and $I^* > 0$, when the latter exists, can be obtained by linearizing the limiting equation about each of those constant solutions. The linearization about any steady state V yields the following characteristic equation

$$1 = \int_0^\infty e^{-\lambda s} \int_0^\infty A(a+s,a) e^{-I^* \int_a^{a+s} \kappa(\sigma)d\sigma} \, da \, ds$$
$$- I^* \int_0^\infty e^{-\lambda s} \int_s^\infty d\tau \int_0^\infty A(a+\tau,a)\kappa(a+\tau-s)e^{-I^* \int_a^{a+\tau} \kappa(\sigma)d\sigma} \, da \, ds. \qquad (4.13)$$

For the trivial equilibrium we get the characteristic equation

$$1 = \int_0^\infty \int_0^\infty e^{-\lambda s} A(a+s,a) \, ds \, da$$
$$= \int_0^\infty \int_0^\infty e^{-\lambda s} \kappa(a) p_\infty(a) e^{-\int_a^{a+s}(\mu(\sigma)+\gamma(\sigma))d\sigma} \, ds \, da. \qquad (4.14)$$

Comparing (4.14) to (4.11) we note that the solutions for λ of (4.14) satisfy $\mathrm{Re}(\lambda) < 0$ if $T < 1$. When $T = 1$, $\lambda = 0$ is a solution of (4.14), and it is easy to see that there are no solutions which are pure imaginary because

$$\left| \int_0^\infty \int_0^\infty A(a+s,a) \cos(ys) \, ds \, da \right| < T = 1,$$

by the positivity of $A(a+s,a)$. When $T > 1$, choosing $\lambda = x$ to be real, we note that the right-hand side of (4.14) is a monotone decreasing function of x which tends to zero as $x \to \infty$ and which equals $T > 1$ at $x = 0$. Hence, there exists a unique positive $x > 0$ such that $\lambda = x$ solves (4.14). This leads to the following result:

Proposition 4.2. *The trivial equilibrium $i(a) \equiv 0$, is locally asymptotically stable when $T < 1$ and unstable when $T > 1$.*

We now consider the stability of the positive equilibrium which exists when $T > 1$. From equation (4.2) it is seen that if $I(t)$ tends to a constant I^*, then $i(a,t)$ tends to the equilibrium solution $i^*(a)$ given by (4.12). Consequently, we only need to consider the local stability of I^*. The characteristic equation in this case is (4.13).

Because of the complexity of (4.13), we have not determined the stability of I^*, however, we have the following partial result:

If $T > 1$, the trivial equilibrium $I^* = 0$ is unstable and the endemic equilibrium $I^* > 0$ exists but its stability has not been determined. If $I^* > 0$ is unstable, any roots of the characteristic equation whose real parts are non-negative have non-zero imaginary parts.

The significance of this remark is that it shows that destabilization of the endemic equilibrium could possibly lead to the bifurcation of time-periodic solutions. In certain simple cases it can be shown that the destabilization of $I^* > 0$ cannot occur and we give one such result.

Theorem 4.2. *Suppose that $\kappa(a) = \kappa = $ constant for $a \geq A$ and $\kappa(a) = 0$ for $a < A$, where $A \geq 0$ is a constant. Then, if $T > 1$, the trivial equilibrium is unstable while the endemic equilibrium $I^* > 0$ is locally asymptotically stable.*

5. Numerical Simulations.

In this section we present numerical results which illustrate the dynamic behavior of the $s \to i \to s$ model with the intracohort, the intercohort and a more general form of the force of infection. For all cases we take the maximal age $A = 60$ with $\mu(a) = 1/|60 - a|$ and $\beta(a) = 84.66 \sin(\pi a/100)$. All figures show the graphs of $I(t) = \int_0^\infty i(a,t)da$ and $S(t) = \int_0^\infty s(a,t)da$.

For the intracohort case we chose $\lambda(a,t) = 1000i(a,t)$, and $\gamma = 0.001$. In Figure 5.1 we use $q = 0$ and, consequently, $T = 0$ (no vertical transmission), and the graph illustrates the extinction of the infective population that is predicted by the analysis. In Figure 5.2 we show the same situation, however, with $q = 0.6$, and the effect of vertical transmission in raising the level of the total infected population $I(t)$ is clearly seen.

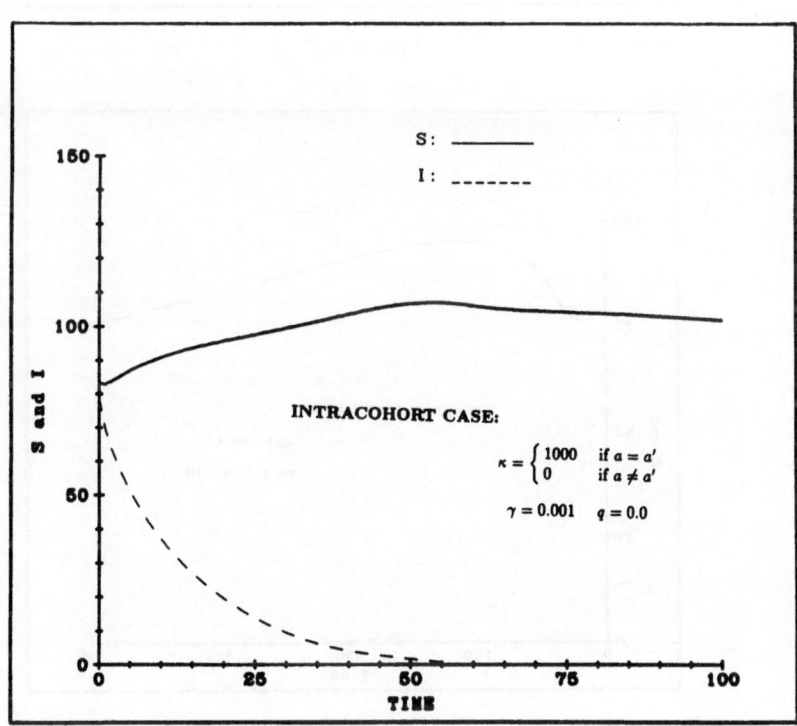

Figure 5.1

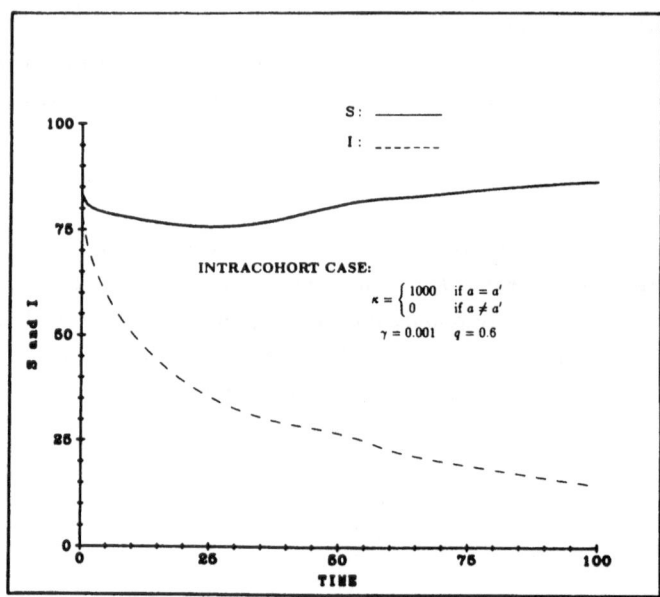

Figure 5.2

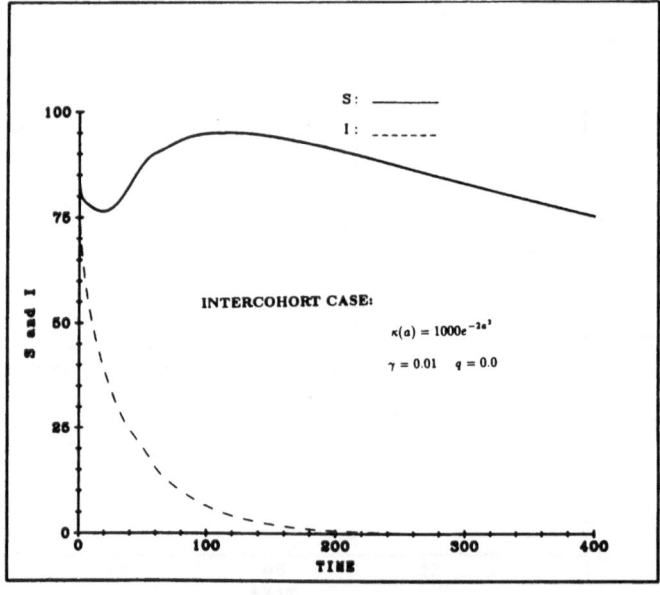

Figure 5.3

Figures 5.3, 5.4 and 5.5 illustrate the behavior of the intercohort model for which we have obtained only partial stability results. We have taken $\lambda(a,t) = k\exp(-2a^2)i(a,t)$, $\gamma = 0.01$, and with $k = 1000$ for Figures 5.3 and 5.4, and $k = 2,300$ for Figure 5.5. The graphs show the difference in effect of changes in both the vertical transmission probability and the force of horizontal transmission. Note that there are no persistent oscillations in these graphs and we also have not found any such oscillations in a fairly large number of such numerical integrations of the intercohort model.

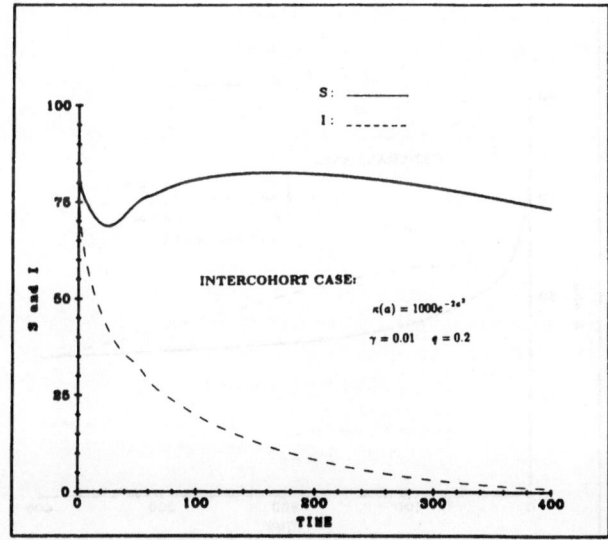

Figure 5.4

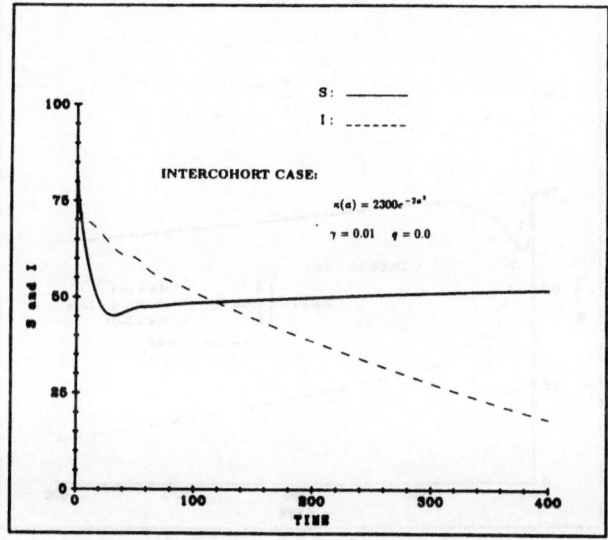

Figure 5.5

In Figures 5.6 and 5.7 we use a general form of the force of horizontal infection and take

$$\kappa(a, a') = \begin{cases} 0 & \text{if } a < \omega/4 \\ ke^{-2(a-a')^2} & \text{if } \omega/4 < a < 3\omega/4 \\ 0 & \text{if } a > 3\omega/4 \end{cases}$$

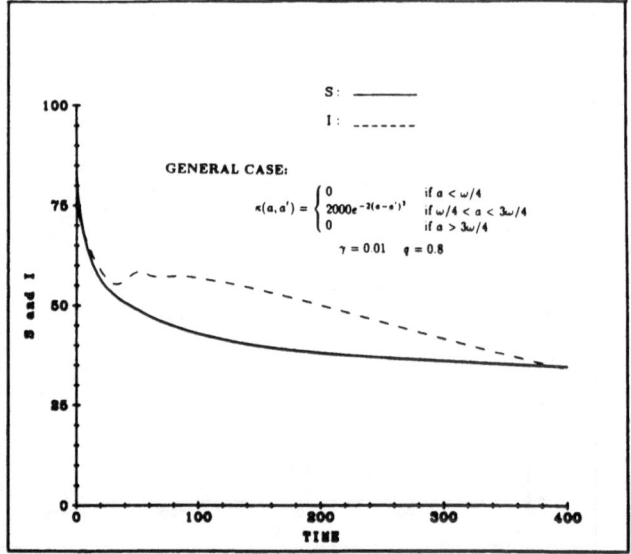

Figure 5.6

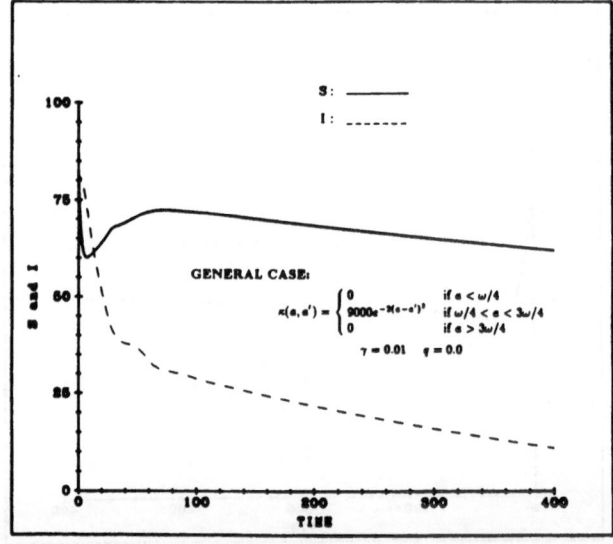

Figure 5.7

with k 2000 and $q = 0.8$ in Figure 5.6, while $k = 9000$ and $q = 0$ in Figure 5.7 and $\gamma = 0.01$.

In both of these graphs the maximum age of the population is assumed to be $\omega = 60$. The relative influences of changes in the horizontal and vertical transmission coefficients are seen as well as the lack of persistent oscillations. Again, our numerical studies of this type of model have not led to any situation with persistent oscillations in the subpopulation levels.

These numerical simulations were obtained using a program developed by S. Busenberg and W. Tsai based on an extension of a method of discretizing the linear McKendrick equations whose convergence and stability is proved in Saints [9].

Acknowledgment. The authors wish to thank Roy Anderson for comments that led to an improvement of Section 2.

REFERENCES

1. Busenberg, S. and Iannelli, M., "Separable models in age-dependent population dynamics", J. Math. Biology **22** (1985) 145-173.

2. Busenberg, S., Cooke, K. and Iannelli, M., "Endemic thresholds and stability in a class of age-structured epidemics", SIAM J. Applied Math., **48** (1988) 1379-1395.

3. Cooke, K. and Busenberg, S., "Vertically transmitted diseases", in *Nonlinear Phenomena in Mathematical Sciences*, V. Lakshmikantham editor, Academic Press, (1982) 189-197.

4. Dietz, K. and Schenzle, D., "Mathematical models for infectious disease statistics", Centenary volume of the International Statistical Institute, (1985) 167-204.

5. Greenhalgh, D., "Analytical results on the stability of age-structured recurrent epidemic models", preprint.

6. Kim, Y. and Aron, J., "On the equality of average age and average expectation of remaining life in a stationary population", preprint.

7. McKendrick, A., "Applications of mathematics to medical problems", Proc. Edin. Math. Soc. **44** (1926) 98-130.

8. Muench, H., *Catalytic Models in Epidemiology*, Harvard U. Press (1959).

9. Saints, K., "Discrete and continuous models of age-structured population dynamics", Harvey Mudd College Senior Research Report 1987-1 (1987).

10. Thieme, H., "Asymptotic estimates of the solutions of nonlinear integral equations and asymptotic speeds for the spread of populations", J. Reine und Angewandte Math. **306** (1979) 94-121.

MULTIPLE TIME SCALES IN THE DYNAMICS OF INFECTIOUS DISEASES

Viggo Andreasen
Department of Mathematics and Physics
Roskilde University, DK-4000 Roskilde, Denmark

Abstract

For an age-structured SIR-model with constant host life span, the dominant eigenvalues of the endemic equilibrium are computed using asymptotic expansions of the stability equation. To the first order the eigenvalues are purely imaginary and the real part of the second order term is negative showing that the endemic equilibrium is always stable. In an age-structured epidemic model with two interacting viral strains that confer partial cross-immunity, purely imaginary eigenvalues exists for some parameter values indicating the possibility of a limit cycle appearing through a Hopf bifurcation.

1. Introduction

Long term persistence of an immunizing infectious disease depends on the course of the infection within the individual host as well as on the demographic processes leading to the introduction of new susceptible hosts. The population biology of infectious diseases thus involves processes that take place on two time scales that differ by 3–4 orders of magnitude, namely the duration of infection (typically a few days), and the life span of a host (for humans about 70 years). This observation has both biological and mathematical implications. Biologically, the importance of the host replacement process means that other factors that change on the same time scale cannot be ignored. Mathematically, the presence of two different time scales implies that large parameter approximations may be possible.

In this paper I consider the effect two aspects that vary over the host life span, namely of age-dependent mortality and changes in virus genetics, and I show how perturbation methods simplify the mathematical analysis. To my knowledge, asymptotic expansions have not previously been used in connection with age-structured SIR-models, though numerical difficulties associated with the simultaneous presence of different time scales have been recognized (Anderson and May, 1984; Castillo-Chavez et al., 1989) This paper focuses on the significance of the two time scales; the mathematical details will be published elsewhere (Andreasen, 1988, ms.).

2. Age-dependent mortality

Soper (1929) was the first to investigate theoretically the mechanisms responsible for the long-term persistence and recurrent outbreaks of infectious diseases. His model combines the host's vital dynamics with the epidemiology of a contagious disease. The host population is divided into 3 classes: susceptibles, S, infectious, I, and recovered and immune, R. Assuming that the per capita mortality rate is μ for all classes, and that births balance deaths, Hethcote (1974) and Dietz (1975) modified Soper's model and obtained the well known SIR-model:

$$\dot{S} = -\beta SI - \mu S + \mu N$$
$$\dot{I} = \beta SI - (\mu + \nu)I \tag{1}$$
$$\dot{R} = \nu I - \mu R.$$

Here ν is the recovery rate, and β is a transmission factor describing the rate at which infectious individuals meet and infect susceptibles (Anderson, 1982). Since the population size N by assumption is constant, the R-equation is redundant.

It is well known (May, 1986) that (1) has an endemic equilibrium if and only if the basic reproductive rate, $R_0 = \beta N/(\mu + \nu)$ exceeds unity. The stability of the endemic fixed point for the S, I-equations is determined by the dominant eigenvalue p

$$p = -\frac{\mu R_0}{2} \pm i\sqrt{\mu(\mu + \nu)(R_0 - 1) - (\frac{1}{2}\mu R_0)^2}$$
$$\approx -\frac{\mu R_0}{2} \pm i\sqrt{\mu\nu(R_0 - 1)}.$$

Since the average life span of a host, $\mu^{-1} = A$, is about 70 years and the basic reproductive rate, R_0, typically is on the order of 2–10 (Anderson, 1982), (1) is weakly damped with an interepidemic period of $T \approx 2\pi\sqrt{AD/(R_0 - 1)}$, where $D = \nu^{-1}$ is the duration of infection.

The Soper model (1) assumes an exponentially decreasing host survivorship, which is unrealistic for human populations, and several authors (Hoppensteadt, 1974; Dietz, 1975; Dietz and Schenzle, 1985; Castillo-Chavez et al., 1989) generalize the model allowing for age-dependence in both mortality and other parameters. For simplicity I assume that all hosts die at the same age, A, and that all parameters except μ are age independent. Such a mortality structure gives a better approximation to the observed mortality in industrialized countries than does a constant mortality (Anderson and May, 1983). With the notation from above, the model becomes

$$\frac{\partial S}{\partial a} + \frac{\partial S}{\partial t} = -\Lambda S$$
$$\frac{\partial I}{\partial a} + \frac{\partial I}{\partial t} = \Lambda S - \nu I \qquad 0 \le a \le A \tag{2}$$
$$\Lambda(t) = \beta \int_0^A I \, d\alpha$$
$$S(0, t) = \varrho \qquad I(0, t) = 0,$$

where Λ denotes the force of infection. New individuals are born susceptible and at a constant rate, ϱ, insuring that the total population size, $N = \varrho A$, is fixed. The equation (2) is valid for $0 \le a \le A$, at age A all individuals die, and $S(a, t) = I(a, t) = 0$. I omit the redundant equation for R, and explicit reference to initial conditions.

Equation (2) can be non-dimensionalized in a way that allows us to see explicitly the effects of the two time scales. By measuring time and age in units of host lifespan, and S and I in fractions of ϱ, we obtain the non-dimensionalized equations

$$\frac{\partial s}{\partial a} + \frac{\partial s}{\partial t} = -\lambda s$$

$$\frac{\partial i}{\partial a} + \frac{\partial i}{\partial t} = \lambda s - ui \qquad 0 \le a \le 1$$

$$\lambda(t) = b \int_0^1 i\, d\alpha \tag{3}$$

$$s(0, t) = 1 \qquad i(0, t) = 0.$$

The model contains only 2 dimensionless parameters, the transmissioncoefficient $b = \beta AN$ and the ratio between the time scales $u = A/D$. For most infectious diseases u is much lager than unity. I now show that the relevant values for b are on the order of u.

Any equilibrium age-distribution (s^*, i^*) for (3) can be found by the method of Dietz (1975): first solve the steady-state equations assuming that the force of infection at equilibrium λ^* is an (unknown) constant, and then determine λ^* implicitly by requiring that the force of infection λ^* equals b times the total number of infectious, i.e.,

$$\lambda^* = b \int_0^1 i^* \, d\alpha = \frac{b\lambda^*}{\lambda^* - u} \left(\frac{1 - e^{-u}}{u} - \frac{1 - e^{-\lambda^*}}{\lambda^*} \right). \tag{4}$$

Since $\lim_{\lambda \to 0}(1 - e^{-\lambda})/\lambda = 1$, equation (4) has a root at $\lambda^* = 0$, corresponding to a disease free equilibrium.

Dietz and Schenzle (1985) show, under more general circumstances, that (4) has exactly one positive root if and only if the threshold condition

$$1 < R_1 = \frac{b}{u^2/(u - 1 + e^{-u})} \approx \frac{b}{u} \tag{5}$$

is satisfied. As in the non-age-structured case, the threshold quantity, R_1, gives the number of secondary cases per infectious individual in a totally susceptible population (Andreasen, 1988; Busenberg et al., 1989). R_1 is therefore the basic reproductive rate of the disease, and consequently R_1 is on the order of 2–10. From equation (5) it follows that b is of the same order of magnitude as u.

The stability of the endemic equilibrium, $(s^*(a), i^*(a))$, can be determined by examination of perturbations of the form

$$\hat{s}(a,t) = s(a)e^{pt}$$
$$\hat{i}(a,t) = i(a)e^{pt}$$
$$\hat{\theta}(t) = \theta e^{pt},$$

where $\hat{s}$ and $\hat{i}$ denote displacements away from the equilibrium values of s and i respectively, while $\hat{\theta}$ is the displacement of λ. The age distributions of the perturbations must to the first order follow the equations

$$\frac{ds}{da} = -\lambda^* s - \theta s^* - ps$$
$$\frac{di}{da} = \lambda^* s + \theta s^* - (p+u)i$$
$$s(0) = i(0) = 0.$$

In order for the perturbation to be consistent with the definition of $\hat{\theta}$, we in addition require (Castillo-Chavez et al., 1989)

$$\theta = b \int_0^1 i(\alpha)\, d\alpha.$$

This yields the stability equation in the eigenvalue p

$$\frac{1}{b} = \frac{p}{\lambda(p+u)(p+\lambda)} + \frac{\lambda e^{-\lambda-p}}{(\lambda-u)p(p+\lambda)} \\ - \frac{(p-\lambda)e^{-\lambda}}{\lambda p(p+u-\lambda)} - \frac{ue^{-u-p}}{(\lambda-u)(p+u)(p+u-\lambda)}. \tag{6}$$

Here and in the following, I omit the $*$ and assume that all quantities are evaluated at the endemic equilibrium.

The stability equation cannot be solved analytically, but the fact that u is much larger than unity allows us to find approximate solutions for the roots near the imaginary axis. These roots give the dominant eigenvalues and hence determine the stability of the model. The term involving e^{-u-p} can only be important when either e^{-u-p} is large or when $p \approx -u, -u+\lambda$. In both cases real part of p is much smaller than -1, so the term is not relevant for determining the stability of (3). Since the remaining part of the equation is essentially of the form $w(p) = u(p)e^{-p}$ where $w(p)$ and $u(p)$ are polynomials of degree 4 and 2 respectively, the equation has at most a finite number of roots with positive real part (Bellman and Cooke, 1963). The equation contains terms of different magnitude (since $u \gg 1$). I multiply through in (6) by the common denominator and expand in orders of u, noticing that by (5), $u/b \approx 1/R_0 \approx 1$:

$$0 = u[-p\lambda u/b + \lambda e^{-\lambda}(1-e^{-p})]$$
$$+ u^0[-p^3 u/b - p^2 2(1-e^{-\lambda}) + p\lambda(\lambda u/b + e^{-\lambda} - 2e^{-p-\lambda}) + \lambda^2 e^{-\lambda-p}] \tag{7}$$
$$+ u^{-1}[-p^4 u/b] + o(u^{-1}p^3).$$

Equation (7) has a root at $p = -\lambda + o(u^{-1})$ which is not a root for (6).

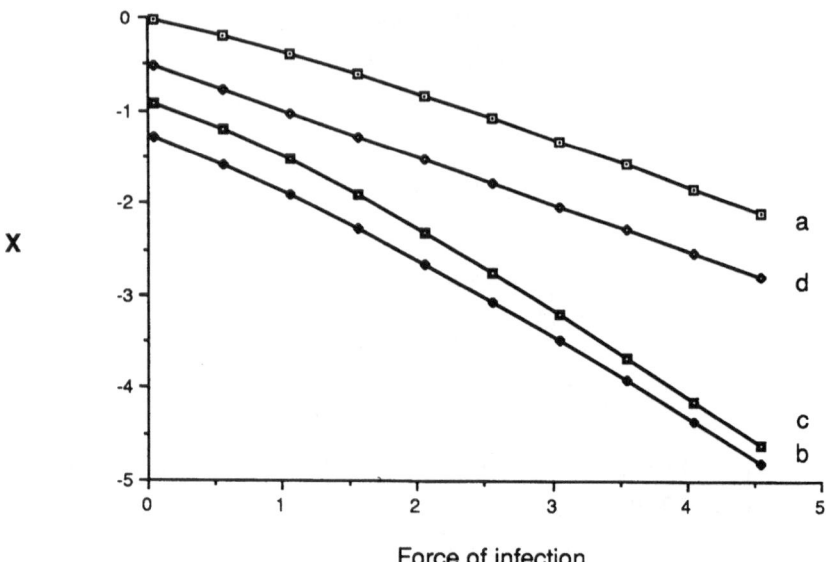

X

Force of infection

Figure 1. Stability of SIR-model with fixed duration of host life span. The curves show bounds of x for the real part of the eigenvaues that determine the stability of the endemic equilibrium of model (3). The two curves (a) and (b) give upper and lower bound for the real part of the eigenvalues, $p = i\sqrt{\lambda u} + o(1)$. The curve (c) is an upper bound for the real part of the roots with imaginary part near 0. The line (d) shows the real part of the eigenvalues of the original Soper model (1).

By trying roots of the form $|p| = cu^{\alpha}$, one sees that there are only two possible types of roots of (7) near the imaginary axis, $|p| \approx c + o(u^{-1})$ and $|p| \approx c\sqrt{u} + o(1)$.

For $|p| = c$ the equation becomes

$$-p\lambda u/b + \lambda e^{-\lambda}(1 - e^{-p}) + o(u^{-1}) = 0.$$

All roots, except the ones at 0 and at $-\lambda$, have real part less than x_1, where x_1 is given implicitly by

$$a^2 e^{-2x_1} = \pi^2/4 + (a - x_1)^2,$$

and $a = \lambda/(e^{\lambda} - 1) < 1$ (Andreasen, 1988). The root $p \approx 0$ has a large second order term so $p \approx 0$ does not correspond to a root for (6).

For $|p| = c\sqrt{u}$, the roots are

$$p = \pm i\sqrt{\lambda u} + o(1),$$

and the stability is determined by the second order terms. It is possible to show that the real part of the second order term is always negative, so that the model is stable. More careful analysis (Andreasen, 1988) shows that the real part of p must be less than a critical value, $x_0(b)$. A heuristic argument indicates that as b — or λ — varies the dominant eigenvalue will fluctuate rapidly, but continuously, within an interval $[x_m; x_0]$. The heuristic lower bound for the real part of the order $\sqrt{u}$-roots, x_m, is always smaller than x_1, upper limit for the real part of the order 1 roots, but numerical solutions of (6) show that the weakly damped, long period oscillations always dominate, (figure 1).

The interepidemic period predicted by the model is therefore $T \approx 2\pi/\sqrt{\lambda u}$ — or in dimensional quantities $T = 2\pi\sqrt{AD/\Lambda A}$, where A is the lifespan of the host and D is the duration of the disease. The age-structured model (3) thus predicts the same interepidemic period as does the standard non-age-structured SIR-model (1), since $\Lambda A = R_0 - 1$.

The real part of the dominant eigenvalue for the model with constant mortality $\mu = A^{-1}$, is in non-dimensional quantities $x_s = -\frac{1}{2}bu/(1+u) \approx -\frac{1}{2}b$. Figure 1 shows that x_s lies between the upper and the lower bound for the real part of the eigenvalue in the age-structured model (3). Numerical solutions of (7) show that in most cases the age-structured model has an eigenvalue closer to the imaginary axis than its non-age-structured counterpart. This explains the numerical observations that age-structured SIR-models tend to be less damped the model (1) (Anderson and May, 1983, 1984; Castillo-Chavez et al., 1989).

3. Changes in virus genetics

Due to the short generation time of virus, the genetics of virus populations can change rapidly compared to the host turnover rate. High mutation rates appear to be quite common both from theoretical considerations (Bremerman, 1980) and emperical observations. Rapid mutation has been observed in measles (Birrer et al., 1981), AIDS (Wong-Staal and Gallo, 1985), myxomatosis (Fenner and Ratcliffe, 1965), and influenza (Webster et al. ,1982; Couch and Kasel, 1983). In some systems, e.g., measles, the virus variation seems to have little consequence for the population dynamics of the disease, but in other systems the effect can be significant, either because of variation in disease induced mortality (myxomatosis) or — as we shall see in more detail — because of changes in the antigenic properties of the viral variants (influenza).

In the human-influenza system the virus genetics affect the virus-host-interaction in a way that appears to be unique to that system (Webster et al., 1982). The influenza virus' surface consists of a membrane densely covered with spikes of two types of glycosylated proteins. The amino acid composition of these proteins can change due to mutations (drift) or recombinations (shift). Since immunity to influenza is induced in response to these proteins, changes in the protein structure give rise to antigenically different strains. Castillo-Chavez et al. (1988) review some studies of influenza immunity and conclude that closely related viral strains may confer partial cross-immunity in such a way that an individual exposed to one viral strain may be less susceptible to related strains.

In order to describe the co-circulation of two related strains, $l = 1, 2$, Castillo-Chavez et al. (1988, 1989) modify the SIR-model (2). They assume that partial cross-immunity can be described as a reduction, σ_l, in the transmission coefficient, β_l, and that susceptibles become infected with strain l ($l = 1, 2$) at the same rate, $\Lambda_l = \beta_l(I_l + V_l)$, as in the case where only strain l is present. Individuals who have already been exposed to one strain get infected at a reduced rate, $\sigma_k \Lambda_k$. With the notation from last section, the model becomes

$$\frac{\partial S}{\partial a} + \frac{\partial S}{\partial t} = -(\Lambda_1 + \Lambda_2)S - \mu(a)S$$

$$\frac{\partial I_1}{\partial a} + \frac{\partial I_1}{\partial t} = \Lambda_1 S - \nu_1 I_1 - \mu(a)I_1$$

$$\frac{\partial I_2}{\partial a} + \frac{\partial I_2}{\partial t} = \Lambda_2 S - \nu_2 I_2 - \mu(a)I_2$$

$$\frac{\partial R_1}{\partial a} + \frac{\partial R_1}{\partial t} = \nu_1 I_1 - \sigma_2 \Lambda_2 R_1 - \mu(a)R_1$$

$$\frac{\partial R_2}{\partial a} + \frac{\partial R_2}{\partial t} = \nu_2 I_2 - \sigma_1 \Lambda_1 R_2 - \mu(a)R_2 \qquad (8)$$

$$\frac{\partial V_1}{\partial a} + \frac{\partial V_1}{\partial t} = \sigma_1 \Lambda_1 R_2 - \nu_1 V_1 - \mu(a)V_1$$

$$\frac{\partial V_2}{\partial a} + \frac{\partial V_2}{\partial t} = \sigma_2 \Lambda_2 R_1 - \nu_2 V_2 - \mu(a)V_2$$

$$\frac{\partial Z}{\partial a} + \frac{\partial Z}{\partial t} = \nu_1 V_1 + \nu_2 V_2 - \mu(a)Z$$

$$\Lambda_1(t) = \beta \int_0^\infty (I_1 + V_1)\, d\alpha \qquad \Lambda_2(t) = \beta \int_0^\infty (I_2 + V_2)\, d\alpha$$

$$S(0, t) = \varrho \qquad I_1 = I_2 = R_1 = R_2 = V_1 = V_2(0, t) = 0.$$

Here I_l denotes previously unexposed individuals, infected with strain l, R_l individuals that have recovered from strain l but have not yet been exposed to the other strain, V_l individuals that are infected with strain l and have already been exposed to the other strain, and Z individuals that have recovered from both strains. In order to maintain a constant population size N, I again require $\varrho = \int_0^\infty \exp(-\mu(\alpha))\, d\alpha$.

If the mortality is constant $\mu(a) = \mu$, (8) reduces to a system of ordinary differential equations and Castillo-Chavez et al. (1989) show by numerical solution of the stability equation that the endemic equilibrium is locally asymptotically stable when it exists.

When the host life span is fixed as in the previous section, Castillo-Chavez et al. (1989) find numerically that the prevalence of each strain exhibits sustained oscillations for some parameter values. In the symmetric case ($\beta_1 = \beta_2 = \beta$ $\sigma_1 = \sigma_2 = \sigma$ $\nu_1 = \nu_2 = \nu$) it is possible to decompose the linearized dynamics near a symmetric endemic equilibrium in a component describing the total disease level and a component describing the relative abundance of the two strains: $(S, I_1 + I_2, R_1 + R_2, V_1 + V_2)$ and $(I_1 - I_2, R_1 - R_2, V_1 - V_2)$.

Asymptotic expansion of the stability equation for the latter part shows that the dominant eigenvalues to highest order are pure imaginary corresponding to oscillations with a period of

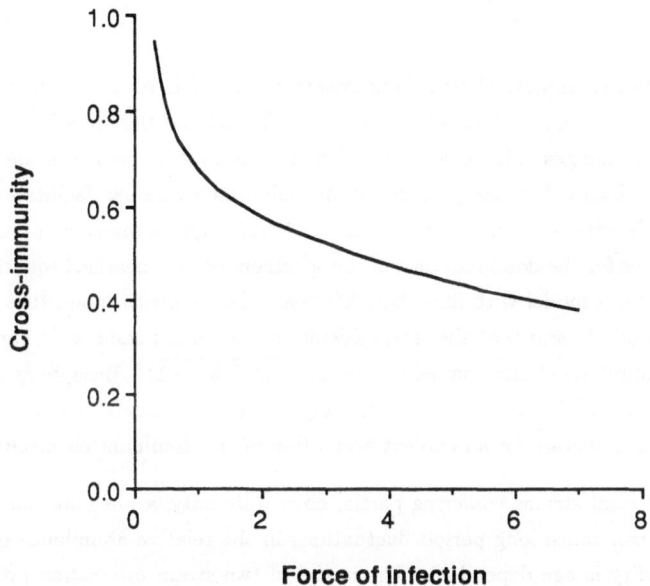

Figure 2. Approximate bifurcation diagram for the two strain model (8). For parameter values above the curve, the endemic equilibrium is stable. There exist parameter values below the curve so that the linearized equations have eigenvalues with positive real part and large imaginary part.

$T = 2\pi\sqrt{AD/\sigma\Lambda A}$, where Λ is the average force of infection at equilibrium,

$$\Lambda = \frac{\beta}{2} \int_0^A (I_1 + I_2 + V_1 + V_2)\, d\alpha.$$

Second order approximation of the dominant roots shows that when the cross-immunity is sufficiently strong (σ small) there exist values of β so that the real part of the roots is positive indicating the possibility of a limit cycle appearing through a Hopf bifurcation (figure 2). For $\sigma \approx 0$ the large parameter approximation breaks down, and one expects some kind of degeneracy due to the assumption that the two strains are *exactly* identical (Andreasen, 1988, ms.).

The presence of multiple strains with partial cross-immunity in conjunction with age structure can thus give rise to sustained oscillations with long period. For type A influena, some evidence suggest that the same viral strain causes recurrent epidemics with long interepidemic period [but the recurrence is probably associated with viral shifts that involve virus populations in wild life hosts (Beveridge, 1977)]. Neither the cross-immunity nor the age-structure alone causes sustained oscillations, so the phenomenon is a consequence of the interaction between these two "slow" processes and the "fast" dynamics of the individual infection.

4. Discussion

In this paper, I give two examples, where it is necessary to explicitly recognize the importance of both the "fast" processes associated with the period of infection and the "slow" processes associated with demographic changes in the host population or evolutionary changes in the virus. In the first part of the paper, I show how the presence of multiple time scales can facilitate the stability analysis of SIR-models with age-dependent mortality. Using large parameter approximations, I give explicit expressions for the dominant part of the spectrum of the linearized equations near the endemic equilibrium for a model with fixed host life span. The analysis shows that the endemic equilibrium is always stable and that the interepidemic period is the same as in the well-known SIR-model with constant mortality, namely $T \approx 2\pi\sqrt{AD/(R_0 - 1)}$. Busenberg et al. (1988) prove of a more general stability result for a restricted class of SIS-models. The advantage of the present method is that it allows for an explicit evaluation of the dominant eigenvalues.

The co-circulation of viral strains confering partial cross-immunity, a phenomenon that may be unique to influenza, can cause long period fluctuations in the relative abundance of the strains when the host mortality is age dependent. The model of two strain interaction provides one of the first examples of a deterministic, autonomous SIR-model exhibiting sustained oscillations. In mathematical terms, the interaction between co-circulating strains thus activates the inherent propensity to oscillations in age-structured SIR-models.

Though the sustained oscillations may be a mathematical phenomenon, the observations in tha last section of this paper indicate that frequency-dependent natural selection in virus populations may be a complex process that must be understood in a non-equilibrium context, again emphasizing the importance of the two time scales in the dynamics of infectious diseases.

5. Acknowledgments

I thank F. Adler, M. Cain, C. Castillo-Chavez, H. Hethcote, and S.A. Levin for their help with this research. I am pleased to acknowledge the support form NSF grant DMS-8406472 to S.A. Levin and from McIntire-Stennis grant NYC-183568 to S.A. Levin. Part of this research was conducted on the Cornell National Supercomputer Facility, Center for Theory and Simulation in Science and Engineering, which is funded, in part, by the National Science Foundation, New York State and IBM Corporation.

REFERENCES

Anderson, R.M. 1982. Directly transmitted viral and bacterial infections of man. In *Population biology of infectious diseases,* R.M. Anderson (ed.), pp. 1–37. Chapman and Hall, London.

Anderson, R.M., and R.M. May. 1983. Vaccination Against Rubella and Measles: Quantitative Investigations of Different Policies. *J. Hyg. Camb.,* 90, 259–325.

Anderson, R.M., and R.M. May. 1984. Spatial, Temporal, and Genetic Heterogeneity in Host Populations and the Design of Immunization Programmes. *IMA J. Math. Appl. in Med. Biol.,* 1, 233–266.

Andreasen, V. 1988. *The dynamics of epidemics in age-structured populations — analysis and simplification* (Ph.D. thesis, Cornell University).

Andreasen, V. ms. The influence of host survivorship on the dynamics of influenza epidemics (manuscript).

Bellman, R., and K.L. Cooke. 1963. *Diffenrential-Difference Equations.* Academic Press, New York.

Beveridge, W.I.B. 1977. *Influenza: The last Great Plague.* Heineman, London.

Birrer, M.J., S. Udem, S. Nathenson, and B.R. Bloom. 1981. Antigenic variants of measles virus. *Nature*, **293**, 67–69.

Bremerman, H.J. 1980. Sex and Polymorphism as Strategies in Host-Pathogen Interactions. *J. Theor. Biol.*, **87**, 671–702.

Busenberg, S., K.L. Cooke, and M. Iannelli. 1988. Endemic Thresholds and Stability in a Class of Age-Structured Epidemics. *SIAM J. Appl. Math.*, **48**, 1379–1395.

Busenberg, S., K.L. Cooke, and M. Iannelli. 1989. Stability and Thresholds in some Age-Structured Epidemics. This volume.

Castillo-Chavez, C., H.W. Hethcote, V. Andreasen, S.A. Levin, and W-m. Liu. 1988. Cross-immunity in the Dynamics of Homogeneous and Heterogeneous Populations. In *Mathematical Ecology, Proc. of the Autumn Course Research Seminars, Trieste 1986,* L.J. Gross, T.G. Hallam, and S.A. Levin (eds.), pp. 303–316. World Scientific Publishing, Singapore.

Castillo-Chavez, C., H.W. Hethcote, V. Andreasen, S.A. Levin, and W-m. Liu. 1989. Epidemiological Models with Age Structure, Proportionate Mixing, and Cross-immunity. *J. Math. Biol.*, in press.

Couch, R.B., and J.A. Kasel. 1983. Immunity to Influenza in Man. *Ann. Rev. Microbiol.*, **37**, 529–549.

Dietz, K. 1975. Transmission and Control of Arbovirus Diseases. In *Epidemiology,* D. Ludwig and K.L. Cooke (eds.), pp. 104–121. Society for Industrial and Applied Mathematics, Philadelphia.

Dietz, K., and D. Schenzle. 1985. Proportionate Mixing Models for Age-Dependent Infection Transmission. *J. Math. Biol.*, **22**, 117–120.

Fenner, F., and R.N. Ratcliffe. 1965. *Myxomatosis.* Cambridge University Press, London.

Hethcote, H.W. 1974. Asymptotic behavior and stability in epidemic models. In *Mathematical Problems in Ecology,* P. van den Driessche (ed.), pp. 83–92. Lecture Notes in Biomathematics 2. Springer, Berlin.

Hoppensteadt, F. 1974. An Age Dependent Epidemic Model. *J. Franklin Inst.*, **297**, 325–333.

May, R.M. 1986. Popuation Biology of Microparasitic Infections. In *Mathematical Ecology,* T.G. Halam and S.A. Levin (eds.), pp. 405–442. Springer, Berlin.

Soper, H.E. 1929. Interpretation of Periodicity in Disease Prevalence. *J. Roy. Stat. Soc.*, **92**, 34–73.

Webster, R.G., W.G. Laver, G.M. Air, and G.C. Schild. 1982. Molecular Mechanisms of Variation in Influenza Viruses. *Nature*, **296**, 115–121.

Wong-Staal, F., and R.C. Gallo. 1985. Human T-Lymphotropic Retroviruses. *Nature*, **317**, 395–403.

A DISTRIBUTED-DELAY MODEL FOR THE LOCAL POPULATION DYNAMICS OF A PARASITOID-HOST SYSTEM

Fred Adler
Center for Applied Mathematics
Cornell University
Ithaca, NY 14853

Lincoln Smith
Dept. of Entomology
Cornell University
Ithaca, NY 14853

Carlos Castillo-Chavez
Biometrics Unit
Dept. of Plant Breeding
and Biometry
Cornell University
Ithaca, NY 14853

Abstract

A data-based model for the population dynamics of the parasitoid-host system involving *Urolepis rufipes* Ashmead and *Musca domestica* L. is introduced and analyzed. The model uses distributed delays to account for variable development times of parasitoid larvae, and assumes that recruitment of susceptible hosts is unaffected by local dynamics. Using an arbitrary form for the functional response by parasitoids to the abundance of parasitized and unparasitized pupae, we compute the range of fly and parasitoid life history parameters for which the parasitoids persist. The model converges to a stable equilibrium whether or not the parasitoids persist, and the equilibrium level of hosts can be forced to arbitrarily low values by extremely efficient parasitoids. The model is intended to describe the dynamics of a single patch in a system where host dispersal greatly exceeds parasitoid dispersal, and may have implications for biological control under these circumstances.

1. Introduction

In several instances, parasitoids have proven to be effective biological control agents of agricultural pests (Caltagirone, 1981; Huffaker and Messenger, 1976; Greathead, 1986). A number of models have been designed to indicate the stability properties of parasitoid-host systems under a variety of assumptions about the parasitoids (Hassell, 1986; Hassell and May, 1973; Murdoch *et al.*, 1985). Our objective is the development of models that will be quantitatively predictive in a controlled environment in order to establish bounds on the effectiveness of this parasitoid over a range of environmental conditions. In this paper we develop a data-based model designed to assess the effectiveness of one species of parasitoid in a laboratory situation. We have therefore chosen to ignore the difficulties associated with measuring and modelling the spatial patterns of searching and foraging which may control the dynamics in the real world (Hassell, 1978; Reeve and Murdoch, 1985).

Smith and Rutz (1985,1986,in press) have recently described the demographic statistics of the parasitoid *Urolepis rufipes* Ashmead (Hymenoptera: Pteromalidae) parasitizing *Musca domestica* L., the house fly, in New York State. To assess the possible value of this parasitoid for biological control, we have constructed a simple model for the dynamics of this parasitoid in a controlled environment. The parameters for this model have been measured at a variety of temperatures (Smith and Rutz, 1986), and analysis of the model with these values will be soon forthcoming.

The life cycle of *U. rufipes* has been described by Smith and Rutz (1985). The following features have formed the basis of our model. First, in contrast to the systems studied by Hassell

(1978), the summer breeding seasons of both flies and parasitoids are characterized by multiple overlapping generations. We have thus modelled using continuous rather than discrete time. Second, this particular parasitoid appears to avoid super-parasitism; adult parasitoids do not lay their eggs in fly pupae which have already been stung. Extensive super-parasitism is characteristic of many species, and could substantially complicate the dynamics. Finally, we assume that the flies disperse over a larger scale than that of the model, whereas local parasitoid densities respond only to local conditions (Danthanarayana, 1986). This formulation is thus most appropriate for modelling local sources, rather than local control, of adult flies which are usually considered the pest.

2. Development of the Model

The dynamics we model follows the flow pictured in Figure 1.

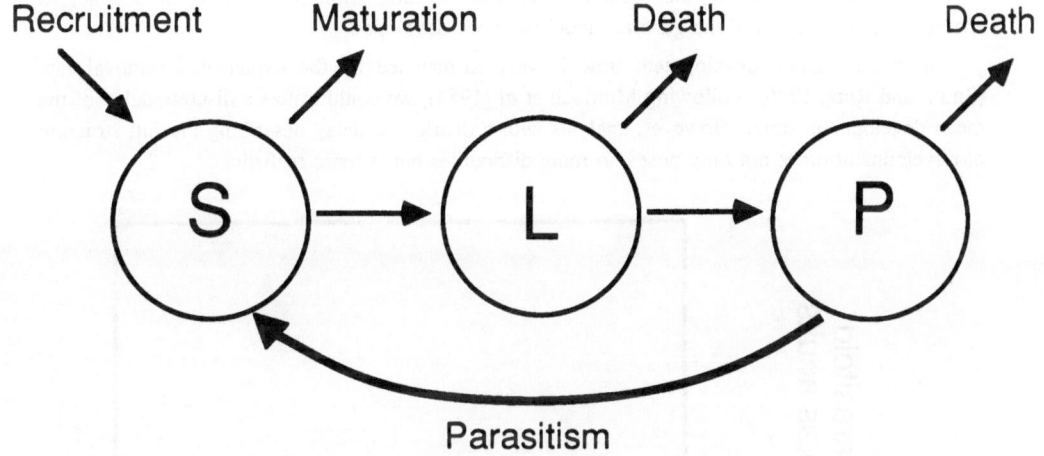

Figure 1. Diagram of the basic model. S - susceptible hosts, L - parasitized hosts (=parasitoid larvae), P - parasitoid adults.

Assuming an exponentially distributed development time for the parasitoid larvae and fly pupae, and ignoring age structure in the larval and adult parasitoids, seasonality in the life history parameters, and changes in the parasitoid sex ratio, we arrive at the following model:

$$S' = \beta - l\,P\,S - v\,S \tag{1a}$$

$$L' = l\,P\,S - m\,L - \delta L \tag{1b}$$

$$P' = m\,L - n\,P \tag{1c}$$

where $'=d/dt$, and the parameters are defined in Table 1

Parameter	Meaning
β	Rate of addition of fly pupae from the environment
l	Some measure of adult parasitoid fecundity
ν	Fly maturation rate
m	Parasitoid larval maturation rate
δ	Parasitoid larval death rate
n	Adult parasitoid death rate

This model, which follows closely the framework of Anderson and May (1981), can be made far more realistic for the system under consideration without becoming intractable, so analysis is postponed until the general model has been developed.

In reality, larval development time is very ill-matched by the exponential removal rate (Smith and Rutz, 1986). Following Murdoch *et al* (1987), we could utilize a discrete delay of the mean development time. However, analysis with a distributed delay describing the full structure of development times not only poses no more difficulties but is more realistic.

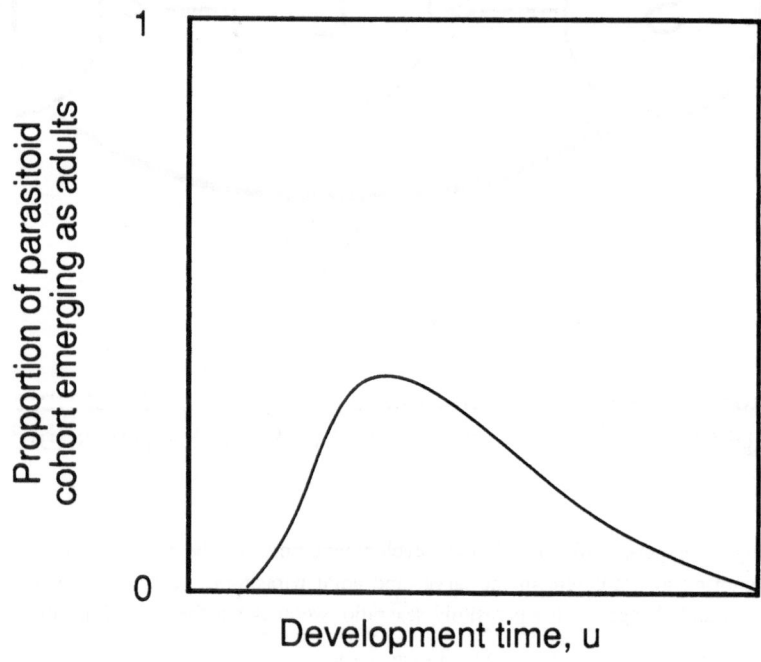

Figure 2. Typical shape of adult emergence as a function of development time.

Some data on the development times of *U. rufipes* are available in Smith and Rutz (1986). We incorporate this into our model by introducing a density of development times $p(u)$. Figure 2 illustrates a typical form of such data. Denoting by q the fraction of parasitoid larvae which complete development yet fail to emerge, we have

$$\int_0^\infty p(u)du = 1-q,$$

q can be quite large at high temperatures (Smith and Rutz, 1986). We observe that $\int_0^\infty p(u)e^{-\delta u}du$ represents the expected fraction of eggs laid which will mature successfully into adult parasitoids. This fraction includes losses due both to larval death during development and to the failure of fully developed larvae to mature, and plays a critical role in determining the location and stability of equilibria in the system. The maturation term mL appearing in equations (1b) and (1c) is then replaced by

$$\int_0^\infty l\,P(t-u)\,S(t-u)\,p(u)\,e^{-\delta u}du.$$

If $p(u)=m\,e^{-mu}$ this model reverts to model (1) under suitable initial conditions (see Appendix A).

The oviposition rate, lPS, analogous with the transmission rate in epidemiological models, fails to correspond very accurately to observed behavior (Hassell, 1986; Holling, 1959). In particular, oviposition rate per adult does not increase indefinitely with increasing availability of fly pupae, since it is limited by the number of eggs or by the handling time required for stinging and oviposition. Furthermore, this term fails to take into account the effect of interference from other adult parasitoids. This interference can take two forms: direct interference due to contact between adult parasitoids, and indirect interference due to the presence of parasitized pupae which take time to recognize as unsuitable hosts. We replace the term lPS by $lP\,r(S,L,P)$ where l is now the potential maximum oviposition rate of an adult parasitoid, and $r(S,L,P)$, the fraction of maximum rate of locating and attacking unparasitized hosts by a female parasitoid, takes on values between 0 and 1. A biologically reasonable r satisfies:

$$r(\infty,0,0)=1 \quad \text{(Potential maximum)}$$

$$r(0,L,P)=0 \quad \text{(No super-parasitism)}$$

$$\frac{\partial r}{\partial S}>0 \quad \text{(Availability of unparasitized hosts)} \tag{2}$$

$$\frac{\partial r}{\partial L}\le 0 \quad \text{(Indirect interference)}$$

$$\frac{\partial r}{\partial P}\le 0 \quad \text{(Direct interference)}$$

This leads to the following distributed delay model.

$$S'=\beta-l\,Pr(S,L,P)-vS \tag{3}$$

$$L'=l\,Pr(S,L,P)-\int_0^\infty l\,P(t-u)\,r(S(t-u),L(t-u),P(t-u))\,p(u)\,e^{-\delta u}du-\delta L$$

$$P' = \int_0^\infty l\, P(t-u)\, r(S(t-u), L(t-u), P(t-u))\, p(u)e^{-\delta u}\, du - n\, P,$$

where the state variables are evaluated at time t if not otherwise indicated, and u is a dummy variable representing development time. Initial conditions for this model must specify all populations from time immemorial to the present (Bellman and Cooke, 1963). Since development times are bounded in practice, one only needs to look back as far as the oldest potential immature parasitoid.

A reasonable form for the function $r(S,L,P)$ can be derived from a simple time budget analysis (as in Arditi, 1983). We set h_S to be the handling time for unparasitized fly pupae, h_L to be the handling time for parasitized pupae, and $\dfrac{t_0}{S+L}$ to be time to find one pupa, assumed here for convenience to decrease linearly with increasing density of parasitized and unparasitized pupae. We assume no direct interference. Then the fraction of time spent stinging will be

$$\frac{h_S \dfrac{S}{S+L}}{h_S \dfrac{S}{S+L} + h_L \dfrac{L}{S+L} + \dfrac{t_0}{S+L}},$$

which can be written in the form

$$r(S,L,P) = \frac{S}{\kappa + S + c\,L}. \tag{4}$$

The assumptions behind this model would not be difficult to check in the laboratory, and the parameters probably not unduly difficult to estimate.

3. Analysis of the Model with no Direct Interference

Throughout this section we will assume that there is no direct interference between the adult parasitoids, i.e. $\dfrac{\partial r}{\partial P} = 0$.

If S^*, L^*, P^* is any equilibrium of system (3), writing

$$\alpha = l\, P^* \frac{\partial r}{\partial S}(S^*, L^*)$$

$$\theta = l\, P^* \frac{\partial r}{\partial L}(S^*, L^*) \tag{5}$$

$$r^* = r(S^*, L^*)$$

$$k(z) = \int_0^\infty e^{-\delta u} e^{-z u} p(u)\, du,$$

and linearizing around the equilibrium, we obtain the characteristic equation of the system as

$$\det \begin{bmatrix} -\alpha - \nu - z & -\theta & -l\,r^* \\ \alpha(1-k(z)) & \theta(1-k(z)) - \delta - z & l\,r^*(1-k(z)) \\ \alpha k(z) & \theta k(z) & l\,r^*k(z) - n - z \end{bmatrix} \tag{6}$$

The roots of the characteristic equation determine the local stability of the equilibrium (Bellman and Cooke, 1963).

The system always has a parasitoid free equilibrium at

$$(S^*, L^*, P^*) = (\frac{\beta}{\nu}, 0, 0).$$

At this equilibrium $\alpha = \theta = 0$, so that the roots of the characteristic equation are determined by the diagonals of the above matrix. Thus, the roots are $-\nu$, $-\delta$ and the roots of

$$f(z) \hat{=} -n - z + l\, r^* k(z) = 0. \tag{7}$$

Define R_0 to be the expected number of adult female offspring produced by an emerging adult female parasitoid when $r = 1$. We may express R_0 as

$$R_0 = (\text{rate of oviposition}) \times (\text{expected lifespan}) \times (\text{expected offspring survivorship})$$

$$= \frac{l}{n} \int_0^\infty e^{-\delta u} p(u)\, du \tag{8}$$

$$= \frac{l}{n} k(0).$$

At an equilibrium (S^*, L^*, P^*), the expected lifetime offspring production R of an emerging adult will be

$$R = R_0 r^* = \frac{l}{n} k(0) r^*.$$

Thus we can rewrite equation (7) as

$$-n - z + n R \frac{k(z)}{k(0)} = 0. \tag{9}$$

Global stability of this equilibrium would imply the extinction of the parasitoids under any initial conditions. Since we are interested in biological control, we need to establish the conditions under which this equilibrium is unstable and the parasitoids can invade the system. We have established the following conclusion, proven in Appendix B, which is weaker than demonstrating global stability. **The parasitoid free equilibrium is locally stable if and only if $R < 1$. If $R > 1$, equation (9) has a positive real root.** Computer simulations indicate that the parasitoid-free equilibrium is globally stable if it is locally stable.

When R crosses 1, the parasitoid-free equilibrium becomes unstable by a transcritical bifurcation, producing an internal equilibrium. This equilibrium is characterized by two conditions: the first indicating that the expected number of adult offspring of an adult parasitoid must be exactly 1, and the second relating equilibrium numbers of parasitoid larvae and fly pupae.

$$r^* = \frac{1}{R_0} \tag{10a}$$

$$L^* = \frac{\beta - \nu S^*(1 - k(0))}{\delta}. \tag{10b}$$

These conditions are sketched in figure 3. Appendix C provides a proof that this sketch is justified. One can see that there can be only one internal equilibrium when there is no direct interference.

At this internal equilibrium, we can write the characteristic equation as

$$(v+z)(\delta+z)(n+z)+\alpha(\delta+z)(n+z)-\frac{n}{k(0)}k(z)(\delta+z)(v+z)-\theta(1-k(z))(n+z)(v+z)=0$$

Evaluating this expression at $z=0$ gives $\alpha\delta n-\theta(1-k(0))nv$ which is always greater than 0. Therefore $z=0$ is never a root. Stability can only be lost through a Hopf bifurcation, if at all.

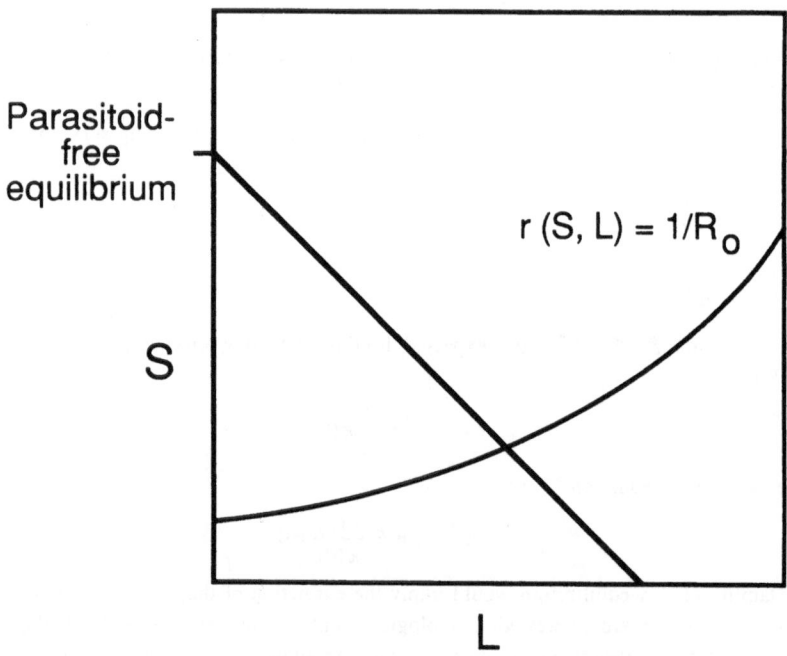

Figure 3. Equilibria of the model when there is no direct interference between adult parasitoids. See text for explanation.

We wish to determine whether predator-prey type oscillations are possible around this equilibrium. We have only been able to successfully analyze the case with no indirect interference, that is $\theta=0$. In this case, where adult parasitoids essentially do not interact, we show in Appendix D that the internal equilibrium is always locally stable.

This is not the most interesting case biologically, since it assumes that adult parasitoids do not waste any time or energy in recognizing parasitized fly pupae or in fighting with each other. Computer simulations indicate that stability is preserved in the case of indirect interference, and a computer algebra proof using MACSYMA has been produced for $p(u)$ of the form $m^2 e^{-mu}$ for an arbitrary biologically reasonable function $r(S,L)$. We conjecture that this stability property holds for arbitrary unimodal $p(u)$.

As shown in Nisbet and Gurney (1976), strong direct interference producing overcompensation is almost certain to produce oscillations. Further experments are needed to determine whether this is a possibility, and what the mechanism for such interference might be.

4. Effects of Parasitoid Parameters on Internal Equilibrium

In Nicholson-Bailey style models (see Beddington *et al.*, 1978), the internal equilibrium number of hosts cannot be driven to zero with increasing predator efficiency. In our model, for some forms of the function r, the equilibrium value S^* can be arbitrarily small. Of course, much of the difference here concerns our assumptions about the dynamics of the fly population; no matter how low the pupal population becomes, there is still a constant supply of fresh pupae.

Using the handling time derivation of r as in equation (4), condition (9a) for an equilibrium becomes

$$S(R_0-1)=\kappa+c\,L,$$

where κ represents a measure of searching efficiency and c is the ratio of handling times of parasitized to unparasitized fly pupae. We can thus solve for S^* as

$$S^* = \frac{\kappa+\beta c\,\Delta}{(R_0-1)+\nu c\,\Delta} \tag{11}$$

where $\Delta = \dfrac{1-k(0)}{\delta}$. Δ has a biological interpretation as the mean amount of time spent in the larval stage, at least in the case where q (the fraction of larvae which mature but fail to emerge) is equal to 0. This is shown in Appendix E.

We also know that the existence of an internal equilibrium requires $R_0 r(\frac{\beta}{\nu},0,0)>1$, or with this r, $R_0>1+\dfrac{\kappa\nu}{\beta}$. We can therefore write

$$R_0=1+\zeta\,\frac{\kappa\nu}{\beta}$$

for $\zeta>1$. Then

$$S^* = \frac{\beta}{\nu}\left[\frac{\kappa+\beta c\,\Delta}{\kappa\zeta+\beta c\,\Delta}\right] \tag{12}$$

We can thus see immediately that if R_0 increases due to changes in the adult parameters l and n, that the equilibrium level of pupae decreases. Also, S^* increases monotonically with increasing larval residence time, as expected, since this indicates a slower rate of natural increase for the parasitoid population. This result exposes another weakness of the model assumptions, in that fly pupae only affect adult parasitoid searching behavior while they contain live fly or parasitoid larvae. This is probably not the case for parasitoids of house fly pupae. We expect experiments to indicate that unoccupied or dead fly pupae have a decreasing interference effect as they age. This component could be incorporated into the model by addition of a class of dead fly pupae.

5. Conclusions and Further Directions

The analysis of the distributed-delay model developed in this paper indicates several ideas regarding biological control. A host-parasitoid model in which the recruitment of susceptible hosts is unaffected by local dynamics, has a tendency to converge to a stable equilibrium whether or not the parasitoids can persist. This equilibrium can have arbitrarily small numbers of susceptible hosts when the parasitoids are extraordinarily efficient. We show that the ability of the parasitoids to persist can be established in terms of basic fly and parasitoid life history parameters.

Even without including the difficulties associated with dispersal and foraging behavior, several modifications may be necessary to use this model in the field. Adult parasitoids sting fly pupae, without laying eggs, in order to drink the hemolymph (Patterson and Rutz, 1986). This direct predation could be easily incorporated into the model if measurements indicate that it takes place at a rate roughly proportional to the rate of oviposition. Super-parasitism occurs in many species of parasitoid, especially when unparasitized hosts are scarce. Tests need to be made of this potentially complicating factor. The sex ratios of parasitoid progeny can vary in response to host abundance and quality, interference, and environmental conditions (Waage, 1986). This could invalidate the total population number approach taken in this model. Also, as noted above, after a fly or parasitoid emerges from a fly pupa, the puparium remains in the environment and may be confusing to foraging parasitoids. Measures of the indirect interference component of the model need to incorporate this effect.

Two plausible and tractable modifications could be incorporated in the model to make the host dynamics more realistic. One could make the local recruitment rate β a function of temperature or time, perhaps fitting it to measured abundance throughout the summer breeding season. Alternatively, by modelling as a closed system, one could estimate the ability of parasitoids to control hosts below their environmental carrying capacity (Andreasen, ms.). Most simply, one could set the local fly recruitment rate to be βS, which assumes that the fly life cycle is significantly shorter than that of the parasitoid.

Experimental tests will show whether this model framework warrants extension to more realistic situations.

Acknowledgments

F. Adler and C. Castillo-Chavez acknowledge the support of McIntire-Stennis grant (NYC-183568) and the National Science Foundation grant DMS-8406472 to S. A. Levin. C. Castillo-Chavez acknowledges also the support of the Office of the Provost, the Center for Applied Mathematics at Cornell University, and a Ford Foundation Postdoctoral fellowship to minorities.

REFERENCES

Anderson, R. M. and R. M. May, "The population dynamics of micro-
    parasites and their invertebrate hosts," Phil. Trans. Roy.
    Soc. B, vol. 291, pp. 451-524, 1981.

Andreasen, V., Disease regulation of age-structured host popula-
    tions. In press

Arditi, R., "A unified model of the functional response of preda-
    tors and parasitoids," J. Anim. Ecol., vol. 52, pp. 293-303,
    1983.

Beddington, J. R., C. A. Free, and J. H. Lawton, "Characteristics
    of successful natural enemies in models of biological con-
    trol of insect pests," Nature, vol. 273, pp. 513-519, 1978.

Bellman, R. and K. Cooke, Differential-difference equations,
    Academic Press, New York, 1963.

Caltagirone, L. E., "Landmark examples in classical biological
    control," Ann. Rev. Entomol., vol. 26, pp. 213-232, 1981.

Danthanarayana, W., Insect flight - dispersal and migration,
    Springer-Verlag, New York, 1986.

Greathead, D., "Parasitoids in Classical Biological Control," in
    Insect Parasitoids, ed. J. Waage and D. Greathead, pp. 290-
    315, Academic Press, New York, 1986.

Hassell, M. P., The Dynamics of Arthropod Predator-Prey Systems,
    Princeton University Press, Princeton, 1978.

Hassell, M. P., "Parasitoids and Population Regulation," in In-
    sect Parasitoids, ed. J. Waage and D. Greathead, pp. 201-
    222, Academic Press, New York, 1986.

Hassell, M. P. and R. M. May, "Stability in insect host-parasite
    models," J. Anim. Ecol., vol. 42, pp. 693-736, 1973.

Holling, C. S., "Some characteristics of simple types of preda-
    tion and parasitism," Can. Ent., vol. 91, pp. 385-398, 1959.

Huffaker, C. B. and P. S. Messenger, Theory and Practice of Bio-
    logical Control, Academic Press, New York, 1976.

Murdoch, W. W., J. Chesson, and P. L. Chesson, "Biological control in theory and practice," Am. Nat., vol. 125, pp. 344-366, 1985.

Murdoch, W. W., R. M. Nisbet, S. Blythe, R. Gurney, and J. D. Reeve, "An invulnerable age class and stability in delay-differential parasitoid-host models," Am. Nat., vol. 129, pp. 263-282, 1987.

Nisbet, R. M. and W. S. G. Gurney, "A simple mechanism for population cycles," Nature, vol. 263, pp. 319-320, 1976.

Patterson, R. S. and D. A. Rutz, Biological control of muscoid flies, Entomol. Soc. Amer., misc. pub. 61, 1986.

Reeve, J.D. and W.W. Murdoch, "Aggregation by parasitoids - the successful control of the California Red Scale: A test of theory," J. Anim. Ecol., vol. 54, pp. 797-816, 1985.

Smith, L. and D. A. Rutz, "The occurence an biology of Urolepis rufipes (Hymenoptera: Pteromalida), a parasitoid of house flies in New York dairies," Environ. Entomol., vol. 14, pp. 365-369, 1985.

Smith, L. and D. A. Rutz, "Devolopment rate adn survivorship of immature Urolepis rufipes (Hymenoptera: Pterolmalidae), a parasitoid of pupal house flies.," Environ. Entomol., vol. 15, pp. 1301-1306, 1986.

Smith, L. and D. A. Rutz, "Reproduction, adult survival and intrinsic rate of growth of Urolepis rufipes (Hymenoptera: Pteromalidae), a pupal parasitoid of house flies, Musca domestica," Entomophaga, vol. 32, In press.

Waage, J. K., "Family Planning in Parasitoids: Adaptive Patterns of Progeny and Sex Allocation," in Insect Parasitoids, ed. J. Waage and D. Greathead, pp. 201-222, Academic Press, New York, 1986.

Part IV.   Acquired Immunodefiency Syndrome   (AIDS)

A MODEL FOR HIV TRANSMISSION AND AIDS

Herbert W. Hethcote[*]
Department of Mathematics
University of Iowa
Iowa City, Iowa 52242

Abstract

The model formulated for transmission of HIV (the AIDS virus) and the subsequent progression to AIDS is a system of nonlinear differential equations. They describe the infection process by interactions within and between risk groups such as homosexual men, bisexual men, female prostitutes, intravenous drug abusers and heterosexually active men and women. The progression to AIDS after infection is modeled by a sequence of stages. A modification of this model is being used to study the transmission of HIV infection and the incidence of AIDS in risk groups in the United States.

1. Introduction

First a basic human immunodeficiency virus (HIV) transmission model is formulated which incorporates transmission by sexual intercourse and by needle–sharing between drug abusers. Then this basic model is modified to account for HIV infectives who pass through different disease stages in the development of acquired immunodeficiency syndrome (AIDS). Other modifications are made to account for HIV infection through blood transfusions, blood products for hemophiliacs, perinatal infections and, finally, transmission differences in racial/ethnic groups.

The basic model is a system of nonlinear ordinary differential equations in which the incidences of infection due to interactions between groups follow mass action laws. Some simplifying assumptions are made about the interactions between groups in order to reduce the number of parameters to be estimated to a tractable number. For example, it is assumed that needle–sharing is governed by proportionate mixing so that a drug abuser sharing a needle is more likely to share with a person from a group which has drug abusers who frequently share needles.

The model presented here or a suitable modification of it will be used to study the transmission of HIV infection and the incidence of AIDS in risk groups in the U.S.A.

[*]Supported by Contract 200–87–0515 from the Centers for Disease Control.

Parameters in the model will be estimated by using all available data from the scientific literature and the Centers for Disease Control. For example, transmission parameters will be estimated from surveillance data on HIV and AIDS incidence and by using experience with transmission rates for other sexually transmitted diseases such as gonorrhea, hepatitis B and syphilis.

Computer programs will be written to simulate the spread of HIV infection and the incidence of AIDS in the risk groups in the population. The inputs will be the initial prevalences in the group
and the parameter values. The output will be the monthly incidence for ten years of HIV infection and AIDS in each group.

The computer simulation program can also be used as an experimental tool. For example, the relative importance of sexual transmission and needle–sharing transmission can be determined. The computer model can be used to compare the outputs of simulations without and with prevention programs such as education, promoting condom use, drug therapies for AIDS, counselling for safe sex and vaccination against HIV infection. For example, the effects of counselling homosexuals so that 50% practice safe sex can be estimated. The decrease in incidence after a 50% reduction in needle–sharing can be calculated.

The relative importance of the parameter estimates can be investigated by running the simulations with different parameter sets and observing the outputs. Identifying the most important parameters may suggest that focused efforts need to be made to obtain better estimates of these parameters.

Before the model is presented, it is useful to describe other modeling work. In their research on modeling the transmission of gonorrhea in a heterosexual population, Hethcote and Yorke (1984) found that a core group of highly efficient, very sexually active people was important for the maintenance of gonorrhea. It seems likely that this core group concept is also important in the epidemiology and modeling of other sexually transmitted diseases such as HIV. The division of a population into a core and noncore or into sexual activity level subpopulations is used in the model presented here; heterogeneity in sexual behavior has also been used in other HIV models. Since HIV transmission occurs not only by sexual contact, but also by needle–sharing between intravenous drug abusers, it is important to further divide the population by creating subpopulations of needle–sharing IV drug abusers. Some HIV models including the model presented here do distinguish sexual and needle–sharing contacts. Because individuals with HIV infections progress through a sequence of stages towards AIDS, the model presented here and some other AIDS models incorporate stages for HIV infectives. Some other HIV models are described briefly below.

Pickering et al. (1986) have used a discrete nonlinear model for homosexuals and gonorrhea incidence data to explore the AIDS outbreak in three cities in the United States. Knox (1986) has developed mathematical and computer simulation models for AIDS with both homosexual and heterosexual classes; he has used the model to predict equilibrium levels for prevalence and incidence of HIV infection in the United Kingdom. Dietz (1987) has considered HIV transmission models involving partnership formation and dissolution and has determined the lifetime number of partners necessary for endemicity. Anderson et al. (1986) have formulated several models for the transmission dynamics of HIV infection in homosexual communities; one of their models incorporates heterogeneity in sexual activity and stages of infection. May and Anderson (1987) have used a mathematical model for HIV transmission in a homosexual population and AIDS incidence data to estimate some basic parameter values. Anderson, May and McLean (1988) have used models to investigate the demographic impact of AIDS in developing countries.

Bongaarts (1987) has presented a computer simulation model for HIV transmission with heterogeneity in both homosexual and heterosexual behavior and with stages of infection. He is primarily interested in the demographic consequences of AIDS in Africa. Hyman and Stanley (1988) have used models to study the effects of variations in the infectivity of HIV and the incubation period. DeGruttola and Mayer (1987) have used an HIV model with two interacting populations: a small population of high risk people such as homosexuals who are rapidly infected and a large population of people who can only be infected heterosexually. Jacquez et al. (1988) have considered the effects in an HIV model of different sexual activity groups and how the type of mixing between the groups influences the development of the epidemic. Castillo–Chavez, Cooke and Levin (1987) have studied the effects of long incubation periods in models of HIV infection and AIDS.

2. Epidemiology of HIV and other STDs

Since AIDS was first recognized in the United States in 1981, the number of cases has increased rapidly. As of December, 1986 there were 28,098 reported cases of AIDS, of whom 15,757 have died (CDC, 1986c). Currently, 73% of the AIDS victims are homosexual or bisexual men (8% of these are also IV drug users), 17% are heterosexual IV drug users, 4% are heterosexual partners of people with AIDS and the remainder are children of infected mothers or recipients of transfused blood or blood products (CDC, 1986c). It is estimated that there are now approximately 1.8 million people in the United States with HIV infection and that within 5 to 10 years, between one–quarter and one–half of these infected individuals will develop AIDS (Sivak and Wormsor, 1985). Since the impact of AIDS on the people and the health care system in the United States is enormous, HIV clearly deserves special study.

HIV has some characteristics similar to other sexually transmitted diseases (STDs) such as gonorrhea and genital herpes, but it also has some different characteristics. Transmission of HIV occurs primarily through sexual intercourse and through needle sharing among drug abusers (Peterman, Drotman and Curran, 1985). The latent period for HIV infection is not known, but is probably a few days so that it is short enough to be ignored in a model. Virus transmission has been documented after exposure in the absence of a detectable antibody response (which takes an average of 3–6 weeks). The infectious period is also unknown, but the virus seems to persist in the host indefinitely since it can be isolated from the blood for many years after infection (Curran et al., 1985). Since there is no evidence to the contrary, it is assumed that HIV infectivity continues for life.

Thus HIV is similar to gonorrhea since both have relatively short latent periods and both do not confer immunity; however, people with gonorrhea return to the susceptible category when an antibiotic has cured the gonococcal infection. HIV is similar to genital herpes since the virus remains in the body forever, but people with herpes are only infectious during outbreaks of the herpes virus.

Since individuals seem to be either susceptible to HIV infection or infectious, HIV should be called an SI (susceptible infectious) disease. In contrast, gonorrhea is an SIS disease since people can recover and return to the susceptible class. Many common diseases such as measles, rubella, mumps, chickenpox, pertussis and poliomyelitis are SIR diseases since individuals have permanent immunity when they move into the resistant class upon recovery. A model for an SI disease is actually a special case of the models for SIS and SIR diseases where the recovery coefficient is zero so infectives never recover. Hence many of the modeling concepts and results developed by the project director and others for SIS diseases such as gonorrhea (Lajmanovich and Yorke, 1976; Yorke, Hethcote and Nold, 1978; Hethcote, Yorke and Nold, 1982; Hethcote and Yorke, 1984) and for SIR diseases (Hethcote, 1978; Hethcote, 1983; Hethcote and Theime, 1985; Hethcote and Van Ark, 1987) will carry over to models developed for HIV.

For gonorrhea and other STDs, only people who are sexually active need to be considered in the models. For HIV infections it is also necessary to consider people who might be infected through needle–sharing, transfusion of blood or blood–products or perinatally. STDs are different from measles and chickenpox since the disease spreads in a much more heterogeneous population. For gonorrhea it was useful to divide the population by sex, level of sexual activity, and whether they were asymptomatic or symptomatic when infectious (Hethcote and Yorke, 1984). For HIV separate risk groups must also be considered.

It is not known what fraction of people with HIV infection will eventually gets AIDS related complex (ARC) or what fraction will get AIDS. Several models for AIDS development seem plausible at this time. It may be determined at the time of infection that a fraction p of HIV infectives will eventually gets AIDS and the fraction 1–p will not. This could be determined by the size or viral subtype of the initial innoculum or by the response or susceptibility of the host (Curran et al., 1985; Peterman et al., 1985). One recent study of subtypes of HIV found much larger genetic differences between subtypes from different people than the differences in a person over time (Hahn et al., 1986). If some HIV subtypes lead to AIDS while others do not, then the model above is appropriate. Epidemiologic data is often presented in a manner consistent with the concept that development of AIDS starts in some fraction p at the time of infection. For example, in the San Francisco Cohort Study of 6875 homosexual and bisexual men, it has been found that about one–third of those with HIV infection have developed AIDS (CDC, 1985a). In a study of transfusion–associated AIDS, the mean incubation period of AIDS was found to be 4.5 years (Liu et al., 1986). If all HIV infectives may eventually get AIDS, then p is one.

It is possible that in a person with HIV infection, the AIDS development process starts at some time after infection when it is triggered by a cofactor. For example, the virus could remain latent in infected, nonreplicating cells until the AIDS process is started by antigenic stimulation by immunization or infection with certain lymphotropic viruses such as cytomegalovirus,

Epstein–Barr virus or hepatitis B virus (Laurence, 1986). Gallo (1987) writes, "Once it is inside a T4 cell, the virus may remain latent until the lymphocyte is immunologically stimulated by a secondary infection".

A six stage system of classification of progression of HIV infected persons towards AIDS has been developed (Redfield et al., 1986). When a group of patients were followed for as long as 36 months, it was found that about 90% of them progressed at least one stage during the study. Thus a suitable model may have HIV infectives move through a sequence of stages towards AIDS and death by opportunistic diseases.

If the cofactor is a lymphotropic virus and homosexuals with HIV are more likely to have a subsequence lymphotropic virus infection than those with HIV acquired from blood transfusion, then the average time until the start of the AIDS process should be shorter among the homosexuals than among the transfusion recipients. It may be possible to test this hypothesis using available data.

3. The Basic HIV Transmission Model

The proposed basic model describes the transmission of HIV infection between fourteen groups in the population by both sexual contact and needle–sharing. The fourteen groups are: 1) homosexual men, 2) bisexual men, 3) female prostitutes, 4) sexually active heterosexual women, 5) sexually active heterosexual men, 6) women who interact with only one man who is bisexual, 7) men who interact with only one woman who is a prostitute, 8) men who interact with only one woman who has other partners, 9) women who interact with only one man who has other partners, 10) women who are drug abusers but are not connected to the primary HIV sexual transmission tree, 11) men who are drug abusers but are not connected to the primary HIV sexual transmission tree, 12) men who interact with only one woman who is a drug abuser, 13) women who interact with only one man who is a drug abuser, and 14) children of women in any of the previous groups. The possible sexual transmission between the 14 groups is shown in figure 1.

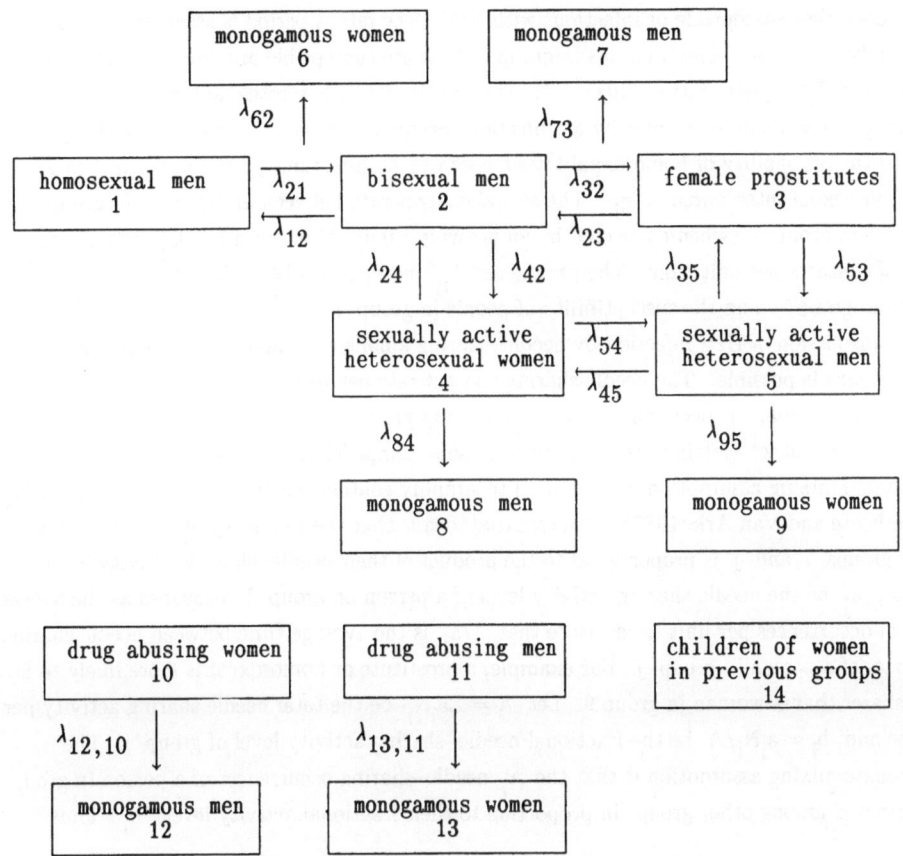

Figure 1. The arrows connecting the risk groups indicate possible directions of sexual transmission of HIV infection. Transmission by needle—sharing between drug abusers can occur between any of the fourteen groups.

The first 5 groups are called primary since they can both give and receive HIV infection. Groups 6–9 are called secondary since they can receive HIV infection from a primary group member by sexual transmission, but cannot give HIV infection since their only sexual partner is the primary group member who is already infected. People in groups 6–9, 12, 13 with only one sexual partner are called monogamous. People in groups 10–14 do not interact sexually with anyone in a primary group so they cannot get HIV infection by sexual transmission, but they can be infected perinatally or by needle—sharing. All primary group members are assumed to be adults; however, monogamous people in groups 6 to 9 could be adults or children who have sexual contact with one primary group member. Group 14 contains children who can only contract HIV infection through perinatal infection.

It is assumed that risk group i has a constant population size N_i and that all people in group i are either susceptible or infectious with HIV. The latent period is assumed to be so small that it can be neglected. The fractions in group i that are susceptible and infectious at time t are $S_i(t)$ and $I_i(t)$ with $S_i(t) + I_i(t) = 1$. The probability of heterosexual transmission of HIV during a new sexual encounter by an infectious person in group i is given by q_i for groups 2 to 5 and the probability of homosexual transmission by an infectious person in group 1 or 2 during homosexual intercourse in q_h. The sexual contact rate between an infective in group j with people in group i (when connected by an arrow in Figure 1) is given by λ_{ij}, the average number of contacts per unit time. The parameter λ_{ij} incorporates both the infectivity of infectives in group j and the susceptibility of people in group i.

Transmission of HIV infection by needle–sharing among drug abusers in any of the fourteen groups is possible. The needle sharing contact rate between an infective in group j with people in group i occurs at a rate η_{ij} sharings per unit time. Since 196 entries in the needle–sharing contact matrix must be estimated, some simplifying assumptions are necessary. A proportionate mixing assumption is often used to simplify contact matrices (Hethcote and Yorke, 1984; Hethcote and Van Ark, 1987). The essential idea is that the frequency of needle sharing between groups i and j is proportional to the product of their needle–sharing activity levels.

Let a_j be the needle sharing activity level of a person on group j measured as the average number of occurrences per unit time. Note that $1/a_j$ is the average time between needle sharing occurrences of a person in group j. For example, a prostitute or homosexual is more likely to be a needle–sharer than a woman in group 9. Let $A = \Sigma a_i N_i$ be the total needle sharing activity per unit time and $b_i = a_i N_i / A$ be the fractional needle–sharing activity level of group i. The proportionate mixing assumption is that the a_j needle–sharing occurrences of a person in group j are distributed among other groups in proportion to their fractional activity levels b_i. Thus

$$\eta_{ij} = \frac{a_j b_i}{A} = \frac{a_j a_i N_i}{A}$$

so that $\eta_{ij} N_j = \eta_{ji} N_i$.

The differential equation model which governs the dynamics of HIV infection is

$$\frac{d(N_i I_i)}{dt} = \sum_{j=1}^{14} (\lambda_{ij} q_j + \eta_{ij}) N_j I_j (1 - I_i)$$

for $i = 1, \cdots, 13$. Note that λ_{ij} is zero except for the sixteen values in figure 1. The q_i is replaced by q_h for homosexual intercourse. The initial conditions at time 0 are given by

$$I_i(0) = I_{i0}$$

for $i = 1, \cdots, 14$.

Before the model above is used to investigate important questions and to predict the incidence of HIV infection, each group and assumption will be reexamined carefully through discussions with CDC personnel and the consultants. It may be desirable to eliminate some groups and to add some other groups.

4. The HIV Model with AIDS

The basic HIV transmission model is now modified to include the progression of some HIV infectives to ARC and to AIDS. Here we formulate a very general model which incorporates both possibilities described earlier. This aspect of the model will also be reexamined carefully before it is used for prediction.

Assume that a fraction p (where p could be 1) of those whose get HIV infection start through a sequence of stages of progression towards AIDS. The waiting time in each stage A_m will have a negative exponential distribution with mean waiting time $1/\gamma_m$. Progression through a sequence of n classes with negative exponential waiting times corresponds to a convolution of exponentials so that the overall waiting time has a gamma distribution with a mean equal to $\Sigma\, 1/\gamma_m$. The dynamics of infection and progression towards AIDS are shown in figure 2.

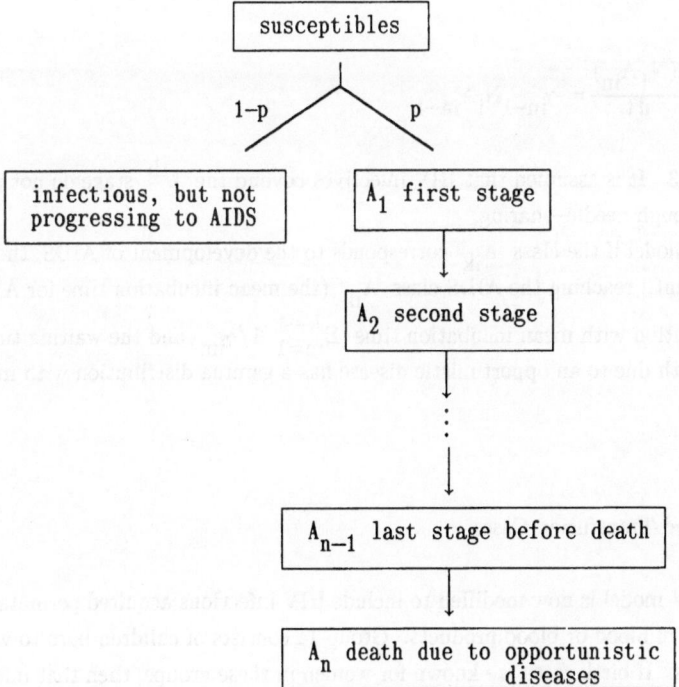

Figure 2. The fraction p of those infected begin with a progression towards AIDS. This fraction p could be 1. The first stage A_1 could correspond to a stage when the HIV virus remains dormant.

Let $I_i(t)$ be the fraction of the population N_i with HIV infection who will not get AIDS and let A_{im} be the fraction of the population N_i that is in the mth stage of progression towards AIDS. The HIV transmission model is now changed to:

$$\frac{d(N_i I_i)}{dt} = (1-p) \sum_{j=1}^{14} (\lambda_{ij}q_j + \eta_{ij})N_j(I_j + \sum_{m=1}^{\ell} A_{j\ell})(1-I_i)$$

$$\frac{d(N_i A_{i1})}{dt} = p \sum_{j=1}^{14} (\lambda_{ij}q_j + \eta_{ij})N_j(I_j + \sum_{m=1}^{\ell} A_{j\ell})(1-I_i) - \gamma_{i1}N_i A_{i1}$$

$$\frac{d(N_i A_{i2})}{dt} = \gamma_{i1}N_i A_{i1} - \gamma_{i2}N_i A_{i2}$$

$$\vdots$$

$$\frac{d(N_i A_{im})}{dt} = \gamma_{im-1}N_i A_{im-1} - \gamma_{im}N_i A_{im}$$

$$\vdots$$

$$\frac{d(N_i A_{in})}{dt} = \gamma_{in-1}N_i A_{in-1}$$

for $i = 1, \cdots, 13$. It is assumed that HIV infectives beyond the ℓth stage do not interact sexually or through needle–sharing.

In this model if the class A_{ik} corresponds to the development of AIDS, then the mean waiting time (until reaching the AIDS class A_{ik} (the mean incubation time for AIDS)) has a gamma distribution with mean incubation time $\Sigma_{m=1}^{k-1} 1/\gamma_{im}$ and the waiting time of those with AIDS until death due to an opportunistic disease has a gamma distribution with mean time $\Sigma_{m=k}^{n} 1/\gamma_{im}$.

5. Perinatal and Transfusion Cases

The HIV model is now modified to include HIV infections acquired perinatally or through the transfusion of blood or blood products. Group 14 consists of children born to women in previous groups. If birth rates are known for women in these groups, then that information will be used. Otherwise, it will be assumed that the probability of an infant being infected by a mother in these groups is proportional to the total fraction of women in these groups who are infected. The differential equation for the number of cases in group 14 are:

$$\frac{d(N_{14}I_{14})}{dt} = (1-p)b \frac{\sum\limits_{\text{women}} N_i (I_i + A_{i1} + \cdots + A_{i\ell})}{\sum\limits_{\text{women}} N_i}$$

$$\frac{d(N_{14}A_{14,1})}{dt} = \frac{\sum\limits_{\text{women}} N_i (I_i + A_{i1} + \cdots + A_{i\ell})}{\sum\limits_{\text{women}} N_i} - \gamma_{14} N_{14} A_{14,1}$$

$$\vdots$$

$$\frac{d(N_{14}A_{14,m})}{dt} = \gamma_{14,m-1} N_{14} A_{14,m-1} - \gamma_{14,m} N_{14} A_{14,m}$$

etc.

If data is available on the fractions of the blood for transfusions and blood products for hemophiliacs that comes from the risk groups, then these fractions will be used. Otherwise, it will be assumed that blood and blood products are taken randomly from the entire population so that the probability of infected blood from group i is $I_i N_i / N$ where N is the entire population size. It will be assumed that blood transfusion and hemophilia treatment are not a significant source of infection in the 14 risk groups. New groups must be added for those who receive blood transfusions and their sex partners and for hemophiliacs and their sex partners. The differential equations for these groups are based on the principles above, but are omitted to save space.

Transfusion acquired HIV infections are now very rare (CDC, 1986d). Since the middle of 1985 all blood for transfusion is screened for antibodies to HIV and positive units are discarded. Therefore the number of infectious units equals the number of false negative test results, which is less than 0.02%. Also blood products for hemophiliacs are now treated to destroy HIV.

6. Racial/Ethnic Groups

About 25% of the AIDS patients are black and 14% are Hispanic even though these groups represent only 12% and 6%, respectively, of the U.S. population (CDC, 1986b). A model incorporating this information is difficult since it requires parameter estimates for contact rates not only within each racial/ethnic category, but also between the categories. Estimates must be made of how much sexual activity and needle–sharing occurs within each category and between the categories. Proportionate mixing assumptions will be useful here to reduce the number of parameters.

Each of the groups previously defined will now be divided into 4 subgroups corresponding to the racial/ethnic categories of black, Hispanic, white and others. Within each race, the structure and parameter estimates found previously will be used, but contact rates will be scaled up for blacks and Hispanics and down for whites until the blacks have 25% of AIDS patients and Hispanics have 14%.

REFERENCES

Anderson R.M., May R.M., and McLean A.R. (1988). Possible demographic consequences of AIDS in developing countries, Nature 332, 228–234.

Anderson R.M., Medley G.F., May R.M. and Johnson A.M. (1986). A preliminary study of the transmission dynamics of the H.I.V., the causative agent of AIDS, IMA J Math Appl in Med and Biol 3, 229–263.

Bailey N.T.J. and Duppenthaler J. (1985). Sensitivity analysis in the modeling of infectious disease dynamics, J Math Biology 10, 113–131.

Bongaarts J. (1987). A model of the spread of HIV infection and the demographic impact of AIDS, preprint, 24 pages.

Castillo–Chavez C., Cooke K., Huang W., and Levin S.A. (1989). On the role of long incubation periods in the dynamics of acquired immunodeficiency syndrome (AIDS). Part 1. Single population models, J Math Biology, in press.

Curran J.W., Laurence D.N., Jaffe H.W. et. al. (1984). Acquired immunodeficiency syndrome (AIDS) associated with transfusion, N Engl J Med 310, 69–75.

Curran J.W., Morgan W.M., Hardy A.M., Jaffe H.W., Darrow W.W. and Dowdle W.R. (1985). The epidemiology of AIDS: Current status and future prospects, Science 229, 1352–1357.

Centers for Disease Control (1984). Declining rates of rectal and pharangeal gonorrhea among males – New York City, MMWR 33, 295–297.

Centers for Disease Control (1985a). Update: AIDS in San Francisco Cohort Study, 1978–1985, MMWR 34, 573–575.

Centers for Disease Control (1985b). Self–reported behavioral change among gay and bisexual men – San Francisco, MMWR 34, 613–615.

Centers for Disease Control (1986a). HTLV III/LAV antibody prevalence in U.S. military recruit applicants, MMWR 35, 421–424.

Centers for Disease Control (1986b). AIDS among blacks and Hispanics – United States, MMWR 35, 655–666.

Centers for Disease Control (1986c). Update: Acquired Immunodeficiency Syndrome – United States, MMWR 35, 757–766.

Centers for Disease Control (1986d). Transfusion–associated HTLV–III/LAV from a seronegative donor – Colorado, MMWR 35, 389–391.

De Gruttola V. and Mayer K.H. (1987). Assessing and modeling heterosexual spread of the HIV in the United States, Reviews of Infectious Diseases 10, #1, 138–150.

Dietz K. (1987). On the transmission dynamics of HIV, preprint, 39 pages.

Gallo R.C. (1987). The AIDS virus, Scientific American, January 1987, 47–56.

Goedert J.J., Biggar R.J., Weiss S.H. (1986). Three–year incidence of AIDS in five cohorts of HTLV–III–infected risk group members, Science 231, 992–995.

Hahn B.H., Shaw G.M., Taylor M.E., Redfield R.R., Markham P.D., Salahuddin S.Z., Wong–Staal F., Gallo R.C., Parks E.S., Parks W.P. (1986). Genetic variation in HTLV–III/LAV over time in patients with AIDS or at risk for AIDS, Science 232, 1548–1553.

Hethcote H.W. (1978). An immunization model for a heterogeneous population, Theor Pop Biol 14, 338–349.

Hethcote H.W. (1983). Measles and rubella in the United States, Am J. Epid 117, 2–13.

Hethcote H.W., Stech H.W. and van den Driessche P. (1981). Periodicity and stability in epidemic models: a survey, in *Differential Equations and Applications in Ecology, Epidemics and Population Problems*, S. Busenberg and K.L. Cooke, Eds. Academic Press, New York, 65–82.

Hethcote H.W. and Theime H. (1985). Stability of the endemic equilibrium in epidemic models with subpopulations, Math Biosci 75, 205–227.

Hethcote H.W. and Van Ark J.W. (1987). Epidemiological models for heterogeneous populations: Proportionate mixing, parameter estimation and immunization programs, Math Biosci 84, 85–118.

Hethcote H.W., Yorke J.A. and Nold A. (1982). Gonorrhea modeling: a comparison of control methods, Math Biosci 58, 93–109.

Hethcote H.W. and Yorke J.A. (1984). *Gonorrhea Transmission Dynamics and Control*, Lecture Notes in Biomathematics 56, Springer–Verlag, Berlin.

Hyman J.M. and Stanley E.A. (1988). Using mathematical models to understand the AIDS epidemic, Math Biosciences, to appear.

Jacquez J.A., Koopman J., Sattenspiel L., Simon C., and Perry T. (1988). Modeling and analysis of HIV transmission: The role of contact patterns, preprint, 48 pages.

Knox E.G. (1986). A transmission model for AIDS, Eur J Epidem 2, #3, 165–177.

Lajmanovich A. and Yorke J.A. (1976). A deterministic model for gonorrhea in a nonhomogeneous population, Math Biosci 28, 221–236.

Laurence J. (1986). AIDS: Definition, epidemiology and etiology, Laboratory Medicine 17, 659–663.

Liu K.–J., Lawrence D.N., Morgan W.M., Peterman T.A., Haverkos H.W., and Bregman D.J. (1986). A model–based approach for estimating the mean incubation period of transfusion–associated acquired immunodeficiency syndrome, Proc Nat Acad Sci 83, 3051–3055.

May R.M. and Anderson R.M. (1987). Transmission dynamics of HIV infection, Nature 326, 137–142.

McKusick L., Horstman W., and Coates T.J. (1985). AIDS and sexual behavior reported by gay men in San Francisco, Am J Public Health 75, 493–496.

Peterman T.A., Drotman D.P. and Curran J.W. (1985). Epidemiology of the acquired immunodeficiency syndrome (AIDS), Epidemiologic Reviews 7, 1–21.

Peterman T.A., Jaffe H.W., Feorino P.M., Getchall J.P., Warfield D.T., Haverkos H.W., Stoneburner R.L., and Curran J.W. (1985). Transfusion–associated acquired immunodeficiency syndrome in the United States, JAMA 254, 2913–2917.

Pickering J., Wiley J.A., Padian N.S., Lieb L.E., Echenberg D.F., and Walker J. (1986). Modeling the incidence of AIDS in San Francisco, Los Angeles and New York, Math Modeling 7, 661–688.

Redfield R.R., Wright D.C. and Tramont E.C. (1986). The Walter Reed staging classification for HTLV–III/LAV infection, N Engl J Med 314, 131–132.

Sivak S.L. and Wormser G.P. (1985). How common is HTLV–III infection in the United States? Correspondence N Engl J Med 313, 1352.

Yorke J.A., Hethcote H.W., and Nold A. (1978). Dynamics and control of the transmission of gonorrhea, Sexually Transmitted Diseases 5, 51–56.

THE ROLE OF LONG PERIODS OF INFECTIOUSNESS IN THE DYNAMICS OF ACQUIRED IMMUNODEFICIENCY SYNDROME (AIDS)

Carlos Castillo-Chavez
Center for Applied Math.
Ecology and Systematics
Biometrics Unit
Cornell University
Ithaca, NY 14853-7801

Kenneth Cooke
Dept. of Mathematics
Pomona College
Claremont, CA 91711

Wenzhang Huang
Claremont Graduate School
Claremont, CA 91711

Simon A. Levin
Cent. for Environ.
Research
Ecol. and Syst.
Cornell University
Ithaca, NY 14853

Abstract.

Single and multiple group models for the spread of HIV (human immunodeficiency virus) are introduced. Partial analytical results for these models are presented for two specific cases. First for a model for which the duration of infectiousness has a negative exponential distribution and second for a model for which all individuals remain infectious for a fixed length of time.

1. Introduction

The discovery by Barre-Sinoussi's and Gallo's groups [1,2,3,4,5] that HIV (human immunodeficiency virus) is the etiological agent for AIDS has brought an unprecedented amount of research on the biology of this retrovirus. At present, however, there is not enough understanding on the consequences of its transmission at the population level. Some routes of HIV transmission are through sex (direct, anal, and oral), through needle sharing, through blood transfusions and through vertical transmission (mother to child at birth). Important epidemiological factors involved in its transmission include: variable infectivity [6,7,8], long periods of infectiousness [9] of eight years or more and cofactors (e.g. whether or not the HIV carrier is infected with another venereal disease). In addition, biological and socio-demographic factors such as sex, age, economic status, race, sexual preference, geographical area of residence, and the nature of the social networks that are particular to each culture, have to be taken into consideration if we are to understand the dynamics of HIV.

This epidemic already has raised many, some perhaps unsolvable, moral, practical, economical and ethical questions regarding the possible implementation of a variety of extreme intervention plans. These include random testing of the population, random testing of specific ethnic groups and the possibility of putting (known) infected individuals in quarantine (see [10] for a commentary). The testing of the possible effectiveness of (such extreme) intervention plans may only make sense in a realistic mathematical framework. However, any mathematical model has its faults. There is always a tradeoff between detail and tractability, and there are inherent limits to predictability. Although mathematical models can suggest possible consequences of intervention plans and assist in thinking about complex issues, we strongly feel that the numerical and mathematical results obtained through their use should not be used to circumvent the moral and ethical questions raised by this epidemic.

In this paper, we report on a series of models that we have developed recently and that are extensions of those of Anderson et al. [11,12]. Our objective has been to identify the role played by the long period of infectiousness associated with HIV on the dynamics of sexually transmitted HIV in homogeneous and heterogeneous populations. We present only a brief description of these models and a partial list of our analytical results. Extensions and proofs of these results can be found in Castillo-Chavez et al. [13-14]. We note that some of our results partially overlap with some of the results obtained simultaneously and independently by Blythe and Anderson [15].

2. Single group models

In this section we describe two models with alternative distributions of the duration of infection. First we assume, as is commonly done in epidemiological models, that individuals are transferred out of the infected class at a constant rate, or equivalently that the duration of infection has a negative exponential distribution (Hethcote et al. [16]). For our second model, we assume that all infected individuals remain infectious for a fixed length of time. This approach allows us to compare the effect of the mean infectious period on the reproductive number (i.e. the number of secondary infections generated by a single infectious individual in a purely susceptible population) and therefore to understand better its role on the dynamics of HIV in a homogeneous population.

a. Model with exponential removal

We divide the population --sexually active male homosexuals with multiple partners-- into five groups: **S** (susceptibles) , **I** (infected that will develop "full-blown" AIDS), **Y** (infected that will not develop full-blown AIDS), **Z** (former Y individuals that are no longer sexually active), and **A** (former I individuals that have developed "full-blown" AIDS). Note that **A** and **Z** are cumulative classes and hence once individuals move into these classes they no longer enter into the dynamics of the disease; however, for bookkeeping purposes, we keep them on record. We do not include a latent class (i.e., those exposed individuals that are not yet infectious). Furthermore, we assume that once an individual develops full-blown AIDS or enters the Z class, he is no longer infectious because he has no sexual contacts. We also assume that all infected individuals become immediately infectious, and that they stop being sexually active or acquire AIDS at the constant rates α_Y and α_I; hence $1/(\mu+\alpha_I)$ denotes the average infectious period and $1/(\mu+\alpha_Y)$ the average sexual longevity of an individual. In addition, we let Λ denote the recruitment rate into the susceptible class (defined to be those individuals who are homosexually active); μ, the natural mortality rate; d, the disease-induced mortality due to AIDS; p, that fraction of the susceptibles that become infectious and will go into the AIDS class; and therefore $(1-p)$, the fraction of susceptible individuals that become infectious and will not develop full blown AIDS. Following Anderson et. al. [11] and using

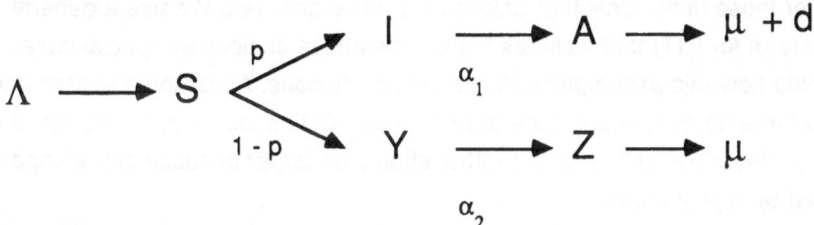

Fig. 1: Flow diagram for a single group model with exponential removal, for details see the text.

Figure 1, we arrive at the following simple epidemiological model with exponential removal:

$$\frac{dS(t)}{dt} = \Lambda - \lambda C(T)(t)\frac{S(t)W(t)}{T(t)} - \mu S(t) \ , \qquad (2.1)$$

$$\frac{dI(t)}{dt} = \lambda p C(T)(t)\frac{S(t)W(t)}{T(t)} - (\alpha_1 + \mu) I(t) \ , \qquad (2.2)$$

$$\frac{dY(t)}{dt} = \lambda(1 - p)C(T)(t)\frac{S(t)W(t)}{T(t)} - (\alpha_Y + \mu) Y(t) \ , \qquad (2.3)$$

$$\frac{dA(t)}{dt} = \alpha_1 I(t) - (d + \mu) A(t) \ , \qquad (2.4)$$

$$\frac{dZ(t)}{dt} = \alpha_Y Y(t) - \mu Z(t) \ , \qquad (2.5)$$

where
$$W = I + Y \text{ and } T = W + S. \qquad (2.6)$$

Here the function $C(T)$ denotes the mean number of sexual partners an average individual has per unit time, given that the population density is T, and λ (a constant) denotes the average sexual risk per partner. More specifically (as in Hyman and Stanley [17]), $\lambda = i\phi$ where i denotes the probability of infection per sexual contact (with an infected individual) and ϕ denotes the average number of contacts per sexual partner. Hence, $\lambda C(T)$ denotes the transmission rate per unit time per infected partner. The factor W/T is the probability that a contact of a susceptible with a randomly selected individual will be with an infectious individual. Since individuals in classes A and Z are not sexually active, $\lambda C(T)SW/T$ denotes the number of newly infected individuals per unit time. $C(T)$ is usually assumed to be approximately linear for small T and to approach a saturation level for a large T. For AIDS, it may be that $C(T)$ should be taken as proportional to T^δ for $0 < \delta \le 1$ for small populations, but treated as a constant for large populations. This is because there is some evidence (Kingsley et al. [18]) that the probability of seroconversion (infection) increases with the number of infected sexual partners for those individuals that practice receptive anal sex. We use a general functional form for $C(T)$ that includes both of the above choices as special cases in order to determine how this assumption affects the conclusions. Anderson and May [12] have shown that in a homogeneous (one-group) model, $C(T)$ should not be the mean number of sexual partners per unit time, but rather should be larger because of the important role played by highly active

individuals who are more likely to acquire infection and are also more likely to transmit it.

Unfortunately, there is evidence that AIDS is actually a progressive disease and that most individuals that have been infected will go on to develop "full-blown" AIDS. If we accept this view, then p is approximately equal to one and equations (2.3) and (2.5) are no longer necessary. In the rest of this article we will report results only for the case $p = 1$: for tho oaoo $0 < p < 1$ tho roador io roforrod to [10-14]. Observe that the dynamics of the classes S and I are governed autonomously, and hence the system (2.1), (2.2), (2.4) can be reduced to

$$\frac{dS}{dt} = \Lambda - \lambda C(T)S\frac{I}{T} - \mu S , \qquad (2.7)$$

$$\frac{dI}{dt} = I\left(\lambda C(T)\frac{S}{T} - \sigma\right) , \qquad (2.8)$$

where $T = S + I$, $\sigma = \mu + \alpha_1$, and where we assume that $C(T)$ is an increasing function of T.

The system (2.7)-(2.8) always has the infection-free state

$$(S,W) = \left(\frac{\Lambda}{\mu},0\right), \qquad (2.9)$$

as an equilibrium.

For this model, the reproductive number R, i.e. the number of secondary infections produced by an infectious individual in a purely susceptible population, is given by

$$R = \lambda C(\frac{\Delta}{\mu}) \frac{1}{\sigma} \ , \tag{2.10}$$

where we observe that $1/\sigma$ denotes the mean infectious period. We note that if $R > 1$ there exists a unique endemic equilibrium given implicitly by the unique positive solution to the system:

$$S = \frac{T}{C(T)} \frac{\sigma}{\lambda} \ , \tag{2.11a}$$

$$I = (\frac{\Delta}{\mu} - S) \frac{\mu}{\sigma} \ . \tag{2.11b}$$

For these equilibria, we have established the following results:

> The system (2.7)-(2.8) has a unique (positive) endemic state if and only if the reproductive number exceeds unity (R>1).The infection-free state (1.9) is globally asymptotically stable (relative to solutions for which S(0) ≥ 0, W(0) ≥ 0) whenever the reproductive number R is less than unity (R<1) and it is unstable when R>1. In addition when R crosses 1 (from below) there is a transcritical bifurcation with the endemic equilibrium becoming globally stable.

b. Models with constant incubation period

As previously, we assume that all individuals become immediately infectious; hence with this formulation the incubation period is again taken to be equal to the infectious period. For the I-infected it is assumed to be constant (ω) and for the Y-infected it is assumed to be a constant τ equal to the average length of their sex-life. Therefore all infected (assumed infectious) individuals remain a fixed length of time (ω) or (τ) in their corresponding classes (more general forms of the model allow (ω) and (τ) to be distributed [13-14]). $I_0(t)$ and $Y_0(t)$ denote those individuals that were in either class I or Y at time $t = 0$, and are still infectious; $Z_0(t)$, those individuals that were in class Z at time $t = 0$, and are still alive; and $A_0(t)$, those individuals that had already developed full-blown AIDS at time $t = 0$, and are still alive. We assume that $Z_0(t)$ and $A_0(t)$ vanish for large enough t, i.e., in mathematical terms that they have **compact support**. Since ω denotes the infectious period and τ the average sex-life of an individual in this population, we assume that $I_0(t) = Y_0(t) = 0$ for $t > \max(\omega,\tau)$. The function $H(x)$ that appears in the following is the Heaviside function, defined as being equal to 1 if $x > 0$ and zero otherwise. The rest of the parameters are defined as in Section 2a. Using these conventions, and with the aid of Figure 2,

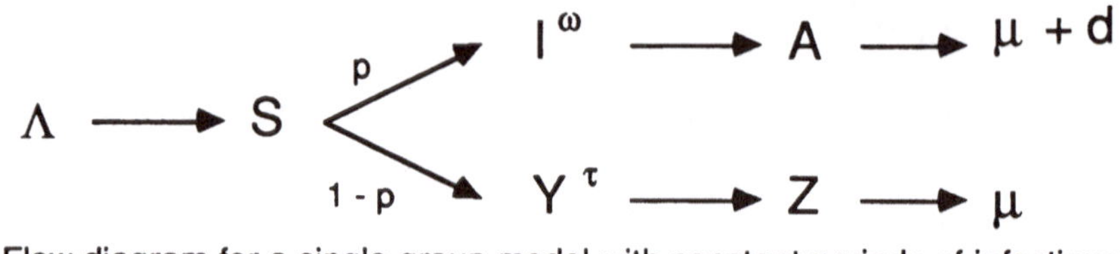

Fig. 2: Flow diagram for a single group model with constant periods of infectiousness, for details see the text.

we obtain the dynamical equations:

$$\frac{dS(t)}{dt} = \Lambda - \lambda C(T)(t)\, \frac{S(t)W(t)}{T(t)} - \mu S(t) \ , \tag{2.12}$$

$$I(t) = I_0(t) + \lambda p \int_{t-\omega}^{t} C[T](x)\, \frac{S(x)W(x)}{T(x)}\, H(x) e^{-\mu(t-x)} dx \ , \tag{2.13}$$

$$Y(t) = Y_0(t) + \lambda(1-p)\int_{t-\tau}^{t} C(T)(x)\, \frac{S(x)W(x)}{T(x)}\, H(x) e^{-\mu(t-x)} dx \ , \tag{2.14}$$

$$A(t) = A_0(t) + \lambda p \int_{0}^{t-\omega} C(T)(x)\, \frac{S(x)W(x)}{T(x)}\, H(x) e^{-\mu(t-x)-d(t-x-\omega)} dx \ , \tag{2.15}$$

$$Z(t) = Z_0(t) + \lambda(1-p)\int_{0}^{t-\tau} C(T)(x)\, \frac{S(x)W(x)}{T(x)}\, H(x) e^{-\mu(t-x)} dx \ , \tag{2.16}$$

where $W(t) = I(t) + Y(t)$, and $T(t) = S(t) + W(t)$ and $C(T)$ is an increasing function of T. Observe that the classes A and Z are completely determined by the classes S, Y, and I. Hence we can restrict our analysis to the system given by (2.12)-(2.14). In addition, the results of Miller [19] and London [20] show that the initial population composition as expressed by, $I_0(t)$, $Y_0(t)$, $Z_0(t)$, and $A_0(t)$ will have a transient effect, but may be neglected for large enough t. The existence , uniqueness and positivity of solutions is established as follows:

First, we specify an appropriate set of initial conditions by setting $S(t) = r(t)$, $I(t) = p(t)$, $Y(t) = m(t)$, on the interval $[-\max(\omega,\tau), 0]$. Moreover, in order to make this system consistent, we assume that $I_0(t) = p(t)$, $Y_0(t) = m(t)$ on the interval $[-\max(\omega,\tau), 0]$. For this set of ordinary delay-differential equations local existence and uniqueness of solutions follows from standard results (see Hale, [21]).

We now assume that $I_0(t) \geq 0$, $Y_0(t) \geq 0$ on $[-\max(\omega,\tau), 0]$, and will show that the solutions remain nonnegative for $t > 0$ as long as they are defined. That is, we will show that the point $(S(t), I(t), Y(t))$ remains in the nonnegative "orthant" in $\mathbb{R}^3$. To correspond to the biological context, we also assume that $S(t) > 0$ on $[-\max(\omega,\tau), 0]$, that at least one of the infectious classes is strictly positive on a subinterval of $[-\max(\omega,\tau), 0]$, and that $\Lambda > 0$.

The trajectory cannot reach a point in the face $S = 0$, since if $S = 0$ then $dS/dt > 0$ in a neighborhood by (2.12). Next we show that a solution cannot reach a face where either $I = 0$ or $Y = 0$. For, if t^* is the first time that $I = 0$ (or $Y = 0$), then from (2.13)

$$I\overset{.}{(t^*)} \geq \lambda\, p \int_{t^*-\omega}^{t^*} C[T](x)\, \frac{S(x)W(x)}{T(x)}\, H(x)\, e^{-\mu(t-x)}\, dx \ ,$$

which is a contradiction. Thus all variables are positive for $t > 0$, under the stated conditions.

As before, we restrict our analysis to the case $p = 1$ (that is we model AIDS as a progressive disease). In this case equation (2.14) is no longer relevant and the study of the steady states reduces to the following set of equations:

$$\frac{dS(t)}{dt} = \Lambda - \lambda C(T)S(t)\, \frac{I(t)}{T(t)} - \mu S(t) \ , \tag{2.17}$$

$$\frac{dI(t)}{dt} = \lambda[C(T(t))S(t)\, \frac{I(t)}{T(t)} - C(T(t-\omega))S(t-\omega)\, \frac{I(t-\omega)}{T(t-\omega)}\, e^{-\mu\omega}] - \mu I(t) \ , \tag{2.18}$$

For this system the infection free-state $(\frac{\Lambda}{\mu}, 0)$ is always an equilibrium, in addition, the

mean infectious period is given by $\dfrac{1 - e^{-\mu\omega}}{\mu}$ and therefore, the reproductive number R,

is given by

$$R = \lambda\, C[\frac{\Lambda}{\mu}]\, (\frac{1 - e^{-\mu\omega}}{\mu}) \ .$$

For this model we have established the following results

The system (2.17)-(2.18) has a unique positive endemic state if and only if the reproductive number exceeds unity (R>1). The infection-free state is globally asymptotically stable whenever the reproductive number R is less than unity (R<1) and it is unstable when R > 1. In addition, the endemic state is locally asymptotically stable whenever R >1. Furthermore, periodic solutions do not arise when one varies parameters from either the endemic state or the infection-free state.

For extensions of these results to the case $0<p<1$ and to the case where distributed (rather than constant) delays are used the reader is referred to [13-14].

3. Multigroup models

In this section we describe two multigroup models: the first assumes that individuals transfer from the infected class at a constant rate (i.e. that the duration of infection has a negative exponential distribution); in the second model we assume that all individuals remain infectious for a fixed length of time. In both models we assume that AIDS is a progressive disease; we relax this assumption in [13-14].

a. Model with exponential removal

A model with three subpopulations ($i = 1,2,3$) with different sexual and social practices is considered (in constructing this model, we follow the approach of Ross [22], and Hethcote and Yorke [23]). Group 1 includes those individuals whose sexual preferences, degree of sexual activity and social practices can facilitate the transmission of HIV. If we assume that the reservoir of the HIV virus is within the (sexually active) homosexual population, then Group 1 could include (sexually active) bisexuals, and perhaps a subgroup of the male and female population of prostitutes. Group 2 includes those heterosexual individuals who have multiple sexual partners, and Group 3 includes those essentially monogamous individuals whose risk of infection arises from social and sexual contact primarily with individuals of Group 2. This classification is somewhat arbitrary, but it is given primarily for the purpose of illustration. We denote by S_i, I_i, and A_i, the corresponding classes for group I as defined in Section 2. In this case $C_i(T) = c_i C(T)$ (c_i appropriate constants for each group), λ_i, Λ_i, α and μ are defined as before but with a subindex to differentiate groups.

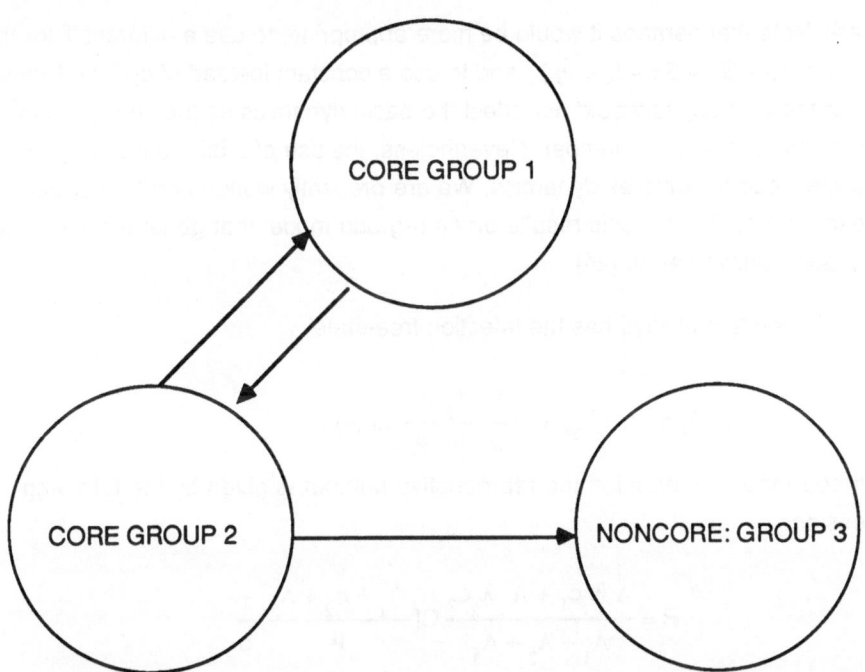

Fig. 3: Three group network, with two core groups and one noncore group, for details see the text.

Proceeding as in Section 2 and guided by Figures 1 and 3 , we arrive at the following model:

$$\frac{dS_1(t)}{dt} = \Lambda_1 - \lambda_1 c_1 C(T(t))S_1(t) \frac{W(t)}{T(t)} - \mu S_1(t) \ , \tag{3.1}$$

$$\frac{dS_2(t)}{dt} = \Lambda_2 - \lambda_2 c_2 C(T(t))S_2(t) \frac{W(t)}{T(t)} - \mu S_2(t) \ , \tag{3.2}$$

$$\frac{dS_3(t)}{dt} = \Lambda_3 - \lambda_3 c_3 C(T(t))S_3(t) \frac{I_2(t)}{T(t)} - \mu S_3(t) \ , \tag{3.3}$$

$$\frac{dI_1(t)}{dt} = \lambda_1 c_1 C(T(t))S_1(t) \frac{W(t)}{T(t)} - \sigma I_1(t) \ , \tag{3.4}$$

$$\frac{dI_2(t)}{dt} = \lambda_2 c_2 C(T(t))S_2(t) \frac{W(t)}{T(t)} - \sigma I_2(t) \ , \tag{3.5}$$

$$\frac{dI_3(t)}{dt} = \lambda_3 c_3 C(T(t))S_3(t) \frac{I_2(t)}{T(t)} - \sigma I_3(t) \ , \tag{3.6}$$

where $\sigma = \mu + \alpha$ and $T = I_3 + S_1 + S_2 + S_3 + W$, $W = I_1 + I_2$.

Remark. Note that perhaps it would be more appropriate to use a different T for the third group (i.e. $T_2 = S_2 + S_3 + I_2 + I_3$), and to use a constant instead of $c_3C(T)$. However, we feel that these changes would not affect the basic dynamics as they have a minimal effect in the reproductive number. Nevertheless, the use of a different C(T) for each group may lead to complex dynamics. We are presently working on further elaborations of these models. For specific results on an n-group model that generalizes this model see Castillo-Chavez et al. [24].

This system always has the infection free-state

$$(S_1,S_2,S_3,I_1,I_2,I_3) = (\frac{\Lambda_1}{\mu},\frac{\Lambda_2}{\mu},\frac{\Lambda_3}{\mu},0,0,0) , \tag{3.7}$$

as an equilibrium. In addition the reproductive number is given by the following expression:

$$R = \frac{\Lambda_1\lambda_1 c_1 + \Lambda_2\lambda_2 c_2}{\Lambda_1 + \Lambda_2 + \Lambda_3} C[\frac{\Lambda_1 + \Lambda_2 + \Lambda_3}{\mu}]\frac{1}{\sigma} ,$$

which is the crucial parameter in the establishment of the following stability result.

The infection-free state is locally asymptotically stable provided its reproductive number R < 1 , and is unstable if R > 1 .

The stability analysis of endemic equilbria for a general C(T) has not yet been fully resolved. In this case there may be more complicated dynamics. For further details on some partial results see [13-14].

b. Model with constant incubation period

If we now modify the previous model by assuming that I-infected individuals remain infected and infectious for a fixed length of time (ω), and ignore transient dynamics (as in Section 2b) we then arrive at the following limiting model (using Figures 2 and 3):

$$\frac{dS_1(t)}{dt} = \Lambda_1 - \lambda_1 c_1 C(T(t))S_1(t)\frac{W(t)}{T(t)} - \mu S_1(t) , \tag{3.8}$$

$$\frac{dS_2(t)}{dt} = \Lambda_2 - \lambda_2 c_2 C(T(t))S_2(t)\frac{W(t)}{T(t)} - \mu S_2(t) , \tag{3.9}$$

$$\frac{dS_3(t)}{dt} = \Lambda_3 - \lambda_3 c_3 C(T(t)) S_3(t) \frac{I_2(t)}{T(t)} - \mu S_3(t) , \qquad (3.10)$$

$$\frac{dI_1(t)}{dt} = \lambda_1 c_1 [C(T(t)) S_1(t) \frac{W(t)}{T(t)} - C(T(t-\omega)) S_1(t-\omega) \frac{W(t-\omega)}{T(t-\omega)} e^{-\mu\omega}] - \mu I_1(t), \qquad (3.11)$$

$$\frac{dI_2(t)}{dt} = \lambda_2 c_2 [C(T(t)) S_2(t) \frac{W(t)}{T(t)} - C(T(t-\omega)) S_2(t-\omega) \frac{W(t-\omega)}{T(t-\omega)} e^{-\mu\omega}] - \mu I_2(t), \qquad (3.12)$$

$$\frac{dI_3(t)}{dt} = \lambda_3 c_3 [C(T(t)) S_3(t) \frac{I_2(t)}{T(t)} - C(T(t-\omega)) S_3(t-\omega) \frac{I_2(t-\omega)}{T(t-\omega)} e^{-\mu\omega}] - \mu I_2(t) , \qquad (3.13)$$

where $T = I_3 + S_1 + S_2 + S_3 + W$, $W = I_1 + I_2$.

This system always has the infection free-state

$$(S_1, S_2, S_3, I_1, I_2, I_3) = (\frac{\Lambda_1}{\mu}, \frac{\Lambda_2}{\mu}, \frac{\Lambda_3}{\mu}, 0, 0, 0) , \qquad (3.14)$$

as an equilibrium. In addition the reproductive number is given by the following expression:

$$R = \frac{\Lambda_1 \lambda_1 c_1 + \Lambda_2 \lambda_2 c_2}{\Lambda_1 + \Lambda_2 + \Lambda_3} C[\frac{\Lambda_1 + \Lambda_2 + \Lambda_3}{\mu}] (\frac{1 - e^{-\mu\omega}}{\mu}) ,$$

which is the crucial parameter in the establishment of the following results:

> **The infection-free state is locally asymptotically stable provided its reproductive number $R < 1$, and unstable if $R > 1$. Furthermore, periodic solutions do not bifurcate from this state when parameters are varied.**

The local stability analysis of endemic equilibria for a general $C(T)$ has not yet been fully resolved. For some partial results in this direction and for partial results for the case $0 < p < 1$ (i.e. when AIDS is not assumed to be a progresive disease), see [13-14, and 24].

Acknowledgments

This work has been partially supported by NSF grants DMS-8406472 to Simon A. Levin, and DMS-8603450 to Kenneth L. Cooke. Carlos Castillo-Chavez' research has been partially supported by The Center for Applied Mathematics, the Office of the

Provost at Cornell University as well as by a Ford Foundation Postdoctoral Fellowship for Minorities. We thank them all.

REFERENCES

[1] Barré-Sinoussi, F., J. -C. Chermann, F. Rey, M. T. Nugeyre, S. Chamaret, J. Gruest, C. Dauguet, C. Axler-Blin, F. Brun-Vézinet, C. Rouzioux, W. Rozenbaum and L. Montagnier. *Isolation of a T-lymphotropic retrovirus from a patient at risk for acquired immune deficiency syndrome (AIDS).* Science 220: 868-70 (1983).

[2] Gallo, R.C., S. Z. Salahuddin, M. Popovic, G. M. Shearer, M. Kaplan, B. F. Haynes, T. J. Palker, R. Redfield, J. Oleske, B. Safai, G. White, P. Foster and P. D. Markhamet. *Frequent detection and isolation of sytopathic retroviruses (HTLV-III) from patients with AIDS and at risk for AIDS.* Science 224: 500-503 (1984).

[3] Gallo, R. C. *The first human retrovirus.* Scientific American 255: 88-98 (1986).

[4] Gallo, R.C. *The AIDS virus.* Scientific American 256: 47-56 (1987).

[5] Wong-Staal, F. and R. C. Gallo. *Human T-lymphotropic retroviruses.* Nature 317: 1985: 395-403.

[6] Francis, D.F., P. M. Feorino, J. R. Broderson, H. M. McClure, J. P. Getchell, C. R. McGrath, B. Swenson, J. S. McDougal, E. L. Palmer, A. K. Harrison, F. Barre-Sinoussi, J. -C. Chermann, L. Montagnier, J. W. Curran, C. D. Cabradilla and V. S. Kalyanaraman. *Infection of chimpanzees with lymphadenopathy-associated virus.* Lancet 2: 1276-77 (1984).

[7] Salahuddin, S.Z., J. E. Groopman, P.D. Markham, M. G. Sarngaharan, R. R. Redfield, M. F. McLane, M. Essex, A. Sliski and R. C. Gallo. *HTLV-III in symptom-free seronegative persons.* Lancet 2: 1418-20 (1984).

[8] Lange, J. M. A., D. A. Paul, H. G. Huisman, et. al. *Persistent HIV antigenaemia and decline of HIV core antibodies associated with transition to AIDS.* Brit. Med. J. 293, 1986: 1459-62.

[9] Medley, G. F., R. M. Anderson, D. R. Cox, and L. Billard. *Incubation period of AIDS in patients infected via blood transfusions.* Nature 328: 719-21(1987).

[10] Gerberding, J. L. and D. K. Henderson. *Design of rational infection control policies for human immunodeficiency virus infection.* J. Infectious Diseases. Vol. 156, No. 6 : 861-864 (1987).

[11] Anderson, R.M., R. M. May, G. F. Medley, and A. Johnson. *A preliminary study of the transmission dynamics of the human immunodeficiency virus (HIV), the causative agent of AIDS.* IMA J. Math. Med. Biol. 3:229-263 (1986).

[12] Anderson, R.M. and R. M. May. *Transmission dynamics of HIV infection.* Nature 326: 137-142 (1987).

[13] Castillo-Chavez, C., K. Cooke, W. Huang and S. A. Levin. *On the role of long incubation periods in the dynamics of acquired immunodeficiency syndrome (AIDS) Part 1. Single population models.* J. of Math. Biol. (in press).

14] Huang, W., C. Castillo-Chavez, K. Cooke and S. A. Levin. *On the role of long incubation periods in the dynamics of acquired immunodeficiency syndrome (AIDS) . Part 2. Multiple group models.* In: *Mathematical and statistical approaches to AIDS epidemiology* (C. Castillo-Chavez, ed). Lecture Notes in Biomathematics, Springer-Verlag (in press).

[15] Blythe, S. P. and R. M. Anderson. *Distributed incubation and infectious periods in models of the transmission dynamics of the human immunodeficiency virus (HIV).* IMA J. Math. Med. Bio. 5:1-19 (1988).

[16] Hethcote, H. W., H. W. Stech and P. van den Driessche. *Periodicity and stability in epidemic models: a survey.* In : Differential Equations and Applications in Ecology, Epidemics and Population Problems. S. Busenberg and K. L. Cooke (eds.) Academic Press, New York: 65-82 (1981).

[17] Hyman, J. M. and E. A. Stanley. *A risk based model for the spread of the AIDS virus.* Math. Biosci. 90:415-473 (1988).

[18] Kingsley, R. A., R. Kaslow, C. R. Rinaldo Jr., K. Detre, N. Odaka, M. VanRaden, R. Detels, B. F. Polk, J. Chimel, S. F. Kersey, D. Ostrow and B. Visscher. *Risk factors for seroconversion to human immunodeficiency virus among male homosexuals.* The Lancet 1:345-348 (1987).

[19] R. K. Miller. *On the linearization of Volterra integral equations.* J. Math. Anal. Appl. 23: 198-208 (1968).

[20] S. O. Londen. *Integral equations of Volterra type.* In: Mathematics of Biology, M. Iannelli (ed.), Liguori Editori, Napoli (1981).

[21] J. K. Hale. Theory of Functional Differential Equations. Springer Verlag. New York (1977).

[22] R. Ross. The Prevention of Malaria. 2nd. Edition, Murray, London (1911).

[23] Hethcote, H. W. and J. A. Yorke. Gonorrhea Transmission Dynamics and Control. Lecture Notes in Biomathematics No. 56, Springer VerlagHeidelberg (1984).

[24] Castillo-Chavez, C., K. Cooke, W. Huang, and S. A. Levin. *Results on the dynamics for models for the sexual transmission of the human immuno-deficiency virus.* Applied Mathematics Letters (in press).

THE EFFECT OF SOCIAL MIXING PATTERNS
ON THE SPREAD OF AIDS

James M. Hyman and E. Ann Stanley
Theoretical Division, MS-B284
Center for Nonlinear Studies
Los Alamos National Laboratory
Los Alamos, NM 87545

Abstract

Mathematical models of the transmission of the AIDS virus can help us better understand the spread of the AIDS epidemic and prepare for the future. Model explorations can indicate which factors the epidemic is most sensitive to and provide guidance in designing interventions, educational programs and social behavior studies. We explore the sensitivity of a transmission model to different social mixing patterns. This model continuously distributes a homosexual community according to sexual partner change rates and can account for infectivity and conversion times that vary with time since infection. An acceptance function determines which partners are acceptable to an individual and defines the mixing between groups with different partner change rates. We find that if people only select partners with very similar behavior the epidemic grows much slower than if they are not as discriminating. Therefore, understanding social mixing patterns may be one of the most urgent tasks if we are to anticipate the future. We also find that the epidemic is sensitive to variable infectivity and conversion times.

I.   INTRODUCTION

Mathematical models for the spread of the Human Immunodeficiency Virus (HIV) that causes AIDS are tools that have the potential to greatly enhance our understanding of the AIDS epidemic. Models provide a framework within which we can study the interactions of social and biological mechanisms that influence the spread of the disease. They allow us to ascertain the relative influence of various factors on the spread of the epidemic, as well as the sensitivity to uncertainties in values.

As a first step in developing a reliable model, we have developed a deterministic model for a homosexual community. This model distributes the population according to the number of sexual partners per year and keeps track of time since infection for infecteds and time since diagnosis for AIDS cases. Susceptible persons are infected through contacts with infected persons, and infected persons develop clinical AIDS (such as Kaposi's sarcoma [KS] or opportunistic infections such as pneumocystis pneumonia [PCP]) at a rate that depends on the length of time since HIV infection. AIDS patients subsequently die at a rate that depends on the length of time since AIDS developed. We assume that an infected person remains infected and infectious for life and that a person maintains the same partner change rates the whole time he remains in the population.

Hyman and Stanley (1988) explored a number of questions with simplified versions of a model similar to the one presented here. They used a model which neglected variations in partner change rates to examine the impact of various plausible shapes for the infectivity as the time since infection varies. These calculations pointed out the importance of measuring the variability of the infectiousness during the disease. They also used a model with no variations with time since infection to show that random partner choice is dramatically different from a strong bias of like prefers like. Other models have also shown that selective partner choice is a crucial determinant of HIV spread (Jacquez et al., in press, Stigum et al., 1988).

Because different mixing patterns can result in radically different epidemics, much more must be known about the interactions between people that lead to AIDS virus spread before it will be possible to accurately predict the AIDS epidemic. The number of sexual partners that people have, the partner-selection process, and the amount and type of contacts between partners must be understood and correlated with sociological information about the partners, such as how many partners your partners have.

In this paper, we further explore the sensitivity of the model to assumptions about partner choice, again using a model which neglects variations in parameters with time since infection. Then we add the distribution of infecteds with time since infection and parameters that vary with time since infection to see what the effects of these variations are.

In our analysis, we focus on the initial growth of the epidemic. If we are to predict where this epidemic is going, we must fully understand its transient dynamics, including the response to changes in the environment of the epidemic. The epidemic will not reach an equilibrium endemic state for a very long time, partly because of the long conversion times from infection to AIDS, during which a person can transmit the virus and partly because medical advances and changes in lifestyle will greatly modify the future of the epidemic. The infectiousness and susceptibility of high-risk individuals in the heterosexual community may be significantly reduced if programs are initiated to quickly identify and treat other STDs. More people are being tested for antibodies to HIV and counseled on the implications of the test results. Treatments are being developed that will prolong the lives of infected persons and perhaps lower their infectivity. A partially effective vaccine may eventually be developed. Models can be used to investigate the effects of each of these programs on the course of the epidemic only if they can capture the transients of the epidemic.

As models are developed they will provide a logical structure for the diverse data that researchers are collecting. Also, new questions and insights will arise to guide investigators in directing their research to add to the general understanding of this epidemic.

II. EPIDEMIOLOGY OF AIDS

The HIV that causes AIDS is primarily transmitted through sexual contact (man-woman, man-man), sharing of hypodermic needles, and exposure to infected blood either perinatally or through blood transfusions. HIV is not transmitted by nonsexual daily contacts, even though the virus has been isolated from almost every body fluid (Fischl et al., 1987). The infection risk to an individual depends both on the behavior of the individual and on the prevalence of infection in the groups with which the individual has sexual contacts or shares needles. This prevalence varies between regions and age groups, as well as between behavioral risk groups. Multiple sexual partners, sexual partners in a high-risk group or from a highly populated area and sharing needles when using drugs all increase risk.

Surveys of risk behaviors in the homosexual communities demonstrate that the variance in the number of sexual partners per year is large. (see Fig. 4.3 in Section IV.D). In this epidemic, it is significant that the people with many partners tend to become infected first and then become carriers who infect less-active people. This distribution can have a marked effect on the course of the epidemic and on which risk group is currently at highest risk of infection.

Risk from sexual activity depends on the probability of choosing an infected partner as well as on the number and type of contacts with an infected partner. The probability of choosing an infected partner depends not only on how many new partners are chosen but also on the manner in which those partners are chosen.

The risk group from which a person chooses partners for sex or needle-sharing is an important social question about which little is known. No large-scale studies specifically aimed at sexual behavior have been conducted in the United States since the Kinsey Studies more than 35 years ago and the sampling procedure for this study decreases its usefulness. The information available from other countries is also poor. However, a number of other studies, such as fertility studies, have included some questions on sexual behavior or have studied specific groups. In addition, NICHD is designing and will implement a nationwide survey of sexual behavior and needle-sharing behavior specifically aimed at gathering information about the transmission of the AIDS virus. Surveys of sexual behavior are being conducted or planned in many countries around the world. For example, a national, randomized survey of 10,000 people has recently been conducted in Norway (Sundet et al., 1988).

Most models for the transmission of venereal diseases (Hethcote and Yorke, 1984; Anderson et al., 1986) have assumed that all partners are picked at random from the pool of available partners. This assumption leads to the proportionate-mixing assumption that the per year probability of someone with i partners per year picking an infected partner with j partners per year is $i \cdot j \cdot P_j / P_T$, where P_j is the number of infected people with j partners per year and P_T is the total number of partners picked per year. These models also assume that the probability of infection per partner is the same. However, it is clear that these assumptions are overly simplistic.

It seems reasonable to assume that there is a tendency for people with fewer partners to have more contacts per partner than do people with many partners. In most communities, there is also a bias of like toward like, so that people with few partners tend to choose partners who also have few partners. This observation led Hethcote and Yorke to use a combination of within-group mixing and random mixing in their 2 risk-level gonhorhea model. Adding these biases into the Anderson et al. model leads to substantially different predictions from their random-mixing model with equal risks (Hyman and Stanley, 1988).

In our model, we assume that an average probability of infection can be assigned to each contact. This assumption may not be sufficiently accurate to predict the spread of HIV and additional factors may need to be included in the model (Peterman et al., 1988). For example, the probability of infection might depend strongly upon the strain of the virus or on the health of the partners.

The infectiousness of a contact probably also depends on the type of contact (man-man, woman-man, man-woman, anal-genital, oral-genital). There is growing evidence that infectiousness may depend as well on other cofactors such as venereal diseases and the use of protective devices (condoms, nonoxynol-9). We need estimates for the prevalence of these cofactors, how frequently protective devices are used, and how much behavior can be influenced by factors such as education, knowledge that a partner or oneself is infected, and fear of infection. As public awareness increases and more people know they are infected, we speculate that the resulting drift toward safer sexual practices will slow the spread of the virus.

The African epidemic demonstrates that the virus can spread quickly through a heterosexual network. Growing evidence suggests that this fast heterosexual spread is partly due to a high prevalence of cofactors, such as genital ulcers caused by chanchroids, which may greatly increase both infectiousness and susceptability. In the developed world, such severe cofactors are virtually nonexistent. However, other cofactors are present, such as gonorrhea, syphilis, and herpes, that may increase transmission rates less dramatically. Without data on infectiousness, with and without cofactors, male-to-female and female-to-male, it is impossible to tell whether or not a self-sustaining heterosexual epidemic will occur in the United States, even though the current heterosexual AIDS cases are primarily due to contacts with homosexuals and IV drug users. A slowly growing heterosexual epidemic could be masked by cases due to contacts with these groups. It is unlikely that models can distinguish between these two possibilities without estimates of transmission probabilities from partner studies (e.g., Fischl et al., 1987; Padian et al., 1987).

The accumulated number of AIDS cases diagnosed in the United States as reported to CDC, A(t), is not growing exponentially but is well approximated by

$$A(t) = 174.6(t - 1981.2)^{3.0} + 340 \pm 2\%$$

(2.1)

for times $t \geq 1982.5$. This polynomial growth is evident in nearly every CDC-defined category including risk behavior, age, region of the country, and ethnic group (Hyman et al., in preparation). The AIDS cases approximated by Eq. (2.1) are based on the pre-June 1987 AIDS definition and do not include dementia and wasting syndrome.

If $C(\tau)$ is the probability that a person infected with HIV at time $t-\tau$ has developed AIDS by time t, and if $I'(t)$ is the number of people infected per year with HIV, then the cumulative AIDS cases reported to CDC satisfies the relationship

$$A(t) = p \int_0^\infty C(\tau) I'(t-\tau) d\tau$$

(2.2a)

or

$$A'(t) = p \int_0^\infty C'(\tau) I'(t-\tau) d\tau$$

(2.2b)

where p is the fraction of infected individuals eventually reported to CDC as AIDS cases. p is the product of the probability that an infection will result in a pre-1987.5 CDC-defined AIDS case (which excludes dementia and slim disease) times the probability it will be reported to CDC. The probability that an AIDS case will be reported to CDC is the product of the probabilities that it will be diagnosed and, once diagnosed, that it will then be reported. Using estimates of $C'(\tau)$, the probability density function for conversion to AIDS, we can solve Eq. (2.2) for $I'(t)$.

As the width of $C'(\tau)$ approaches zero (that is, a delta-function), then the solution of Eq. (2.2) approaches

$$I(t) = p^{-1}A(t + \tau_A) .$$

(2.3)

This estimate can be used as a rough approximation for I(t), even for fairly wide distributions $C'(\tau)$ (see Hyman and Stanley, 1988). This approximation can be used to estimate the number of infected individuals in January 1988. For example, if we assume that 80% of the infected individuals develop CDC-defined AIDS and that 80% of these are reported to the CDC, then p = 0.64. If τ_A = 9 years and the number of AIDS cases in 1997 (= 1988 + τ_A) is 85% of the extrapolated cubic approximation (4.2), then the current cumulated number of infected individuals is

$$I(1988) \simeq \left\lfloor \frac{0.85}{0.64} \right\rfloor \left| 174.6(1988.0 + 9 - 1981.2)^3 + 340 \right| \simeq 915,000 \ . \qquad (2.4)$$

We remark that if only 40% of the infected individuals develop CDC-defined AIDS (as was thought a few years ago) then, even though the predicted AIDS cases are the same, this approximation estimates that there would be 1,830,000 people infected with HIV in the United States.

The cubic polynomial growth can be explained by a wave of infection progressing from populations with high-risk behavior into populations with lower-risk behavior. For example, if individuals with risk behavior r (proportional to the number of sexual partners or needles shared) are infected through interactions with people of similar behavior and if the population is distributed as a decreasing function of risk behavior [e.g., $N(r) \simeq N_0(1 + ar)^{-4}$, where $N(r)$ is the number of individual with risk r], then the highest-risk population is quickly infected, giving rise to an initial transient exponential growth. This growth quickly becomes polynomial as the saturation wave of infection moves into lower-risk (but still high-risk) behavior and finally slows to an $\exp(1/t)$ growth rate (See Sec. V). The polynomial growth is analyzed in more detail in Colgate et al. (1988).

III. MODEL DESCRIPTION

A complete model of the spread of the AIDS virus in a sexually active and IV-drug-using community must account for the complicated interactions between people. However, one must begin by understanding the behavior of simple models before going on to explore more complex ones. In risk-based models, such as the one we use here, the population is stratified according to the amount of risk individuals incur. These models do not model the risk (or protection) of longer-term relations as well as the partnership models of Dietz (1987 and 1988) in which individuals are continually forming and breaking partnerships and the infection is passed only when one individual in the partnership is infected and the other is not. However, in the partnership models it is difficult to account for the wide variations in risk behavior that occur. Because we are primarily concerned with modeling HIV spread in high-risk populations, we use the risk-based approach and account for partnership duration by allowing a variable number of contacts in each partnership.

For modeling purposes, we divide the at-risk community into uninfected susceptibles, infecteds without AIDS, and diagnosed AIDS cases. To model variations in risk behavior within this community, we suppose that the population can be distributed according to their numbers of new sexual partners per year. People mature into a fixed risk group and leave it only upon becoming sexually inactive (and ceasing to share needles). Before the introduction of the AIDS virus, there was a balance between a constant maturation and migration rate into each risk group in the community and a constant rate per individual of retirement or death out of it; these processes continue in the presence of AIDS. Susceptibles become infected through sexual contacts or IV needle-sharing with infected partners. Infected individuals eventually develop AIDS, becoming sexually (or needle-sharing) inactive, and die at an accelerated rate

We further stratify the non-AIDS-infecteds and AIDS cases according to time since infection or AIDS. This allows us to model both a variable infectivity and the distributions of times from infection to AIDS and of times from AIDS to death. Defining

t	:	time,
τ	:	time since becoming infected or developing AIDS
r	:	number of new sexual partners per year,
$S(t,r)$	:	distribution of susceptibles according to the number of partners per year,
$I(t,r,\tau)$	:	distribution of infecteds according to the number of partners per year and the time since infection,
$A(t,\tau)$	:	distribution of AIDS cases according to time since AIDS began,
$i(\tau)$	:	probability of infection from a contact with a person infected τ years ago,
$\gamma(\tau)$	:	rate at which infecteds develop AIDS at a time τ after infection,
$\delta(\tau)$	:	death rate at time τ after AIDS starts,
$A_T(t)$	:	accumulated number of AIDS cases,
$N(t)$	:	number of susceptible and infected individuals without AIDS,
μ	:	death rate of individuals without AIDS,
$c(r,r')$	:	total number of contacts in a partnership between people with r and r' partners per year, and
$S_0(r)$	:	density of people with r new partners per year before the AIDS virus was introduced.

Note that $S(t,r)$ and $S_0(r)$ have the units people-time/partners and $I(t,\tau,r)$ has the units people/partners. The resulting model is

$$\frac{\partial S(t,r)}{\partial t} = \mu(S_0(r) - S(t,r)) - \lambda(t,r)S(t,r) \; ,$$

$$I(t,0,r) = \lambda(t,r)S(t,r) \; ,$$

$$\frac{\partial I(t,\tau,r)}{\partial t} + \frac{\partial I(t,\tau,r)}{\partial \tau} = -(\gamma(\tau) + \mu)I(t,\tau,r) \; ,$$

$$A(t,0) = \int_0^\infty \int_0^\infty \gamma(\tau) \, I(t,\tau,r) \, d\tau dr \; , \qquad (3.1)$$

$$\frac{\partial A(t,\tau)}{\partial t} + \frac{\partial A(t,\tau)}{\partial \tau} = -\delta(\tau)A(t) \; ,$$

$$\frac{dA_T}{dt} = \int_0^\infty \int_0^\infty \gamma(\tau) \, I(t,\tau,r) \, d\tau dr \; ,$$

$$<rN(t)> = \int_0^\infty rN(t,r) \, dr$$

and

$$N(t,r) = S(t,r) + \int_0^\infty I(t,\tau,r) \, d\tau$$

This model portrays a community in which people mature or migrate into the susceptible community with risk r at a constant rate $\mu S_0(r)$. People without AIDS die (or become inactive) at a constant rate, with μ^{-1} their average life expectancy. Infection occurs through sexual contact with an infected partner.

There may be a wide variation in infectiousness as the disease progresses. A constant rate of progressing to AIDS would impose an exponentially decaying distribution of times to AIDS. However, cohort studies have found that the probability of getting AIDS increases with time since infection for at least the first 7 years (see Section IV.A).

The infectivity, $i(\tau)$, is an average over all individuals infected at time τ and is discussed in more detail in Section IV.B.

We must still define $\lambda(t,r)$. We discuss below some possible choices: random partner choice, a bias of people towards partners like themselves, and a combination of the two.

Defining $\lambda(t,r)$

We assume that the average r-r' partnership is sufficiently short and infectivity is sufficiently low that the probability that a person has already become infected in the partnership is small, i.e.,

$$max_{t} \, i(t)c(r,r') \ll 1 \; .$$

Furthermore, the epidemic cannot grow so fast that the chance that a partner is infected becomes significantly different during the course of the partnership from an unmatched person from the same risk group.

Under these assumptions , $\lambda(t,r)$ can be approximated by

$$\lambda(t,r) = r \int_{0}^{\infty} F(t,r,r') \, k(t,r,r') dr'$$

$$(3.2)$$

$$k(t,r,r') = c(r,r') \int_{0}^{\infty} i(t) \frac{I(t,\tau,r')}{N(t,r')} \, d\tau \; .$$

Here $k(t,r,r')$ is the probability of being infected by a partner of risk r'. $F(t,r,r')$ is the fraction of partners of people with risk r that have risk r'. For random partner choice, this is

$$F_{random}(r,r') = r'N(t,r')[<rN(t)>]^{-1} \; .$$

$$(3.3)$$

If we assume that partners are chosen at random from the entire population, then $\lambda(t,r)$ is given by

$$\lambda_{random}(t,r) = \frac{r}{<rN(t)>} \int_{0}^{\infty} c(r,r')r' \int_{0}^{\infty} i(t) \, I(t,\tau,r') d\tau \, dr' \; .$$

$$(3.4)$$

A version of this model with no differences in partnership durations and no variability in infectiousness ($c(r,r')$ and $i(\tau)$ constant) was first proposed by Anderson et al. (1986).

The $\lambda(t,r)$ given by Eq. (3.4) does not account for the fact that people do not choose partners at random from all groups but instead prefer partners of a certain type and choose them when available. Ideally, the partner selection in any model should be based on sociological data. This question will be discussed in more detail in a later report; as a first step towards addressing this question we present below a model which allows a wide range of partner choices to be specified.

To account for partnership biasing, $F(r,r')$ is determined by the fraction of partnerships from r' that are both available and acceptable. Thus, if partners of risk r' are accepted by people with risk r with a frequency $f(r,r')$ then the fraction of partnerships available and acceptable to a person of risk r is

$$F(r,r') = f(r, r')r'N(t,r') \left[\int_0^\infty r'' f(r,r'')N(t,r'')dr'' \right]^{-1}. \tag{3.5}$$

There are, however, constraints on F(t,r,r'): the total rate that r-r' partnerships form, rN(t,r)F(t,r,r'), must be equal to the total rate that r'-r partnerships form. We would also like to ensure that a person from r has r partners/year. There is no unique way to do this. However, if we let the person from the lower risk group always be the one which decides on the acceptability of the partnership, then

$$F(t,r,r') = \begin{cases} \left(1 - \int_0^r F(t,r,x)dx\right) \cdot \dfrac{f(r,r')r'N(t,r')}{\displaystyle\int_r^\infty xf(r,x)\,N(t,x)d'x} , & \text{for } r < r' , \\[4ex] F(t,r',r)\,\dfrac{r'N(t,r')}{rN(t,r)} , & \text{for } r > r' \end{cases} \tag{3.6}$$

is a reasonable choice. $f(r,r') = 1$ gives random mixing (3.4). Substituting Eq. (3.6) into Eq. (3.2) defines $\lambda(t,r)$.

The system in Eq. (3.1) with different choices of $\lambda(t,r)$ allows the implications of a wide variety of partner-selection mechanisms to be investigated.

If mixing occurred only with people from the same risk group, then the virus could not spread between groups, $\lambda(t,r)$ would be equal to $k(t,r,r)$, and the system in Eq. (3.1) would describe separate epidemics for each value of r. However, this perfect isolation is unrealistic. The mixing between people of similar, but not identical, risk behavior leads to diffusion of the virus from one group to another. Using

$$f(r,r') = exp\left[-(1/2\varepsilon)(r-r')^2/(r+a)^2\right] \tag{3.7}$$

$f(r,r') = exp[-(1/2\varepsilon)(r-r')^2/(r+a)^2]$ to define F and letting $\varepsilon \to 0$ in Eq. 3.2 gives

$$\lambda(t,r) = r\left[k(t,r,r) + \frac{\varepsilon}{2(r+a)rN(t,r)} \frac{\partial}{\partial x}\left((x+a)^2 xN(t,x)\frac{\partial k(t,r,x)}{\partial x}\right)\right]_{at\ x=r} \tag{3.8}$$

to $O(\varepsilon)$.

With this λ, Eq. (3.1) becomes a partial differential equation. Although this model is only an approximation to the full system, it shows that the complex integro-differential equations of (3.1), (3.2), and (3.6) model a diffusive process. Exploring this model (and other limiting models) can help us understand much of what is occurring in the numerical simulations of the full model.

Even within the male homosexual and the IV needle-sharing communities, behavior patterns are not this simple. Depending on the community of interest, there may be a very different mixing pattern from the ones described here. An individual's behavior will change over time, and people with many partners one year may have only a few the next, or vice versa. Social groups within which mixing is strong, and between which it is weak, may cause low-activity people in one group to be infected before high-activity people in another group.

The social/nonsocial mixing behaviors modeled by Sattenspiel (1987) and Sattenspiel and Simon (1988) may also play an important role in the spread of this disease. Models with a variety of mixing assumptions need to be developed and compared, both with each other and with behavioral and serological studies, to ascertain what complexities are necessary for modeling HIV spread.

IV. MODEL PARAMETERS

The models discussed in the previous section contain a number of parameters that must be estimated in order to make calculations. Some of these parameters can be estimated fairly well (μ, γ, or $\delta(\tau)$), but for most of them only partial information is known. It is important to explore the effects of parameter changes, within plausible ranges, on the solution of these models.

A. RATE OF DEVELOPING AIDS

HIV causes a slow decline in the immune system. This picture of progressive immune-system decline indicates that most infected individuals eventually die from HIV-induced illness and that the probability that an individual will develop AIDS depends on how long he has been infected. The time from infection to diagnosis of AIDS is extremely variable. HIV-infected adults have developed AIDS in less than 2 years and some have remained well for more than 9 years. The distribution of times between infection and clinical AIDS is only partially known because of the long times involved. In studies of patients for whom an estimate of date of infection can be made (such as hemophiliacs), the percentages developing AIDS in any given year after infection are either still increasing or are remaining roughly constant, which leads to an estimate of an average time to AIDS of at least 8 years.

We have chosen to use the Weibull distribution of Medley et al. (1987)

$$C'(\tau) = pq^p\tau^{p-1}e^{-(q\tau)^p} ,$$

(4.1)

with p = 2.4, q = 0.11 for the times from infection to AIDS, primarily because it agrees well with the first 7 years of estimates from the portion of the San Francisco Hepatitis B cohort for whom the date of infection can be estimated (George Lemp, personal communication). This distribution, shown in Fig. 4.1, has a maximum at 7.5 years, a median value of 8 years and a mean of 8.9 years. This function is chosen such that all infected persons eventually get AIDS. If less than 100% of the infected people get AIDS, the tail of the distribution should be reduced, but the first 7 years should be left unchanged.

The rate $\gamma(\tau)$ of getting AIDS at time τ after infection is the conditional probability density given that the person has not yet developed AIDS

$$\gamma(\iota) = C'(\iota)[1 - C(\iota)]^{-1} \ , \ \ C(\iota) = \int_0^{\iota} C'(\iota_a)d\iota_a \ . \tag{4.2}$$

$\gamma(\tau)$ is shown in Fig. 4.1 for the Weibull of Eq. (4.1).

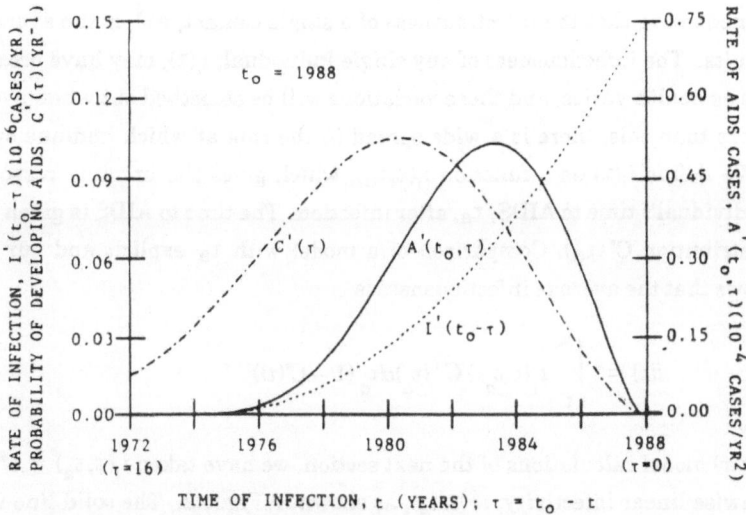

Fig. 4.1. Conversion from infection to AIDS as given by Eq. (4.1) with $p = 2.4$, $q = 0.11$. Here $C(\tau)$ is the probability of developing AIDS by τ years after infection, $C'(\tau)$ is the probability density of developing AIDS at τ years after infection, and $\gamma(\tau)$ is the conditional probability density of first developing AIDS at time τ.

B.   INFECTIVITY

The infectivity may be related to the amount of free virus in the circulatory system of an infected individual. Studies indicate that the amount of free virus goes up in the first few weeks after infection (Francis et al., 1984; Sulahuddin et al., 1984) and then goes down as antibody response occurs, remaining at very low levels for years. As the immune system collapses in the year or so before AIDS develops, viral counts return to high levels (Lange et al., 1986).

Information on average per contact infectivity is only good enough to make estimates on its order of magnitude. Padian et al. (1987) have used partner studies to estimate an average per contact infectivity from man to woman of 0.001 when no other venereal diseases are present. Grant et al. (1987) have used seroprevalence estimates to estimate a per partner infectivity for man-to-man transmission (with receptive and insertive intercourse) of $i_p = 0.10$, but they had no information on numbers of contacts between partners. They also make some estimates for per contact infectivity assuming a fixed number of contacts per month and get a range of 0.004 for 8 contacts to 0.03 for 1 contact per month. Only a study with information about the number of contacts between partners and the clinical status of the partner can give actual numbers, but these data indicate that the average infectivity of a sexual contact probably lies between 0.001 and 0.03.

We assumed above that the infectiousness of a single contact, $i(\tau)$, is the average for all infected adults. The infectiousness of any single individual, $i_i(\tau)$, may have occasional ups and downs as health varies, and these variations will be smoothed out when averages are taken. More than this, there is a wide spread in the rate at which immune systems deteriorate. We define $i_i(\tau)$ as a function $i_i(\tau,\tau_a)$, which gives the immune response in terms of the individual's time to AIDS, τ_a, after infection. The time to AIDS is given by the probability distribution $C'(\tau_a)$. Comparison of a model with τ_a explicit and our model without τ_a shows that the average infectiousness is

$$i(\tau) = \int_\tau^\infty i_i(\tau,\tau_a) C'(\tau_a) d\tau_a (1 - C'(\tau))^{-1}. \tag{4.3}$$

For the (τ,r) model calculations of the next section, we have taken $i_i(\tau,\tau_a) = i^*(\tau/\tau_a)$. We use a piecewise linear infectivity, $i^*(\tau/\tau_a)$, as shown in Fig. 4.2. The solid line in Fig. 4.2 shows the effect of applying Eq. (4.3) to the Weibull of Fig. (4.1) and the $i^*(\tau/\tau_a)$ shown as $i_i(\tau,8)$.

C.   DEATH RATES

The death rates μ and $\delta(\tau)$ are the model parameters for which the best data exist. If we take μ to represent the rate of attrition out of the at-risk community, a μ^{-1} of 30-50 years is reasonable. In our calculations, we use $\mu = 0.02$.

The probability of death once AIDS symptoms appear can be estimated from CDC mortality data, where deaths are recorded according to diagnosis date. The rate of death is high at first and gradually decreases. An exponentially decreasing probability density for death as a function of time since AIDS, which gives a constant death rate, fits adequately. A slightly better fit is found by taking the density function to be

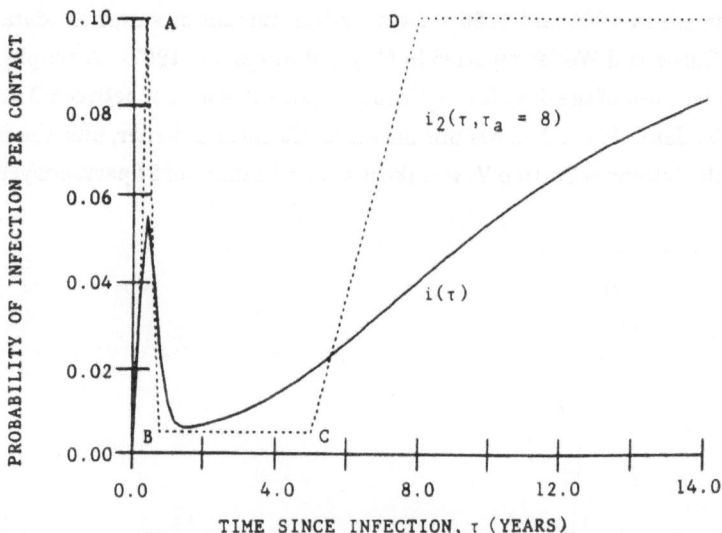

Fig. 4.2 The infectivity of an average person infected at time τ is a smeared version of the infectivity of an individual. We have postulated an individual infectivity $i_i(\tau,\tau_a) = i^*(\tau/\tau_a)$. The dotted line shows $i_i(\tau,\tau_a)$ for $\tau_a = 8$ years, and the solid line shows the average infectivity, $i(\tau)$, given by Eq. (4.3) with $C'(\tau)$ as in Fig 4.1.

$$D'(\tau) = d_1 exp[-d_2\tau(1+d_3\tau)^{-1}] , \tag{4.4}$$

where τ is the time since AIDS symptoms appear and $D_1 \simeq 1$ is chosen to normalize the area to 1 at $\tau = 20$ years. Now we get the rate of death to be decreasing with τ:

$$\delta(\tau) = D'(\tau)[1 - D(\tau)]^{-1} , \quad D(\tau) = \int_0^\tau D'(\tau_d) d\tau_d . \tag{4.5}$$

$d_2 = 0.075$ and $d_3 = 0.05$ give reasonably good fits to the CDC data, with 48% dead in 1 year and 90% dead about 5 years later. A recent follow-up of AIDS cases found that deaths were severely under reported (Hardy et al., 1987). Thus, this distribution might underestimate the true death rate due to AIDS. This underestimate will be somewhat less severe than it might have been because of the widespread use of AZT.

D.   DISTRIBUTION OF RISKS

Sexual activity data from studies of homosexual men show that there is an enormous variation between individuals in the numbers of partners and the amount and type of contacts. Participants in the Multicenter AIDS Cohort Study (MACS), who were questioned between April 1984 and March 1985, reported between 1 and 500 male partners in the previous 6 months, with a mean of between 5 and 10 (Kingsley et al., 1987). The San Francisco Men's Health Study (Winklestein et al., 1987) and homosexual men surveyed in London in 1984 and 1986 show a similar amount of variation (data from T. McManus and Carne and Weller reported in May and Anderson, 1987). A simple function that approximates most of the data is $(n+1)(n\mu)^{n+1}(n\mu+r)^{-n}$ with n between 3 and 4, and μ matched to the data. Fig. 4.3 shows this data from Carne and Weller, plus the fit with n = 4. For the calculations of Section V, we take n = 4 and a mean of 24 partners/year.

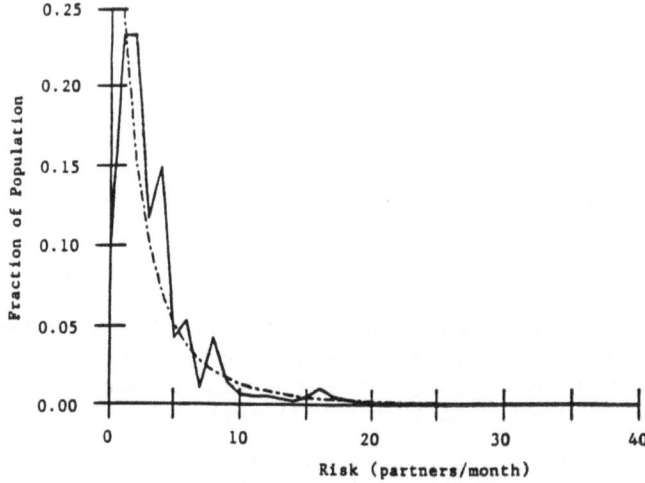

Fig.4.3. The distribution of homosexual men attending STD clinics in London, according to the number of partners per month from Carne and Weller . The dotted line shows the inverse quartic with the same mean as the data. (Data reported in May and Anderson, 1987).

Information on the number of contacts between different types of partners (long term, casual, prostitutes) is scarce, even for these homosexual cohorts. This critical information is beginning to be collected (Joseph et al., 1987). Because transmissibility through different types of contacts may be different, the frequency of each type of contact needs to be quantified. Without such knowledge, the best that we can do is to make some reasonable assumptions and explore various possibilities.

The assumptions that we use are that people with large numbers of partners have one contact with each partner and that people have more contacts with each partner when both partners have fewer partners. For the calculations in this paper we use the contact function $c(r,r') = 1+(c_1-1)\exp[-c_2(r+r')]$ with $c_1 = 11$ and $c_2 = 0.1$.

E. INITIAL CONDITIONS

In order to solve system 3.1, we need to specify initial conditions for $S(0,r)$, $I(0,\tau,r)$ and $A(0,\tau,r)$. For these conditions to be consistent with the epidemic we must define infections and AIDS cases as a function of τ according to what they were at some given time.

For the calculations with no τ dependence, we take the initial infected population to be a Gaussian distribution of 1000 people, with height 100, centered at a risk behavior of 175 partners per year. For the (τ,r) model, the initial infected population should be consistent with the past history of the epidemic, as well as being consistent with the risk-model calculations. For this to hold, we define I(0,τ,r) to be such that

$$\int_0^\infty I(0,\tau,r)dr \approx (1 - C(\tau))\, \Gamma'(t-\tau) \; .$$

where Γ'(t) is defined by Eq. (2.3): $\Gamma'(t) = a(t-t_0)^2$. The integral of I(0,τ,r) over τ is defined to have the same distribution about r = 175 partners per year as in the risk model.

V. SAMPLE CALCULATIONS

In this section we examine some of the qualitative features of the epidemic by comparing the predicted spread of HIV and AIDS cases as we vary the parameters. We focus on early growth because it is important to understand how the epidemic moves into new populations and which interactions are important in its transient dynamics. For the risk-based models, we examine the number of infecteds versus risk and show that different mixing assumptions result in substantial differences in predictions for the growth of the epidemic.

The solutions were integrated in time with an explicit Adams-Bashford-Moulton method to an accuracy of 10^{-6} per unit time. The τ-derivatives were calculated with fourth-order finite differences and the solution was approximated on a uniform grid of between 61 and 201 mesh points in both τ and r. The grid spacing and error tolerance were varied to check convergence of the solutions. We emphasize that these models are too simplistic to give accurate predictions of the AIDS epidemic and that the following calculations are meant only to illustrate the behavior of the models.

A. RISK-BASED CALCULATIONS

For our first set of calculations, we took i(τ), γ(τ), and δ(τ) to be their average values ($i(\tau) = 0.025$, $\gamma(\tau) = 0.133$ years^{-1}, and $\delta(\tau) = 0.5$ years^{-1}). This allows us to reduce Eqs. 3.1 and obtain a model where the non-AIDS infecteds and the AIDS cases are distributed only according to partner change rates and not according to τ. We use this collapsed model to examine the sensitivity of the model to different choices of the acceptance functions, f(r,r').

The initial susceptible population is distributed in risk as an inverse quartic $S(0,r) = S_0\, 3(2m)^3(2m+r)^{-4}$, with total population $\int S(0,r)dr = 10$ million, mean $m = \int rS(0,r)dr$ (10 million)$^{-1}$ = 24 partners/year. There is migration into all risk categories with migration rate equal to the natural death rate, $\mu = 0.02$ times $S_0(r) = S(0,r)$. Initially, there is a Gaussian distribution of 0.001 million infected individuals, centered at risk $r = 175$, with height 0.0001 million - years/partner, and no AIDS cases.

First we compare acceptance functions where the width of the mixing regions are similar and increases linearly with r. These functions differ in the amount of mixing between dissimilar groups. For this purpose, we use the Gaussian shown in Eq. (3.7) and

$$f(r,r') = \left[1 + \frac{(r - r')^n}{\varepsilon(r + r_m)^n} \right]^{-1} \qquad (5.1)$$

with $n = 2$ and $n = 4$. Here $r_m = 10$ partners and ε is chosen so the width of the acceptance function is approximately the same for each of the three functions [$f(r,r \pm 24) \approx 0.1$ at $r = 75$ partners/year)]. These three functions are shown in Fig. 5.1.

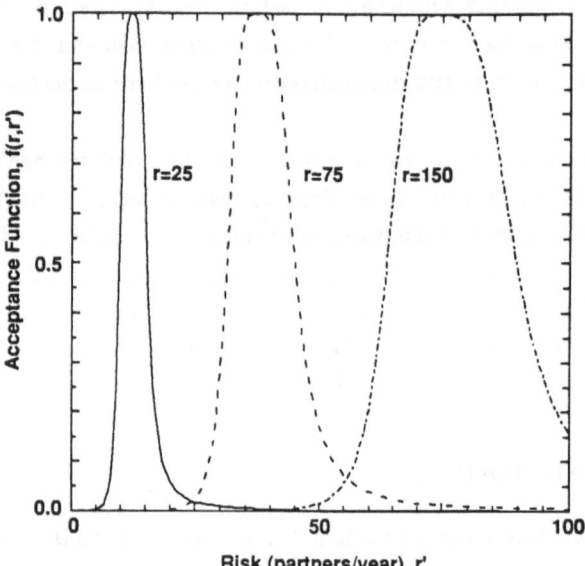

Fig. 5.1a. The inverse quartic function $f(r,r')$ of Eq. (5.1) with $n = 4$, $r_m = 10$ partners/year and $\varepsilon = 0.00065$, is shown for $r = 25, 75$ and 150 partners/year.

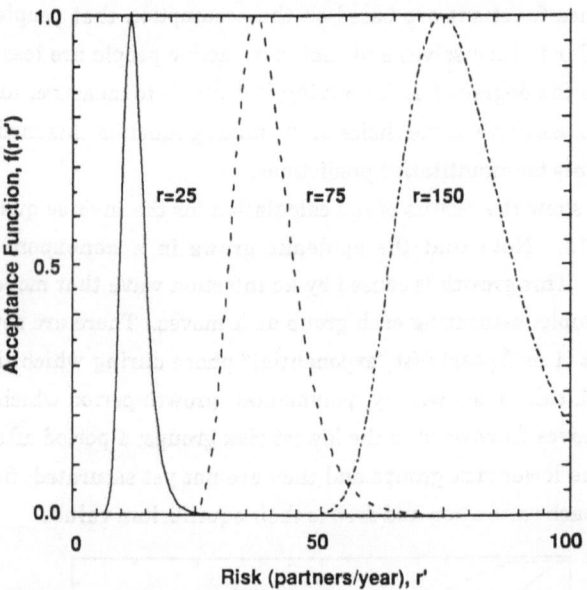

Fig. 5.1b. The Gaussian function f(r,r') of Eq. (3.7) with ε = 0.017 is shown, for r = 25, 75 and 150 partners/year.

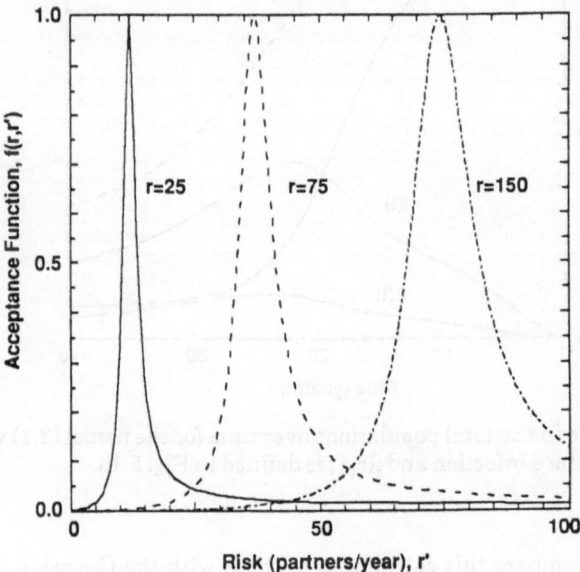

Fig. 5.1c. The inverse quadratic function f(r,r') of Eq. (5.1) with n = 2, ε = 0.0085 is shown for r = 25, 75 and 150 partners/year.

These acceptance functions are based on the assumption that people preferentially mix with those similar to themselves and that more active people are less picky than the less active. Because the degree of social mixing is difficult to measure, unless the model solutions are fairly insensitive to the choice of the mixing function this modeling approach can not be used reliably for quantitative predictions.

In Fig. 5.2, we show the results of the calculation for the inverse quartic acceptance function of Fig. 5.1a. Note that the epidemic grows in a nonexponential, roughly polynomial, fashion. This growth is caused by an infection wave that moves from highest risk to lower risk people, saturating each group as it moves. There are several phases to this growth: a short (1 or 2 year) fast "exponential" phase during which the highest risk groups are saturated; this is followed by "polynomial" growth period which lasts about 10 years as the wave moves downward to the lowest risk groups; a period after the epidemic wave has reached the lower risk groups and they are not yet saturated; finally, even the lowest risk groups reach saturation and drop to their equilibrium values.

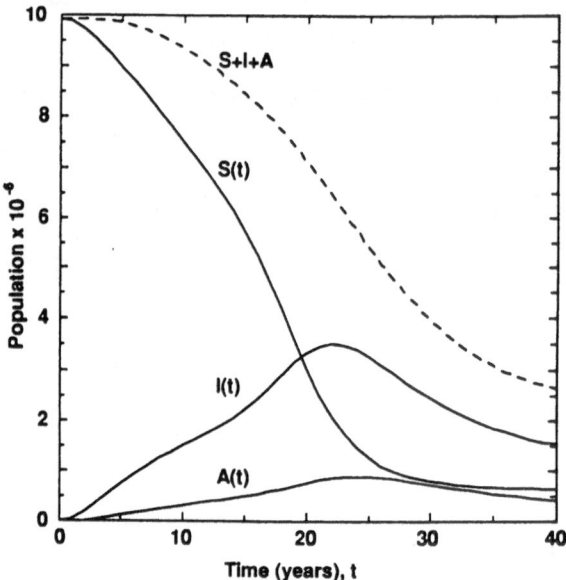

Fig. 5.2a. The change in the total populations over time for the model (3.1) when there is no dependence on time since infection and f(r,r') is defined in Fig. 5.1a.

In Fig. 5.3 we compare this calculation to those with the Gaussian and the inverse quadratic functions of Figs. 5.1b and 5.1c. We see that there is very little difference between the behavior of the exponential function and the inverse quartic. In both cases, the number infected have two inflection points before reaching a maximum and agree quantitatively. However, the quadratic function gives a faster epidemic that is more uniform in behavior and reaches low-risk groups earlier. This occurs because, although the

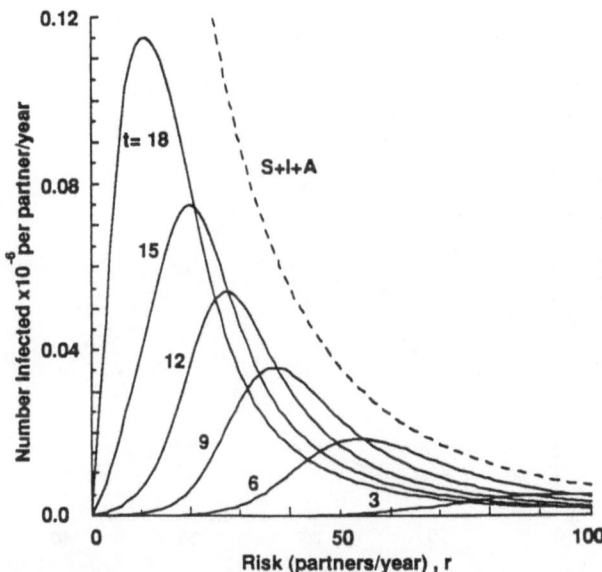

Fig. 5.2b. An infection wave moves from high risk to low risk. Shown is the infection profile at 3, 6, ..., 18 years for the calculation of Fig 5.2a.

local mixing has a similar width, the long tails of the quadratic allow some mixing between high and low risk people. Thus, we see that the epidemic is fairly sensitive to even a small amount of nonself selective mixing.

To determine the sensitivity of the epidemic to the width of the mixing region, we compared the results from the inverse quartic of Figs. 5.1 and 5.2 with those for the same function when it is twice ($\varepsilon = 0.01$) and four times ($\varepsilon = 0.17$) as wide. In Fig. 5.4a we see that as the acceptance function gets wider, the initial epidemic grows faster. That is, the less discriminating people are in selecting partners similar to themselves, the faster the epidemic grows and spreads into the lower risk populations. The wave of infection for the wider acceptance function ($\varepsilon = 0.01$) is not as sharp as in Fig. 5.2. When $\varepsilon = 0.17$, Fig. 5.4b, the wave almost disappears and the infection quickly begins growing in the lower risk groups, as in the random mixing model ($f(r,r') = 1$) (Hyman and Stanley, 1988).

This epidemic, with $\varepsilon = 0.17$, is only slightly faster than for the quadratic of Fig. 5.1c, showing that the width of local mixing is not as important to know as the amount of mixing between very dissimilar groups. Because most of the susceptibles have low risk behavior (small r), the less selective partner choice is assumed to be, the more the partners of high-risk behavior people have low risk and the more the high-risk group acts as a pool of infection for the lower-risk group, causing the lower-risk populations to become infected more quickly. The early AIDS cases had, on average, a large number of partners, indicating that mixing was probably fairly self-selective.

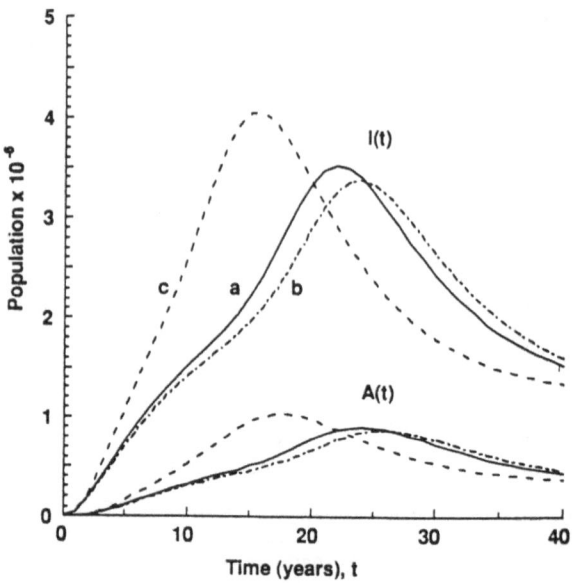

Fig. 5.3a. The total number infected and total AIDS cases for the three functions in Fig. 5.1. The letters a, b, c correspond to the acceptance functions in those of the Figures 5.1a, b, c. The small amount of mixing between low and high risk people allowed by the inverse quadratic (c) gives a faster epidemic than the inverse quartic (a) or the exponential (b), which are very similar.

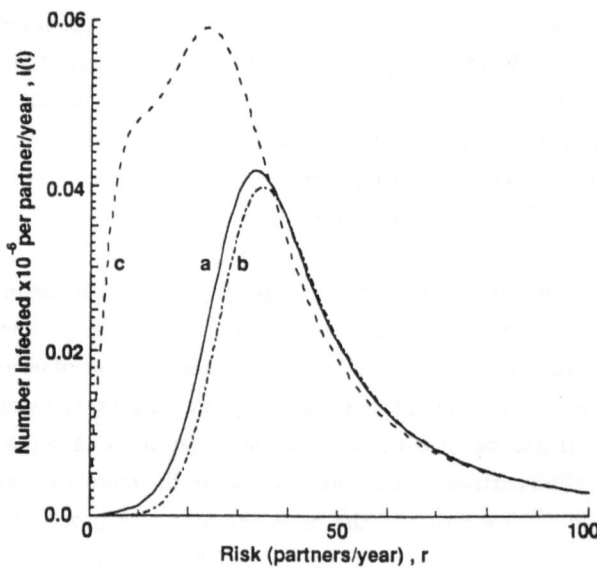

Fig. 5.3b. The distribution of infection according to risk at 10 years for the three functions of Fig. 5.1. The infection has already reached low risk groups for the inverse quartic.

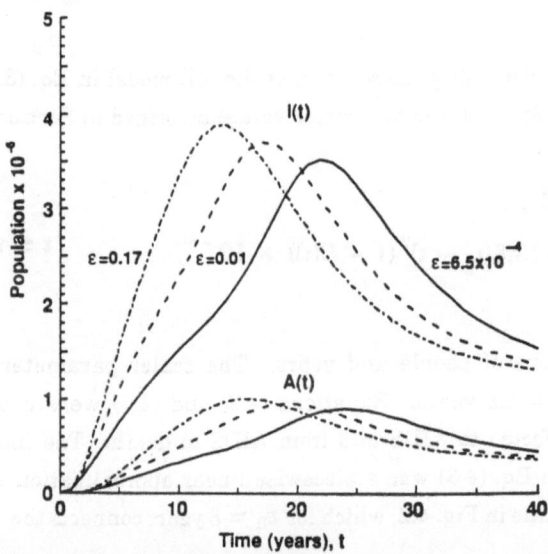

Fig. 5.4a. As the width of the acceptance function is increased, the epidemic spreads faster and saturates the population earlier. Infections and AIDS cases are shown for the calculation of Fig. 5.2 with the inverse quartic of Fig. 5.1a, an inverse quartic twice ($\varepsilon = 0.01$) and four times ($\varepsilon = 0.17$) as wide as in Fig. 5.1a.

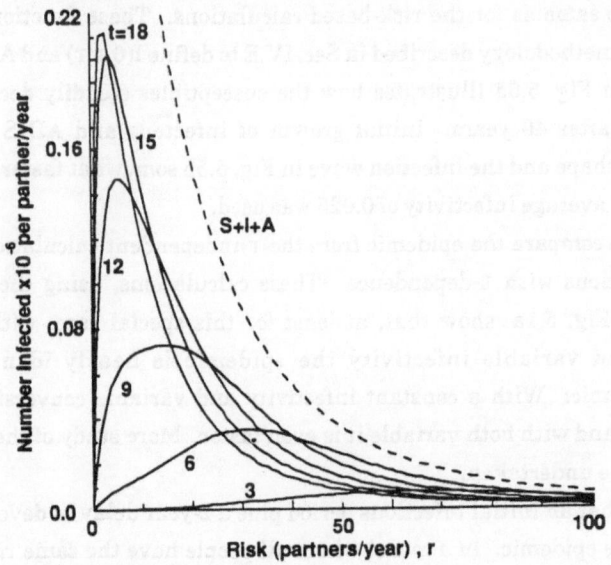

Fig. 5.4b. When the width of the acceptance function is quadrupled, the infection wave moves quickly into the lower-risk population. The distribution of infecteds versus risk is shown for $\varepsilon = 0.17$ of Fig. 5.4a at 3, 6, ..., 18 years. As the acceptance function becomes wider, the behavior approaches that of random mixing, with the wave-front behavior disappearing and there are more infected individuals with low-risk behavior than with high-risk behavior early in the epidemic.

B. (τ,r) MODEL

We finish by calculating the solution of the full model in Eq. (3.1), using the same f(r,r') in λ(t,r) as for Fig. 5.2, the parameter values described in Section IV and the initial conditions

$$\int S(0,r)dr = 10 ,$$

$$\int I(0,\tau,r)dr = 523.8(\tau_0 - \tau)^2(1 - C(\tau)) \times 10^{-6}, \qquad \tau \leq \tau_0 , and \qquad (5.2)$$

$$A(0,\tau,r) = 0 .$$

The units are millions of people and years. The scaler parameters used were μ = 0.02 year^{-1} and τ_0 = 1.8 years. Equations (4.2) and (4.5) were used for the rates of progression from infected to AIDS and from AIDS to death. The individual infectivity $i_i(\tau,\tau_a) = i^*(\tau/\tau_a)$ in Eq. (4.3) was a piecewise linear approximation $L[(\tau_1,i_1), (\tau_2,i_2)...]$ shown as the dotted line in Fig. 4.2, which for τ_a = 8 years connects the (τ,i) data points

$$i_i(\tau,\tau_a) = L[(0,0), (0.1,0), (0.4,0.1), (0.7,0.005), (5.0,0.005), (8.0,0.1)] . \qquad (5.3)$$

This distribution and the resulting i(τ) are shown in Fig. 4.2. Note that each individual has an average per contact infectivity of 0.024. The initial conditions for $\int I(0,\tau,r)d\tau$ and $\int A(0,\tau,r)d\tau$ were the same as for the risk-based calculations. These functions were then combined using the methodology described in Sec. IV.E to define I(0,τ,r) and A(0,τ,r).

The solution in Fig. 5.5a illustrates how the susceptibles steadily decline to near-equilibrium values after 40 years. Initial growth of infecteds and AIDS cases has a somewhat different shape and the infection wave in Fig. 5.5b somewhat faster than the one in Fig. 5.2 where the average infectivity of 0.025 was used.

In Fig. 5.5d we compare the epidemic from the τ-independent calculation of Fig. 5.2 and several calculations with τ-dependence. These calculations, using the highly self-selective mixing of Fig. 5.1a, show that, at least for this special case, with a constant conversion rate and variable infectivity the epidemic is nearly identical to the τ-independent epidemic. With a constant infectivity and variable conversion rate it is significantly faster, and with both variable it is even faster. More study of the full model's sensitivites need to be undertaken.

This confirms that an initial infectious period plus a 2-year delay in developing AIDS can greatly speed the epidemic. In a model where all people have the same risk behavior, we can dramatically change the rate at which the susceptible population is infected by varying the infectivity profile, even when the average infectiousness of an individual, $\int_0^1 i^*(x)dx$, is the same (Hyman and Stanley, 1988). The shape of the initial peak in

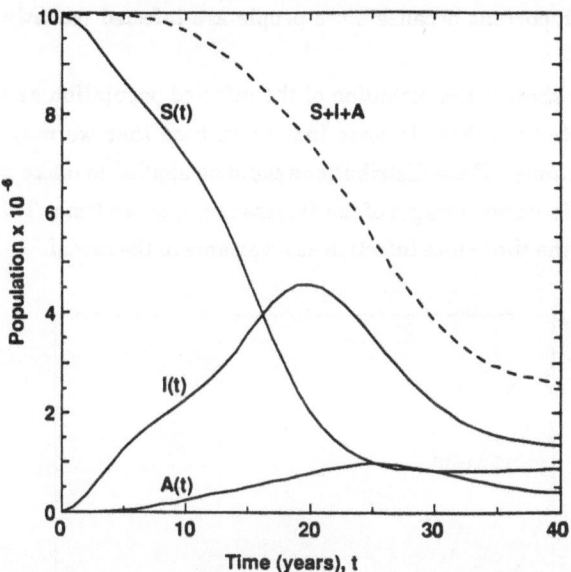

Fig. 5.5a. The solution of the model in Eqs. (3.1), (3.4) with the initial conditions $\tau_o = 1.8$ years and infectivity as in Eq. (4.3) and $f(r,r')$ as in Fig. 5.1a. Here $S(t) = \int S(t,r)dr$, $I(t) = \int\int I(t,\tau,r)d\tau dr$ and $A(t) = \int\int A(t,\tau,r)d\tau dr$.

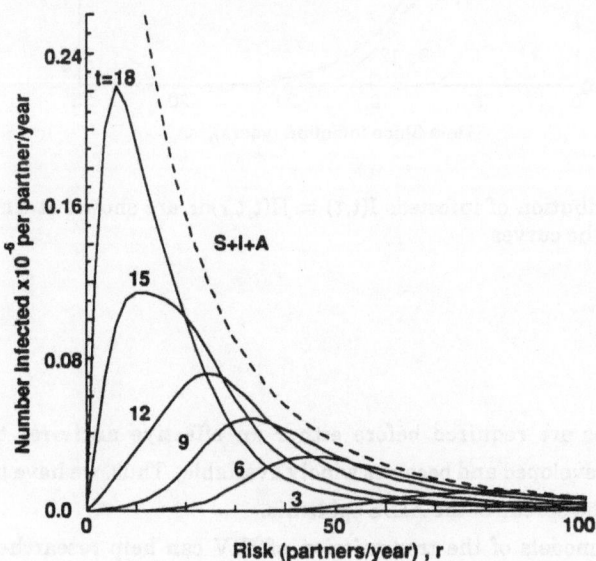

Fig. 5.5b. The infected population forms a wave that sweeps from high-risk behavior groups into lower-risk groups. The distribution of infecteds is shown every 3 years, at the times marked on the curves.

infectivity is most important because more people are infected recently (low τ) than 5-7 years ago (high τ).

In Fig. 5.5c we show the distribution of the infected population as a function of time since infection. Note that there is some indication here that we may not have chosen optimal initial conditions. These distributions could be applied to make predictions of how many people will be in various stages of the disease at any given time. This is an additional benefit in including the time since infection as a variable in the model.

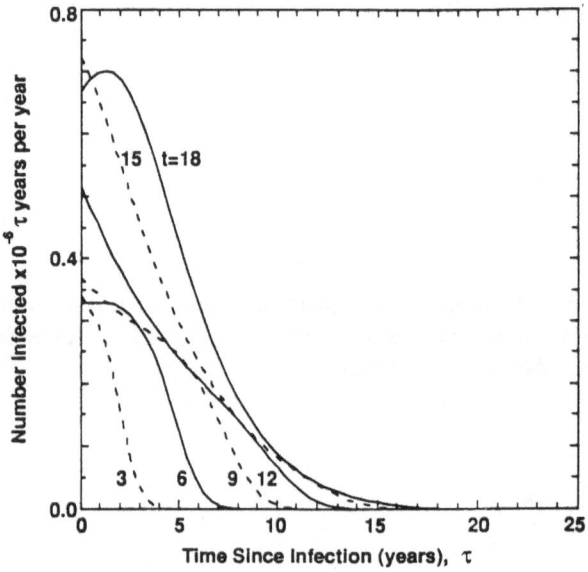

Fig. 5.5c. The distribution of infecteds $I(t,\tau) = \int I(t,\tau,r)dr$ are shown at times 3, 6, ..., 18 years, indicated on the curves.

VI. SUMMARY

Major advances are required before either an effective antiviral therapy or an effective vaccine is developed and becomes widely available. Thus, we have to prepare for a long battle against the spread of the AIDS epidemic.

Mathematical models of the transmission of HIV can help researchers develop an understanding of the complex interactions that lead to the epidemic's spread. The complexity results from the long asymptomatic period after infection with the human immunodeficiency virus (HIV) that causes AIDS, the social behavior of human populations, and changes in the environment of viral transmission. These models can show how the early infection of high-risk groups, behavioral changes, and future medical advances such

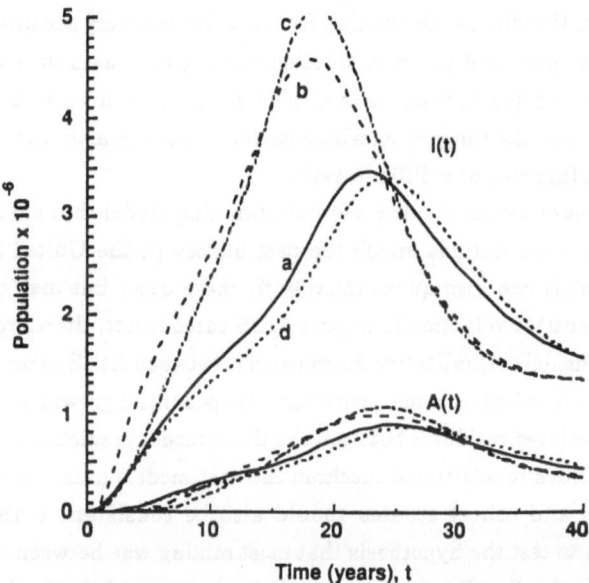

Fig. 5.5d. The infected population and AIDS cases from (a) Fig. 5.2b with no τ-dependence and i = 0.0244 (b) Fig. 5.5a with τ-dependence; (c) τ-dependence with i(τ) = 0.0244; and (d) τ-dependence with γ(τ) = 0.1333 and i(τ) as in (b). Having an initial viremic peak and no one developing AIDS for 2 years after infection greatly speeds the epidemic.

as treatments and vaccines will affect the future course of this epidemic. The effects will be highly nonlinear functions of the parameter values and at times may even lead to changes that are counter to both intuition and simple extrapolated predictions. The mathematical model predictions of these counterintuitive mechanisms may greatly improve our understanding of the observations.

In our computer models, the amount of sexual contact and needle-sharing between high-activity and lower-activity individuals determines both who gets infected and the speed with which the epidemic progresses. If there is little mixing between these groups, then the individuals in high-risk groups are nearly all infected before the infection moves into lower-risk groups. However, if mixing is large, many more lower-risk individuals will be infected in the early stages of the epidemic. The epidemic moves much faster when mixing is large because there are many more low-risk individuals than high-risk ones. In a model where partners are chosen with little regard for their partner-change rate, the total number of infected low-risk individuals quickly exceeds the number of infected high-risk individuals. This result is contrary to experience (Darrow et al., 1987; Goedert et al., 1984; Auerbach et al., 1984) and reflects the urgent need to collect and analyze the information on mixing patterns to estimate critical model parameters.

This sensitivity raises an obvious question: is it possible to measure mixing sufficiently accurately to predict the spread of the epidemic? We have seen that the model is not very sensitive to the shape of the mixing function, but it is very sensitive to its width. Thus, although we do not need to know whether mixing decreases in a Gaussian or a polynomial fashion as people become more dissimilar, we may need to estimate within better than a factor of two the range from which partners are primarily chosen. Even with the best possible data this may be a difficult task.

We can choose parameters in our preferential-mixing model that ensure that AIDS cases in the numerical simulations match the past history in the United States. Many other reasonable models can also quantitatively fit these cases but may predict a very different future. Quantitatively matching past AIDS cases is not, therefore, sufficient to distinguish between models. Qualitative discrepancies between AIDS cases and the model need to be explained; for example, models with initial exponential growth do not fit the U.S. AIDS case data. Correlated residuals between the fitted model predictions and AIDS data may give important clues to additional mechanisms that models must incorporate. Data from seroprevalence and cohort studies should also be consistent with the model's predictions. We plan to test the hypothesis that most mixing was between men of similar risk behavior by analyzing San Francisco data on behavior versus infection before behavior changes became widespread.

Although it is unlikely that any model will provide accurate long-term predictions of the numbers of AIDS cases, transmission models could eventually allow investigators to answer many questions. For example, one can assume increased condom use by people in a targeted age group and region and then determine how much that increased use will slow the local course of the epidemic. This predictive ability would then help authorities decide if it is more effective to encourage condom use in that group than to use another strategy, such as stressing the importance of having fewer partners or reducing the incidence of other sexually transmitted diseases, to lower the probability of infection for some population groups. As another example, a partially effective vaccine with potentially harmful side effects might be developed. Somehow it must be ascertained which persons should be vaccinated. The model would be used to understand how vaccinating each group affects the spread of the epidemic.

ACKNOWLEDGMENTS

The authors express their appreciation to C. Castillo-Chavez, S. A. Colgate, K.W. Dietz, J. S. Koopman, S. P. Layne, L. Sattenspeil, J. Wiley and C. R. Qualls for their comments and advice. This research was supported by the Department of Energy under contracts W-7405-ENG-36, KC-07-01-01, and HA-02-02-02.

REFERENCES

Anderson, R. M., May, R. M., Medley, G. F., and Johnson, A. 1986. A Preliminary Study of the Transmission Dynamics of the Human Immunodeficiency Virus (HIV), the Causative Agent of AIDS. *IMA J. Math. Appl. Med. Biol.* 3, 229-263.

Auerbach, D. M., Darrow, W. W., Jaffe, H. W., and Curran, J. W. 1984. Cluster of Cases of the Acquired Immune Deficiency Syndrome. Patients Linked by Sexual Contact. *Am. J. Med.* 76, 487-492.

Colgate, S. A., Stanley, E. A., Hyman, J. M., Layne, S. P., and Qualls, C. 1988. *A Behavior Based Model of the Initial Growth of AIDS in the United States.* Los Alamos National Laboratory report.

Darrow, W. W., Echenberg, D. F., and Jaffe, H. W. 1987. Risk Factors for Human Immunodeficiency Virus Infections in Homosexual Men. *AJPH* 77, 479-483.

Dietz, K. 1987. The Dynamics of Spread of HIV Infection in the Heterosexual Population unpublished report.

Dietz, K. 1988. On the Transmission Dynamics of HIV. Proceedings of the 1987 CNLS Workshop on Nonlinearity in Medicine and Biology, *Math. Bio.* 90, 397-414.

Fischl, M. A., Dickinson, G. M., Scott, G. B., Klimas, N., Fletcher, M. A., and Parks, W. 1987. Evaluation of Heterosexual Partners, Children, and Household Contacts of Adults with AIDS. *JAMA* 257, 640-644.

Francis, D. P., Feorino, P. M., Broderson, J. R., McClure, H. M., Getchell, J. P., McGrath, C. R., Swenson, B., McDougal, J. S., Palmer, E. L., Harrison, A. K., Barre-Sinoussi, F., Chermann, J.-C., Montagnier, L., Curran, J. W., Cabradilla, C. D., and Kalyanaraman, V. S. 1984. Infection of Chimpanzees with Lymphadenopathy-associated Virus. *Lancet* 1276-1277.

Goedert, J. J., Biggar, R. J., and Winn, D. M. 1984. Determinants of Retrovirus (HTLV-III) Antibody and Immunodeficiency Conditions in Homosexual Men. *Lancet* 711-716.

Grant, R., Wiley, J., and Winklestein, W. 1987. The Infectivity of the Human Immunodeficiency Virus: Estimates from a Prospective Study of a Cohort of Homosexual Men. *J. Inf. Dis.* 156, 189-193.

Hardy, A. M., Starcher, E. T., Morgan, W. M., Druker, J., Kristal, A., Day, J., Kelly, C., Ewing, E., and Curran, J. 1987. Review of Death Certificates to Assess Completeness of AIDS Case Reporting. *Pub. Hlth. Rep.* 102, 386-391.

Hethcote, H. W. and Yorke, F. A. 1984. *Gonorrhea: Transmission Dynamics and Control. Lecture Notes in Biomathematics* 56, 1-105.

Hyman, J. M., Qualls, C. R., and Stanley, E. A. (in preparation). *Analysis of CDC AIDS Data*. Los Alamos National Laboratory report.

Hyman, J. M. and Stanley, E. A. 1988. Using Mathematical Models to Understand the AIDS Epidemic. *Mathematical Biosciences* 90, 415-474.

Jaquez, J., Koopman, J., Simon, C., Sattenspiel, L., and Perry, T. 1988. Modeling and Analyzing HIV Transmission: The Effect of Contact Patterns. *Mathematical Biosciences* 92, 119-199.

Joseph, J. G., Montgomery, S., Kessler, R. C., Ostrow, D. G., Emmons, C.A., and Phair, J. P. Two-Year Longitudinal Study of Behavioral Risk Reduction in a Cohort of Homosexual Men. Presented June 1, 1987 at the III International Conference on AIDS in Washington, DC.

Kingsley, L. A., Kaslow, R., Rinaldo, C. R., Jr., Detre, K., Odaka, N., Van Raden, M., Detels, R., Polk, B. F., Chmiel, J., Kelsey, S. F., Ostrow, D., and Visscher, B. 1987. Risk Factors for Seroconversion to Human Immunodeficiency Virus Among Male Homosexuals. *Lancet*, 345-349.

Lange, J.M.A., Paul, D. A., Huisman, H. G., de Wolf, F., van der Berg, H., Coutinho, R., Danner, S. A., van der Noordaa, J., and Goudsmit, J. 1986. Persistent HIV Antigenemia and Decline of HIV Core Antibodies Associated with Transition to AIDS. *Br. Med. J.* 293, 1459-1462.

May, R. M. and Anderson, R. M. 1987. Transmission Dynamics of HIV Infection. *Nature* 326, 137-142.

Medley, G. F., Anderson, R. M., Cox, D. R., and Billard, L. 1987. Incubation Period of AIDS in Patients Infected via Blood Transfusion. *Nature* 238, 719-721.

Padian, N., Wiley, J., and Winkelstein, W. Male to Female Transmission of Human Immunodeficiency Virus: Current Results, Infectivity Rates, and San Francisco Population Seroprevalence Estimates. Presented June 4, 1987, at the III International Conference on AIDS in Washington, DC.

Peterman, T. A., Stoneburner, R. L., Allen, J. R. Jaffe, H. W., and Curran, J. 1988. Risk of Human Immunodeficiency Virus Transmission from Heterosexual Adults with Transfusion-Associated Infections. *JAMA*, 259, 55-58.

Sattenspiel, L. 1987. Population Structure and the Spread of Disease. *Human Biology* 59.

Sattenspiel, L. and Simon, C. 1988. The Spread and Persistence of Infectious Diseases in Structured Populations. Proceedings of the 1987 CNLS Workshop on Nonlinearity in Medicine and Biology, *Math. Bio.* 90, 341-366.

Stigum, H., Groennesby, J. K., Magnus, P., Sundet, J. M., and Bakketeig, L. S. 1988. The Effect of Selective Partner Choice on the Spread of HIV. Presented at the Conference on The Global Impact of AIDS, London.

Sulahuddin, S. Z., Markham, P. D., Redfield, R. R., Essex, M., Groopman, J. E., Sarngadharan, M. G., McLane, M. F., Sliski, A., and Gallo, R. C. 1984. HTLV-III in Symptom-Free Seronegative Persons. *Lancet*, 1418-1420.

Sundet, J., Kralem, I., Magnus, P., and Bakketeeig, L. 1988. Prevalence of Risk Prone Sexual Behavior in the General Population of Norway. Presented at the Conference on The Global Impact of AIDS, London.

Winklestein, W., Lyman, D. M., Padian, N. S., Grant, R., Samuel, M., Wiley, J. A., Anderson, R. E., Lang, W., Riggs, J., and Levy, J. A. 1987. Sexual Practices and Risk of Infection by the Human Immunodeficiency Virus. *JAMA* 257, 321-325.

POSSIBLE DEMOGRAPHIC CONSEQUENCES OF HIV/AIDS EPIDEMICS: II, ASSUMING HIV INFECTION DOES NOT NECESSARILY LEAD TO AIDS

Robert M. May,
Department of Biology
Princeton University
Princeton, N.J. 08544
USA

Roy M. Anderson and Angela R. McLean
Department of Pure and Applied Biology
Imperial College
London University
London, England SW7 2BB

1. INTRODUCTION

It seems likely that mortality associated with HIV/AIDS infections, transmitted horizontally by heterosexual contacts among adults and vertically to the offspring of infected mothers, will have significant demographic effects in Africa, and possibly in other developing countries. Such demographic effects may, indeed, also be significant in the long run among subgroups (such IV-drug abusers), and possibly more generally, in developed countries.

We therefore combine simple epidemiological models for the transmission dynamics of HIV/AIDS with conventional demographic models, to get a qualitative understanding of the way HIV/AIDS may affect overall rates of population growth, the age structure of the population, and other such things. The models are deliberately oversimplified: thus infected individuals have a constant infectiousness, and move at some constant rate either to develop AIDS disease or to become uninfectious (in reality, there is a more complex distribution of incubation times, infection may depend on time since first infected, and it is not known whether some fraction may remain asymptomatic carriers indefinitely); new sexual partners are assumed to be acquired at some average rate (in reality, there are significant heterogeneities in contact rates); and contacts are assumed to be homogeneously mixed among all sexually-active age groups (in reality, there seems likely to be significant age-structure in the "pairing probability"). Most of these realistic refinements have been discussed more fully, and included to one degree or another, in numerical studies of epidemiological models (Anderson et al., 1986; Anderson and May, 1986; May and Anderson, 1987; Medley

et al., 1987; Hyman and Stanley, 1988; Dietz, 1988; Castillo-Chavez et al., 1988).
Essentially all this work, however, retains the assumption -- pervasive in conventional mathematical epidemiology -- that the host population is constant.

The main focus of the present paper is on the interplay between epidemiological
and demographic effects. That is, we seek a qualitative understanding of how long-
term rates of overall population growth and asymptotic age-profiles may depend on
epidemiological and demographic parameters such as birth rates, death rates from AIDS
and from other causes, fraction of HIV infectees who die, fraction of offspring of
infected mothers who will be infected and die, transmission rates, and so on. We
believe the largely analytic study of these deliberately simplified models is a useful preliminary to the necessarily numerical exploration of much more complicated and
realistic models, giving insight into the nature of the dynamics and magnitude of
time-scales on which various demographic effects manifest themselves.

In a companion paper to this one (May et al., 1988, hereafter referred to as
MAMcL), we present such an analysis of the properties of combined epidemiological and
demographic models in the limiting case when all those infected with HIV eventually
go on to develop AIDS ($f = 1$ in the notation defined below). In this limit, the
basic dynamical system is a 2-dimensional one, which makes for many simplifications. The present paper treats the more general case when some finite fraction, f,
of HIV infectees go on to develop AIDS, while the remaining fraction, $1 - f$, do not;
the basic dynamical system is now 3-dimensional and the analysis is correspondingly
more complicated.

In Section 2 the basic model is defined, and its properties are elucidated (by
phase plane techniques and asymptotic analysis). Numerical results for the way population sizes and the fraction infected change over time are given for a variety of
representative parameter values, along with the criterion for long-term rates of
population growth to become negative. Section 3 moves beyond analysis of total population sizes, to explore the effects of HIV/AIDS upon the age-structure of the population. Using these results, Section 4 gives asymptotic results for the effect of
HIV/AIDS upon economic and social indicators such as "child dependency ratios",
defined as the fraction of the total population under the age of 15 years. The main
conclusions are summarized in Section 5.

2. BASIC MODEL COMBINING DEMOGRAPHY WITH EPIDEMIOLOGY

2.1   Definition of the Model.

Following closely along the lines laid down in MAMcL, we first consider the transmission of HIV within a single-sex population. Let the total population $N(t)$ at time t be subdivided into $X(t)$ susceptibles, $Y(t)$ infected (assumed to be also infectious), and $Z(t)$ who have been infected and infectious, but are no longer infectious and who are now leading normal lives (in reality, this class may be empty). It is assumed that a fraction $(1 - f)$ of infecteds move at a constant rate, v (that is, after an average incubation time $1/v$), into the "recovered" class $Z(t)$, while the remaining fraction, f, move at the same rate, v, to develop full-blown AIDS (at which point they are regarded for the purpose of this model as effectively dead, which they will in fact be in about a year or so). Deaths from all other causes occur at a constant rate µ.

This simple system is described by the set of first-order differential equations:

$$dX/dt = B - (\lambda + \mu)X, \qquad (2.1)$$

$$dY/dt = \lambda X - (\mu + v)Y, \qquad (2.2)$$

$$dZ/dt = (1 - f)vY - \mu Z. \qquad (2.3)$$

The dynamics of the total population, $N = X + Y + Z$, thus obeys

$$dN/dt = B - \mu N - fvY. \qquad (2.4)$$

Here B is the net rate which new recruits appear in this population, and λ is the usual "force of infection", representing the probability per unit time that a given susceptible will become infected. As discussed more fully elsewhere (Anderson et al., 1986), for a sexually transmitted disease, such as HIV, we may, to a simplest first approximation, write

$$\lambda = \beta c Y/N. \qquad (2.5)$$

Here β is the probability to acquire infection from any one infected partner, Y/N is the probability that a randomly-chosen partner will be infected, and c is the average rate at which partners are acquired.

The above equations are for HIV transmission in a single-sex population. But our main interest is <u>heterosexual</u> transmission of HIV, and its demographic consequences. In general, this means we must deal with two distinct populations, N_1 of males and N_2

of females, with females acquiring infection only from males (at a rate characterized by the parameter combination $\beta_1 c_1$) and vice versa for female-to-male transmission (characterized by $\beta_2 c_2$). If, however, we make the simplifying assumption that male-to-female and female-to-male transmission takes place at the same intrinsic rate, so that $\beta_1 c_1 = \beta_2 c_2 = \beta c$, then we have a symmetrical situation in which the two-sex set of equations effectively collapses back into the simple set of eqs(2.1)-(2.5); $N(t)$ is the total population size, and sex ratios are assumed to remain around 1:1 at all times in this symmetrical situation. The fact that the ratio of cases of AIDS among females to cases among males is around unity in Africa today suggests that -- as a broad average -- this symmetry assumption may be not too bad for HIV transmission in Africa at present (May and Anderson, 1987). This system of equations is now closed by the additional demographic assumption that the input of susceptibles, B, is given by the net birth rate:

$$B = \nu \left[N - (1 - \epsilon)Y \right]. \tag{2.6}$$

Here ν is the per capita birth rate (females per female, or equivalently offspring per capita for a 1:1 sex ratio) in the absence of infection. We assume that a fraction ϵ of all offspring born to infected mothers survive, while a fraction $(1 - \epsilon)$ die effectively at birth. Thus the net birth rate is diminished below νN by deaths at the rate $\nu(1 - \epsilon)Y$, resulting from vertical transmission.

As outlined in the Introduction, and discussed more fully by many authors (MAMcL; Anderson et al., 1986; May and Anderson, 1987; Hyman and Stanley, 1988; Dietz, 1988; Castillo-Chavez, 1988), the above assumptions are unrealistically simple in many ways. They serve, however, as a basis for qualitative insights and as a point of departure for more realistic models. A few points deserve emphasis. First, it can be seen that the "average rate of acquiring partners", c, is not simply the mean number but rather is closer to the ratio of mean-square number to the mean, averaged over the distribution of rates of partner acquisition: $c \approx \langle i^2 \rangle / \langle i \rangle \approx m + \sigma^2/m$ (where m is the mean, and σ^2 the variance, of the relevant distribution). Second, if the symmetry assumption $\beta_1 c_1 = \beta_2 c_2$ is removed, we must work with separate sets of equations for the male population (infected by females) and the female population (infected by males); the relevant analysis is sketched in MAMcL, where it is shown

that the main conclusions of the symmetric model remain largely intact provided that
"βc" is interpreted as the geometric mean of male-to-female and female-to-male
transmission processes (βc = $[\beta_1 c_1 \beta_2 c_2]^{\frac{1}{2}}$) and that the rate at which HIV infection is
increasing is significantly faster than other epidemiological and demographic rate
processes (βc significantly greater than μ, ν, v). Third, eq(2.5) for λ implicitly
assumes contacts occur homogeneously over all age groups. MAMcL have explored the
opposite extreme where heterosexual contacts are essentially confined to within
age-classes; they find results that are qualitatively similar to those based on homogeneous mixing of all age-classes (eq(2.5)). For further discussion of these and
other points, see MAMcL.

Eqs(2.1)-(2.6) give a complete description of the dynamical behavior of this
model system, under the interplay of intrinsic demographic factors (births and
disease-free deaths, characterized by the parameters ν and μ) and epidemiological
factors (horizontal and vertical transmission and resulting deaths, characterized by
the parameters βc, v, f, and ε). Eqs(2.1)-(2.6) may, after some trivial manipulation, be brought to three equations for the variables N(t), Y(t) and Z(t):

$$dY/dt = Y[\beta c \{1 - (Y/N) - (Z/N)\} - (\mu + v)], \tag{2.7}$$

$$dN/dt = N[(\nu - \mu) - \{fv + (1 - \epsilon)\nu\}(Y/N)]. \tag{2.8}$$

The third equation, for dZ/dt, is simply eq(2.3).

Rather than using the number infected and "recovered", Y(t) and Z(t), respectively, we may use the corresponding fractions:

$$X = Y/N, \tag{2.9}$$

$$\varsigma = Z/N. \tag{2.10}$$

Noticing that $dX/dt = X[(dY/dt)/Y - (dN/dt)/N]$, with a similar expression for
$d\varsigma/dt$, we arrive at a set of three first-order differential equations, for the
three variables N(t), X (t), ς (t):

$$dN/dt = N[r - \theta X], \tag{2.11}$$

$$dX/dt = X[(\wedge - r) - (\beta c - \theta)X - \beta c \varsigma], \tag{2.12}$$

$$d\varsigma/dt = (1 - f)vX - \varsigma[\nu - \theta X]. \tag{2.13}$$

In writing the equations in this form, we have defined some biologically-significant
combinations of parameters. Specifically, we define r to be the AIDS-free growth
rate of the population, $\wedge$ to be the initial exponential growth rate of the infec-

tion (from very low values) within the population, and θ to represent the additional mortality rates associated with infection (both from the direct effects of horizontally transmitted infection, fv, and from the effects of vertical transmission which can depress effective birth rates, $\nu(1 - \epsilon)$):

$$r = \nu - \mu, \tag{2.14}$$

$$\wedge = \beta c - (\mu + v), \tag{2.15}$$

$$\theta = fv + \nu(1 - \epsilon). \tag{2.16}$$

We now outline the qualitative and asymptotic properties of this dynamical system, and give numerical results for $N(t)$ and $\chi(t)$, as functions of time, for representative values of the parameters. Notice, incidentally, that if $f = 1$ (all HIV infectees die of AIDS) then $\zeta = 0$ and we have the simpler, 2-dimensional system of eqs(2.11) and (2.12); MAMcL is confined to this limiting case.

2.2   Phase Plane Analysis

As emphasized by Getz and Pickering (1983), for most sexually transmitted diseases (STDs) the overall transmission rate tends to depend only on the ratio of infected individuals to the total population size -- as in the above model -- and therefore not upon overall population density. It follows that, in these simplest models for the transmission dynamics of STDs, threshold criteria for the epidemic spread or endemic maintenance of STDs tend to depend on transmission parameters (such as per capita contact rates) and not on population density; that is, the models do not lead to the possibility of population regulation by the infection. In such models, STDs can obviously lead to altered population growth rates, and even can depress these growth rates to negative values, but they cannot, by themselves, hold populations around some steady, equilibrium value.

These dynamical properties can easily be seen for the system of eqs(2.11)-(2.13). Consider first the pair of eqs(2.12) and (2.13), for the fractions infected and recovered, χ and ζ, respectively. The dynamical behavior of these two differential equations can be made plain by a phase plane analysis, as shown in Fig. 1. The isocline along which the fraction infected, χ, is unchanging is (from eq(2.12) given by the straight line

$$(\beta c - \theta)\chi + \beta c\zeta = (\wedge - r). \tag{2.17}$$

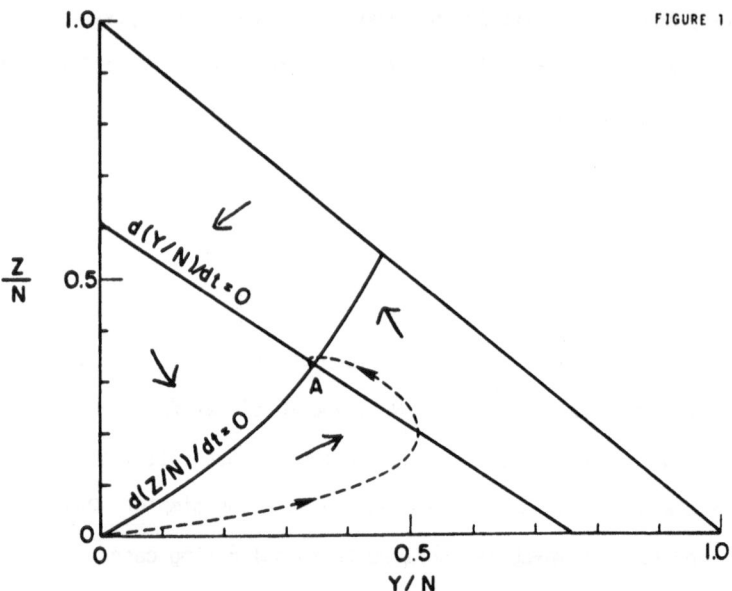

FIGURE 1

Fig. 1: This figure shows the phase plane of Y/N (or X) versus Z/N (or ζ), for dynamical trajectories defined by eqs(2.12) and (2.13); note that (Y/N) + (Z/N) ≤ 1. The isoclines along which d(Y/N)/dt = 0 and d(Z/N)/dt = 0 are shown. These isoclines divide the phase plane into 4 sectors, in each of which trajectories move in the direction shown by the arrows. A stability analysis shows that trajectories spiral in to the fixed point A; the dashed line shows one typical such trajectory. (Specifically, this figure is drawn for $\wedge$ = 0.2, v = 0.06, μ = 0.02, r = 0.03 (all in yr^{-1}), ε = 0.5 and f = 0.5).

X is increasing for values of (X, ζ) below this line, and decreasing above it.

Similarily, from eq(2.13), the isocline for ζ is the upwardly-curving line

$$\zeta = (1 - f)vX/(\nu - \theta X). \qquad (2.18)$$

ζ increases for values of (X, ζ) below this line, and decreases above it. It

follows that trajectories move in the directions shown by the arrows in the four

separate regions defined by these two isoclines, resulting in trajectories that

spiral around the point, A, defined by the intersection of the two isoclines.

A routine analysis of the linearized stability properties of this system (see,

e.g., May, 1974) shows that A is a locally stable equilibrium point provided

$\nu - \theta X^* > 0$ and $\beta c - \theta > 0$ (here X^* is the value of X at the equilibrium point

A). From eq(2.18), the first of these two conditions must hold if ζ^* is to be posi-

tive, and the second condition can be seen to follow from the earlier requirement
that $\Lambda > r$. Moreover, eqs(2.12) and (2.13) are sufficiently simple that A is surely
globally attracting if is locally attracting (although we have not proved this). In
short, the phase plane analysis suggests that χ and ζ exhibit damped oscillations
as they tend asymptotically to the steady values χ^* and ζ^* given by eq(2.17) and
(2.18), provided $\Lambda > r$.

With the dynamical behavior of $\chi(t)$ and $\zeta(t)$ resolved, we return to eq(2.11)
for N(t), and observe that asymptotically the total population will grow at the exponential rate ρ, given by

$$\rho = r - \theta \chi^*. \tag{2.19}$$

As before, χ^* is found by solving the pair of eqs(2.17) and (2.18) for χ^* and ζ^*.

2.3 Asymptotic Behavior

We have just seen that, as $t \to \infty$, the total population will settle to grow or
decline exponentially as $N(t) \to N(0)\exp(\rho t)$, with ρ defined by eq(2.19). We also saw
that ρ differs from the disease-free growth rate r ($\chi^* \ne 0$) only if $\Lambda > r$. That
is, the exponential rate at which the infection initially grows must exceed the
overall population growth rate, otherwise a decreasing fraction will experience
infection even though the absolute number infected is growing.

By solving eqs(2.17) and (2.18) for χ^*, we can get an explicit expression for
the growth rate ρ:

$$\rho = - (\mu + v) + \nu \left(\frac{\beta c}{\beta c - \theta} \right) \left(\varepsilon + \frac{[1-f]\,v}{\rho + \mu} \right). \tag{2.20}$$

Eq(2.20) is a quadratic equation for ρ.

We have chosen to write eq(2.20) in this slightly eccentric way in order to facilitate comparison with the earlier results for f = 1 in MAMcL, and to make clear some
of the limiting properties. First, observe that if f = 1 the second term in the
brackets on the right of eq(2.20)disappears, and the expression for ρ reduces to the
simple linear relation obtained in MAMcL; in the limit as ε → 0 (all vertically
infected offspring die), we have the overall population decreasing at the rate
$-(\mu + v)$. Second, observe that in the more general case when f ≠ 1, ρ is never less
than -μ. All this has an intutive explanation in biological terms: if all infected
individuals die (f = 1) and few offspring survive, the population will tend to

decline at the average per capita death rate, which in this case is $(\mu + v)$; but if some fraction, $1 - f$, of infecteds never go on to develop AIDS, then it is the death rate of this fraction, μ, that will asymptotically dominate. More specifically, for $f \neq 1$ there will be initial phases when HIV/AIDS first spreads and subsequently causes substantial mortality among a fraction, f, of all infecteds, but eventually this muchdiminished population will settle into a final phase of asymptotic decrease in which the death rate is essentially set by the "recovered" fraction of the adult population. The numerical examples in the next sub-section will make this plainer. A more full and formal discussion is given in Appendix A.

Up to this point, we have treated v as simply the overall average birth rate per capita. But, in fact, HIV spreads predominately among sexually active adults, and births occur only from sexually mature females, so that it is more realistic to interpret $N(t)$ as the population of sexually active adults (subdivided into $X(t)$, $Y(t)$, $Z(t)$, as before), rather than the total population. If we retain the previous assumption about symmetry between transmission from males to females and females to males, the basic eqs(2.1)-(2.5) remain true for the re-interpreted (adult population) variables $N(t)$, $X(t)$, $Y(t)$ and $Z(t)$. The expression for input of new susceptibles by births, $B(t)$ from eq(2.6), must however be modified in three ways. First, the recruitment of adults at time t depends on births that occurred at the earlier time $t - \tau$, when τ is the time taken to attain sexual maturity (something like 15 years in the present context). Second, the average birth rate is now expressed per adult, $\hat{v}$ (contrasting with the overall average rate, v, per member of the population). Third, this birth rate must obviously be discounted by a factor s, representing the probability of surviving the first τ years of life (if the death rate is μ, independent of age, then $s = \exp(-\mu\tau)$; but we can treat s as a more general parameter). Eq(2.6) is thus replaced by

$$B(t) = \hat{v}\, s[N(t - \tau) - (1 - \epsilon)Y(t - \tau)]. \tag{2.21}$$

Although these time-delays complicate numerical solution or phase-plane analysis of these new versions of eqs(2.11)-(2.13), the asymptotic results are easily generalized to include the complications. As discussed in more detail in MAMcL, the population still behaves asymptotically as $\exp(\rho t)$, with ρ given by eq(2.20). The only difference is that now, in eq(2.20) and elsewhere, v is to be interpreted as

$$v = \hat{v}\, s e^{-\rho\tau}. \tag{2.22}$$

This replacement must, of course, be made in eqs(2.14) and (2.16) which define r
and e, respectively, in terms of ν, and elsewhere. We will usually know the exponential growth rate of the disease-free population, r, rather than the intrinsic
birth rate per adult, $\hat{\nu}$, or the childhood survival probability, s: in the disease-
free population, these quantities are related by $\nu_0 = r + \mu = \hat{\nu} s e^{-r\tau}$. Thus, in
terms of the parameters we can most usually estimate directly, it may be best to say
that " ν " in eq(2.20), and elsewhere, should be interpreted as

$$\nu = (r + \mu)e^{(r-\rho)\tau}. \tag{2.23}$$

Here r is, as always, the growth rate of the population just before the advent of
HIV/AIDS. These intuitive considerations will be put on a more rigorous basis when
we turn to a fully age-structured analysis in Section 3.

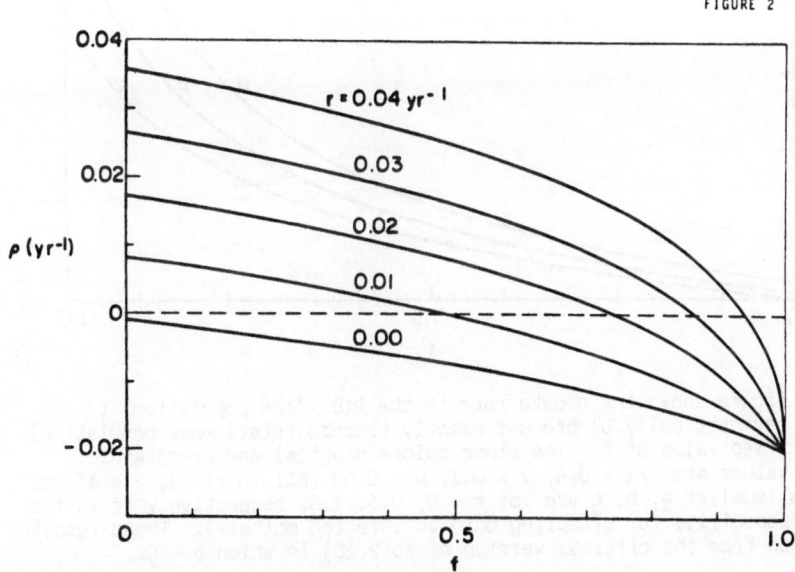

FIGURE 2

Fig. 2: This figure shows the asymptotic rate of exponential population
growth, ρ, as a function of f (the fraction of HIV infectees who die from
AIDS), for several values of the growth rate per annum in the AIDS-free
population, r, as shown; see eq(2.20). It is evident that the population
can exhibit an asymptotic pattern of decline, even for relatively large
r-values, provided f is big enough. In this figure, the other demographic
and epidemiological parameters have the values $\Lambda = 0.4$, $v = 0.1$, $\mu = 0.02$
(all in yr^{-1}), $\epsilon = 0.5$, and $\tau = 15$ yr.

Fig. 2 shows the asymptotic rate, ρ, at which the population is changing exponentially, as a function of f (the fraction of HIV infectees who go on to die from AIDS),
for several different values of the initial, disease-free rate of population growth,
r. The figure is based on the assumption that it is the birth rate per adult female
(rather than some overall average birth rate) that is unchanging, as just discussed;
that is, the curves in Fig. 3 are calculated from eq(2.20) with ν defined in terms
of r, μ, and τ from eq(2.23).

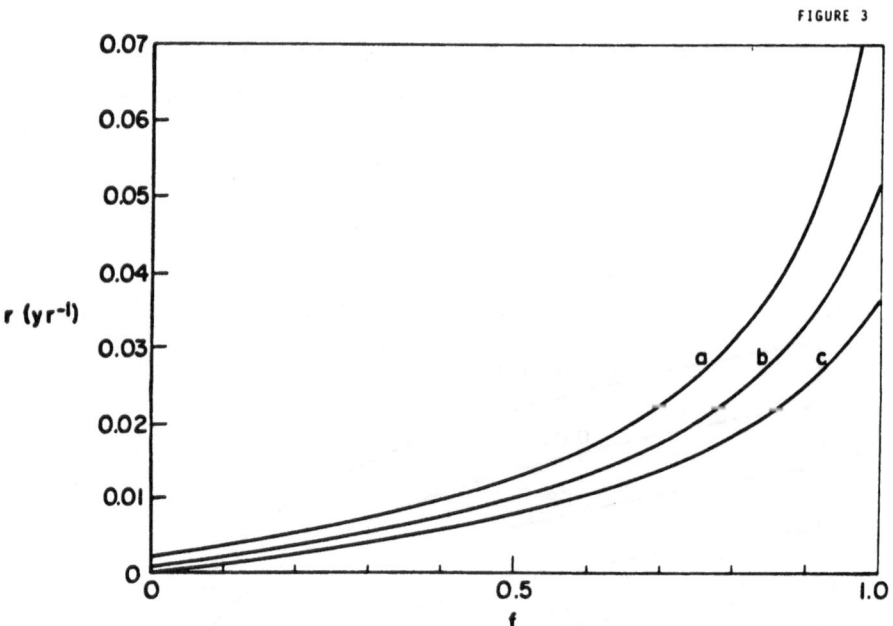

FIGURE 3

Fig. 3: This figure shows the growth rate in the AIDS-free population, r,
which can asymptotically be brought exactly to zero (stationary population)
for a specified value of f. The other epidemiological and demographic
parameter values are $\wedge$ = 0.4, v = 0.1, μ = 0.02 (all in yr⁻¹), τ = 15 yr;
the curves labelled a, b, c are for ϵ = 0, 0.5, 1.0, respectively (ϵ is the
survival probability for offspring born to infected mothers). These results
are obtained from the critical version of eq(2.20) in which ρ = 0.

The combination of values of epidemiological and demographic parameters that
exactly divide population growth from population decline can be found simply by
putting ρ = 0 in eq(2.20). The critical relations between r (disease-free growth
rate) and f that will lead asymptotically to zero growth rates are shown in Fig. 3,
for several values of ϵ (survival probability for offspring infected by vertical
transmission). Figs. 2 and 3 essentially summarize the above discussion by making it

clear that plausible values of the epidemiological and demographic parameters can
result in population sizes eventually declining under the impact of HIV/AIDS.

2.4   Numerical Illustrations.

Figs. 4 and 5 show population trajectories for N(t) as a function of time, t, for
various assumptions about the magnitudes of the demographic and epidemiological parameters. It can be seen that eventual decline in population size is possible, once f
(the fraction of those infected with HIV who go on to develop AIDS, if they do not
die from other causes first) attains substantial values; such relatively high values
of f seem increasingly likely to be realistic.

Notice that these results for N(t) have the features foreshadowed above, with a
preliminary phase in which population growth slows slightly as infection becomes
established, followed often by a relatively fast decline as AIDS mortality asserts

FIGURE 4

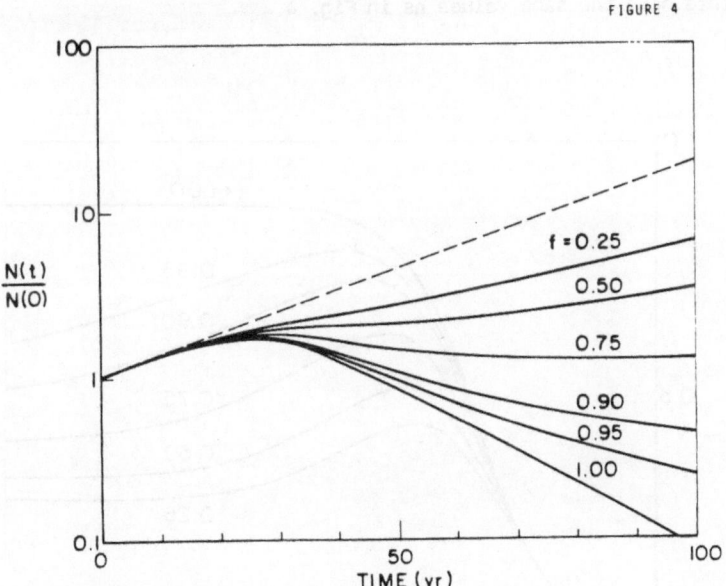

Fig. 4: The magnitude of the total population is plotted (on a logarithmic
scale, as a ratio to the initial magnitude) as a function of time, t,
for a variety of f-values, as shown. The dashed curve illustrates the
AIDS-free rate of purely exponential growth; the other curves have the
features discussed in the text. The epidemiological and demographic
parameters have the values $\bigwedge$ = 0.2, v = 0.06, μ = 0.02, r = 0.03 (all
in yr^{-1}), ε = 0.5; at t = 0, Y(0)/N(0) = 0.01.

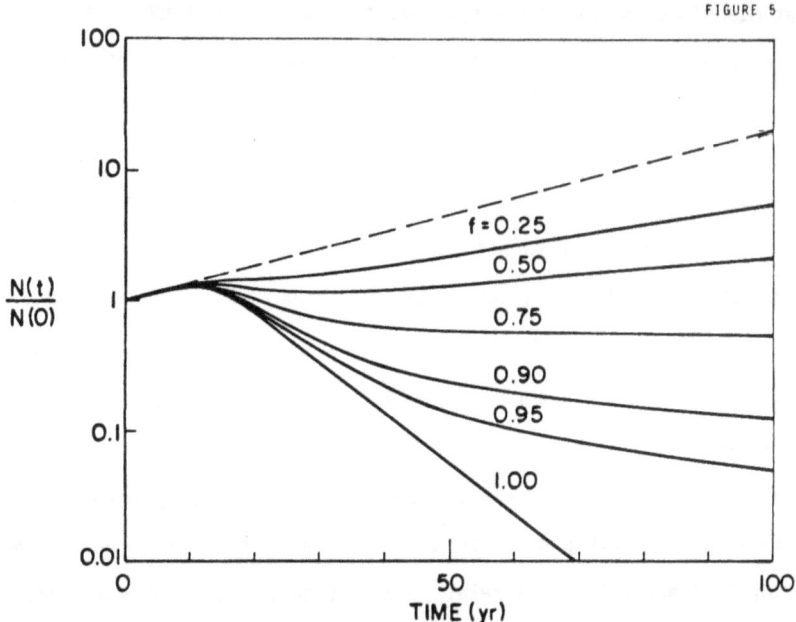

Fig. 5: As for Fig. 4, except here $\wedge$ = 0.4 yr^{-1} and v = 0.1 yr^{-1}; all other parameters have the same values as in Fig. 4.

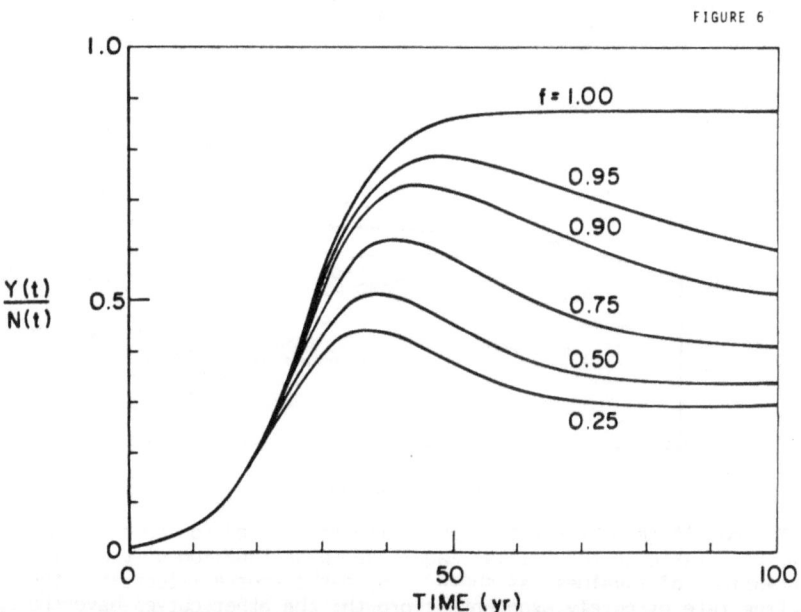

Fig. 6: This figure shows the fraction of the population who are infected with HIV, Y/N, as a function of time, t, for the same range of f-values shown in Fig. 4. All parameters have the same values as in Fig. 4.

itself, and leading eventually to a phase of slower asymptotic decrease (in which the
death rate, μ, of the Z-class of "recovered" individuals may predominate).

Of particular note are the relatively long time-scales that characterize the
dynamics. As dicussed much more fully in MAMcL, it is likely that 30 years or more
may elapse before populations begin to decline, even though long-term rates of
decline may be marked.

Fig. 6 shows the fraction infected, χ (t), as a function of time, t, for the
populations whose overall dynamics are illustrated in Fig. 4. Notice that, as anticipated from the phase plane analysis in Sec. 2.2, these fractions tend to make an
oscillatory approach -- of a kind richly familiar in other epidemiological
contexts -- to their asymptotically steady values.

3. EFFECTS OF HIV/AIDS ON THE AGE-STRUCTURE OF THE POPULATION

3.1   Definition of the Basic Model.

We now extend the basic models to take account of the age-structure of the population. This enables us, among other things, to deal explicitly with the effects of
recruitment lags. The basic demographic and epidemiological framework of the age-
structured analysis retains the simplfying assumptions made in Section 2: constant
infectiousness; constant rate of movement from HIV infection to AIDS disease, in that
fraction, f, who do develop AIDS; Type II mortality from other causes (i.e., μ =
constant); new sexual partners acquired at some average rate (ignoring the substantial variability in such rates that often pertain); and overall transmission of HIV
male-to-female assumed equal to that for female-to-male ($\beta_1 c_1 = \beta_2 c_2 = \beta c$). As
discussed more fully in Section 2 and in MAMcL, such deliberately simplified models
give insight into the dynamical nature of the changes brought about by HIV/AIDS, and
into the magnitude of the time-scales on which different things happen. These
simplified studies thus serve as a guide for the numerical exploration of more
realistic and more complicated models.

We define N(a,t) to be the total number of individuals of age a, at time t. These
total numbers may, as before, be subdivided into susceptible, infected-and-
infectious and no-longer-infectious categories, X(a,t), Y(a,t) and Z(a,t), respectively. The propability, per unit time, that a given susceptible will acquire
infection is λ(a,t), which now depends explicitly on age. In the first instance, we

assume that death from causes other than AIDS occurs at the age-dependent rate $\mu(a)$, and births at the rate $m(a)$; all other rate processes are as in Section 2. The dynamical behavior of this system is described by the following set of partial differential equations:

$$\partial X/\partial t + \partial X/\partial a = -\left[\lambda(a,t) + \mu(a)\right]X(a,t), \tag{3.1}$$

$$\partial Y/\partial t + \partial Y/\partial a = \lambda X - \left[v + \mu(a)\right]Y(a,t), \tag{3.2}$$

$$\partial Z/\partial t + \partial Z/\partial a = (1 - f)vY - \mu(a)Z(a,t). \tag{3.3}$$

The total number, $N = X + Y + Z$, thus obeys

$$\partial N/\partial t + \partial N/\partial a = -\mu(a)N(a,t) - fvY(a,t). \tag{3.4}$$

These equations have as one boundary condition the requirement $X(0,t) = N(0,t) = B(t)$ and $Y(0,t) = Z(0,t) = 0$, where the birth rate, $B(t)$, is

$$B(t) = \int m(a)\left[N(a,t) - (1 - \epsilon)Y(a,t)\right]da. \tag{3.5}$$

The other boundary condition is given by specifying $X(a,0)$, $Y(a,0)$, $Z(a,0)$, $N(a,0)$; that is, by specifying the age-specific numbers in each category at some initial time, $t = 0$.

To complete the description of this system, we generalize eq(2.5) to define $\lambda(a,t)$ as

$$\lambda(a,t) = \beta c \int p(a,a')Y(a',t)da' / \int p(a,a')N(a',t)da'. \tag{3.6}$$

Here, as before, β and c are the transmission probability and mean rate of acquiring new sexual partners. The ratio of integrals in eq(3.6) gives the probability that any one partner will be infected; $p(a,a')$ is the probability that a susceptible of age a will choose a partner of age a'. In MAMCL (which was restricted to the limiting case $f = 1$), we explored the asymptotic properties of this system of equations under the two extreme assumptions that all ages mix homogeneously ($p(a,a')$ = constant, independent of a and a' so long as both ages lie in the sexually-active range) and that partners are restricted to be from the same age cohort ($p(a,a')$ = $\delta(a-a')$). These two assumptions tend to represent opposite extremes. We are encouraged to find that the results in MAMCL are qualitatively similar under these two extreme assumptions, which seem likely to bracket reality. In detail, the age-specific prevalence of HIV infection understandably tends to rise more slowly for the age-restricted model than for the homogeneously mixed one, resulting in HIV/AIDS having somewhat less demographic impact -- other things being equal -- if sexual pairings are age-restricted rather than homogeneously mixed. In what follows, we restrict our attention to the homogeneously mixed case; for a more general analysis, see MAMcL.

This system of age-structured, partial differential equations can now be solved numerically. Starting with some specified set of age profiles for $N(a,0)$, $X(a,0)$, $Y(a,0)$ and $Z(a,0)$ at $t = 0$, we compute the initial birth rate, $B(0)$, and force of infection, $\lambda(a,0)$. Eqs(3.1)-(3.4) then give the age profiles one time step later, and so on. This numerical procedure can be extended in a straightforward way to embrace a variety of the realistic complications mentioned above.

3.2 Aysmptotic Properties.

After a sufficient length of time has elapsed, $N(a,t)$, $X(a,t)$, $Y(a,t)$ and $Z(a,t)$ will in general tend to exhibit stable age-profiles, whose shapes do not change even though the total numbers increase or decrease exponentially (at some rate ρ). We now analyze this asymptotic behavior, and present numerical results for the asymptotically stable age-profiles and for social and economic indicators (such as "child dependency ratios") that can be derived from them.

As in the simpler models without age-structure, as $t \to \infty$ the time dependence in $N(a,t)$ and other such quantities can be factored out as $\exp(\rho t)$. As explained in MAMcL, this is essentially because $\lambda(a,t)$ depends on the ratio of infecteds to total numbers, and thus becomes independent of time as $t \to \infty$. The asymptotic age-profiles can then be obtained by first defining:

$$N(a,t) = n(a)G(a)e^{\rho t}, \tag{3.7}$$

$$X(a,t) = x(a)G(a)e^{\rho t}, \tag{3.8}$$

$$Y(a,t) = y(a)G(a)e^{\rho t}. \tag{3.9}$$

Here the function $G(a)$ is the disease-free stable age profile, except that ρ replaces the disease-free population growth rate r:

$$G(a) = \exp\left[-\rho a - \int_0^a \mu(a')da'\right]. \tag{3.10}$$

The functions $n(a)$, $x(a)$ and $y(a)$ describe the demographic effects of HIV/AIDS upon the shapes of the age-profiles, as distinct from its effects upon the overall growth rate (and thus upon $G(a)$). Substituting eqs(3.7)-(3.10) into (3.1)-(3.4), we see that these functions obey the set of ordinary differential equations

$$dn/da = - fvy, \tag{3.11}$$

$$dx/da = - \lambda x, \tag{3.12}$$

$$dy/da = \lambda x - vy. \tag{3.13}$$

As discussed above, we assume that sexually active adults mix homogeneously: $p(a,a') =$ constant, provided both a and a' have values above some lower limit, τ

(taken to be 15 years in the numerical illustrations), and below some upper limit, γ (which is disregarded, in the sense that it is taken to be infinite, in the numerical illustrations). Under this assumption, the force of infection, $\lambda(a,t)$, is always zero if $a < \tau$, while for $\gamma > a > \tau$ asymptotically it has the constant value $\lambda(a) = \lambda$, with λ given by

$$\lambda = \beta c \int_{\tau}^{\gamma} y(a)G(a)da \ / \ \int_{\tau}^{\gamma} n(a)G(a)da. \tag{3.14}$$

The boundary conditions for the three first-order differential eqs(3.11)-(3.13) are $n(0) = x(0) = B$ and $y(0) = 0$, with the constant B defined by

$$B = \int m(a) [n(a) - (1 - \epsilon)y(a)] G(a)da. \tag{3.15}$$

In the absence of HIV infection, $\lambda = 0$ and thus $n(a) = B$. It follows that the asymptotic age-profile is given by $G(a)$ of eq(3.10), with ρ having the disease-free value $\rho = r$, as stated above. This disease-free rate of population growth, $\rho = r$, follows immediately from the appropriately simplified form of eq(3.15), which gives the demographers' standard Euler equation:

$$1 = \int m(a)exp \left[- ra - \int_{0}^{a} \mu(a')da' \right] da. \tag{3.16}$$

3.3   Asymptotic Age Profiles.

For the numerical illustrations in Figs. 7-13, we assume that mortality is Type II ($\mu(a) = \mu$, where μ is constant), and that the birth rate has some constant value in the age-range in which individuals are sexually active, as just defined ($m(a) = \hat{\nu}$ for $\gamma > a > \tau$, $m(a) = 0$ otherwise). Finally, we take $\gamma \to \infty$ to keep the presentation relatively simple. The corresponding expressions for finite values of γ are straightforward; the numerical results for "child dependency ratios" and other such quantities obtained with $\gamma \to \infty$ are essentially identical with those using $\gamma = 50$ years (because there are very few people in the older age-classes).

Eq(3.12) for $x(a)$ can now be integrated, to get

$$x(a) = 1 \qquad\qquad ; \quad a < \tau, \qquad\qquad\qquad (3.17a)$$

$$x(a) = \exp[-\lambda(a - \tau)] \qquad ; \quad a > \tau. \qquad\qquad\qquad (3.17b)$$

Here we have, without loss of generality, put the constant $B = 1$. Eq(3.13) for $y(a)$ and eq(3.11) for $n(a)$ can now be integrated in turn, to get

$$y(a) = 0 \qquad\qquad\qquad\qquad ; \quad a < \tau, \quad (3.18a)$$

$$y(a) = \frac{\lambda}{\lambda - \nu} \left\{ \exp[-\nu(a - \tau)] - \exp[-\lambda(a - \tau)] \right\} \qquad ; \quad a > \tau, \quad (3.18b)$$

$$n(a) = 1 \qquad\qquad\qquad\qquad\qquad\qquad ; \quad a < \tau, \quad (3.19a)$$

$$n(a) = (1-f) + f\left\{ \lambda \exp[-\nu(a-\tau)] - \nu \exp[-\lambda(a-\tau)] \right\} /(\lambda-\nu) \quad ; \quad a > \tau. \quad (3.19b)$$

These expressions give $n(a)$ and $y(a)$ in terms of defined epidemiological and demographic parameters, along with the quantity λ. The overall age-profile, $N(a)$, and the age-specific numbers of HIV infectees, $Y(a)$, are obtained by multiplying $n(a)$ and $y(a)$, respectively, by the factor $G(a)$ (which for Type II mortality is $G(a) = \exp[-(\rho + \mu)a]$). To complete the calculations, therefore, we need to express λ and ρ in terms of the defined parameters, using eqs(3.14) and (3.15).

At this point it is convenient to define the integrals

$$I = \int_{\tau}^{\gamma} n(a)G(a)da, \qquad\qquad\qquad\qquad (3.20)$$

$$J = \int_{\tau}^{\gamma} y(a)G(a)da, \qquad\qquad\qquad\qquad (3.21)$$

$$K = \int_{0}^{\tau} n(a)G(a)da. \qquad\qquad\qquad\qquad (3.22)$$

In this notation, eq(3.14) takes the form

$$\lambda = \beta c J/I. \qquad\qquad\qquad\qquad (3.23)$$

Remembering that we have put $B = 1$, eq(3.15) becomes

$$1 = \hat{\nu} \left[I - (1 - \epsilon)J \right]. \qquad\qquad\qquad\qquad (3.24)$$

The integrals I, J, K can be evaluated either by routine integration of eqs(3.18) and (3.19), or directly from eqs(3.1)-(3.4) (noting that, asymptotically, $\partial F/\partial t \to \rho F$ and then integrating with respect to a). Either way, we obtain (in the limit $\gamma \to \infty$)

$$I = \frac{N(\tau)[(\rho + \mu + \nu)(\rho + \mu + \lambda) - f\nu\lambda]}{(\rho + \mu)(\rho + \mu + \nu)(\rho + \mu + \lambda)} \qquad\qquad (3.25)$$

$$J = N(\tau)\lambda / [(\rho + \mu + \nu)(\rho + \mu + \lambda)], \qquad\qquad (3.26)$$

$$K = [1 - N(\tau)]/(\rho + \mu). \qquad\qquad\qquad\qquad (3.27)$$

Here $N(\tau)$ is equal to $G(\tau)$; that is, $N(\tau) = \exp[-(\rho + \mu)\tau]$.

Substituting eqs(3.25) and (3.26) into eq(3.23), and performing some algebraic manipulations, we obtain an expression for λ in terms of ρ and the defined epidemiological and demographic parameters:

$$\lambda = [(\rho + \mu)(\wedge - \rho)] / [\rho + \mu + (1 - f)v]. \tag{3.28}$$

Using this relation in conjunction with eqs(3.25) and (3.26) in eq(3.24), we eventually obtain an expression relating the asymptotic rate of population growth, ρ, to the fundamental epidemiological and demographic parameters. This expression is identical with eq(2.20), except that now the definition of $\mathcal{V}$ emerges explicitly from the age-structured analysis as

$$\mathcal{V} = \hat{\mathcal{V}} \exp[- (\rho + \mu)\tau]. \tag{3.29}$$

This is the same as the more intuitively-justified eq(2.22). As pointed out in Section 2, it is more usual to have a direct estimate of the disease-free rate of population growth, r, than of the birth rate per adult, $\hat{\mathcal{V}}$; eq(2.23) translates eq(2.22) into a definition of $\mathcal{V}$ in terms of r and ρ.

We are now in a position to compute the asymptotic age profiles of the population as a whole, N(a), and of the numbers infected, Y(a). Once the basic demographic parameters (μ, τ, and the disease-free r) and epidemiological parameters ($\wedge$, v, ε, f) are specified, the asymptotic force of infection, λ, and rate of population growth, ρ, are calculated from eqs(3.28) and (2.20), respectively. Then N(a) and Y(a) are obtained by multiplying the functions n(a) and y(a) of eqs(3.19) and (3.18), respectively, by the function G(a) = $\exp[- (\rho + \mu)a]$.

Figures 7 and 8 show such age-profiles, for two respresentative sets of values of the demographic and epidemiological parameters. The figures are displayed in a conventional way, plotting the relative numbers in the different age-classes on the horizontal axis, with age itself being the vertical axis; our symmetry assumption implies that male and female age-profiles are identical (more generally, the female age-profile might be shown as the positive x-coordinate and the male age-profile as the negative x-coordinate, resulting in asymmetric profiles). In each of the two figures, the dependence of the asymptotic age-structure upon f is shown by computing the profile for three different values of f (f = 0.4, 0.7, 1.0). In all cases, the stable age-profiles which are shaped by the impact of HIV/AIDS are contrasted with the original age-profiles, before the impact of the disease.

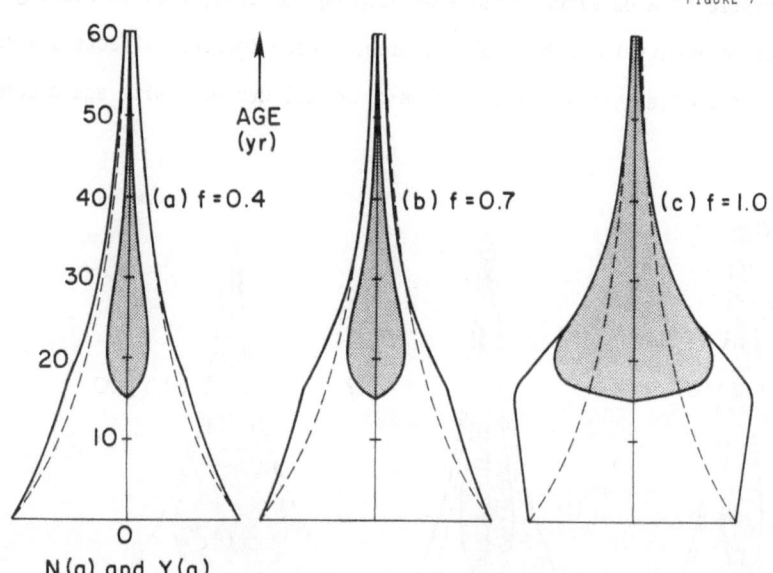

FIGURE 7

Fig. 7: These three figures show the asymptotic proportions of the total population (the solid curves) and of the total numbers infected (the hatched region), in the different age classes (with age, from 0 to 60 years, being the vertical axis). These numbers, N(a) and Y(a) (measured on the horizontal axis), are shown in arbitrary units, because we are concerned here only with relative numbers as a function of age. The dashed lines show the corresponding age-profiles in the pristine populations, before the advent of HIV/AIDS. As discussed much more fully in the text, HIV/AIDS tends to reduce overall rates of population growth, resulting in less steep age-profiles among younger age classes; the superimposed effects of adult deaths from AIDS, however, then reduce the numbers in older age classes. The net ratio of those below age 15 years to those above age 15 years, compared with this ratio in the AIDS-free population, is not easily intuited; the figures suggest that the two main effects tend to cancel, leaving these ratios roughly unchanged. As shown, these three figures are for: (a) f = 0.4, (b) f = 0.7, (c) f = 1.0. The other epidemiological and demographic parameters have the values $\wedge$ = 0.4, v = 0.1, μ = 0.02, r = 0.03 (all in yr^{-1}), ε = 0.5, and τ = 15 yr.

As shown in Figs. 7 and 8, HIV/AIDS has two different kinds of effects upon asymptotic age-profiles. On the one hand, the direct effects of adult mortality from AIDS tend to decrease the ratio of numbers of adults to numbers of children. This direct effect is captured by the function n(a), which is simply a constant, independent of age, in the pristine population. On the other hand, the increased adult mortality (from horizontal transmission) along with the effectively decreased birth rate (resulting from vertical transmission) cause the overall rate of population growth to decrease; this has the indirect effect of tending to produce less steep age-profiles

(more characteristic of a developed than a developing country), thus increasing the
ratio of numbers of adults to numbers of children. These indirect effects are captured in the function G(a), whose pattern of exponential decrease with age becomes

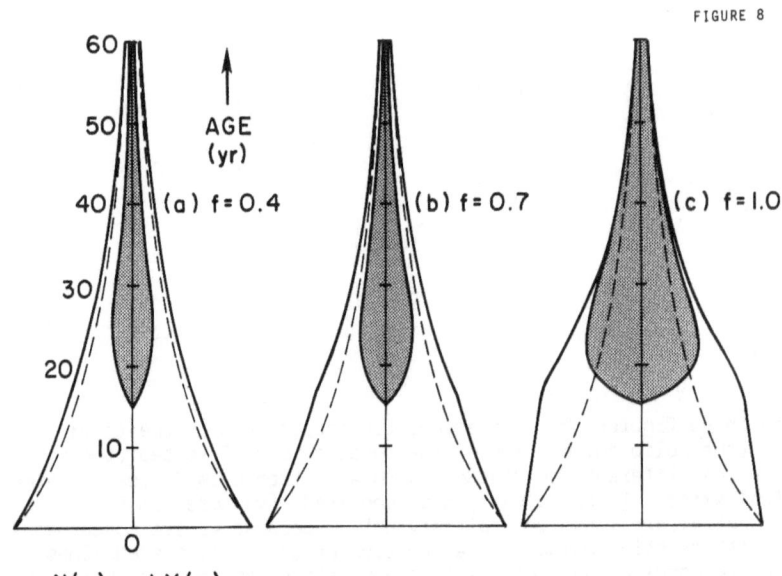

FIGURE 8

N(a) and Y(a)

Fig. 8: As for Fig. 7, except here the epidemiological parameters $\wedge$ and v
have the values $\wedge$ = 0.2 yr^{-1} and v = 0.06 yr^{-1}; all other parameters
are exactly as in Figs. 7a,b,c. For these values of $\wedge$ and v, HIV spreads
more slowly and AIDS kills after a longer incubation period, than for the
parameter values used in Fig. 7. As a result, the asymptotic age-profiles
shown here tend to be less different from the corresponding AIDS-free
age-profiles, than is the case in Fig. 7. As in Fig. 7, however, the
asymptotic ratio of those below age 15 years to those above 15 years is
roughly unchanged by HIV/AIDS, for the reasons discussed more fully in the
text.

less and less steep as the asymptotic rate of population growth, ρ, becomes smaller.
Figures 7 and 8 thus confirm the observations made in MAMcL for the special case when
f = 1.

The net result of these two countervailing effects -- one tending to decrease the
ratio of adults to children and the other tending to increase this ratio -- cannot be
derived intuitively, and must be calculated for any specific set of demographic and
epidemiological assumptions. We now present numerical results pertaining to such
ratios, which have obvious social and economic implications.

4. CHILD DEPENDENCY RATIOS

One rough measure of, as it were, the ratio of tax consumers to tax producers is the so-called "child dependency ratio", CDR, defined as the fraction of the total population who are below the age of 15 years. Using the notation established in the previous section, the asymptotic value of this ratio is defined in terms of the quantities I and K of eqs(3.20) and (3.22) (in the limit $\gamma \to \infty$) as

$$CDR = K/(K + I). \qquad (4.1)$$

Figures 9-11 show this ratio as a function of f, in each case for pristine growth rates in the population before HIV/AIDS of $r = 0.01, 0.02, 0.03, 0.04$ yr^{-1}. In Figs. 9, 10, 11 the survival probabilities for offspring born to infected mothers are $\varepsilon = 0, 0.5, 1.0$, respectively. The other parameters have the values $\wedge = 0.4$ yr^{-1} (corresponding roughly to a doubling time of 2.5 years for HIV prevalence in the early stages of the epidemic), $v = 0.1$ yr^{-1} (corresponding roughly to a 10 year incu-

FIGURE 9

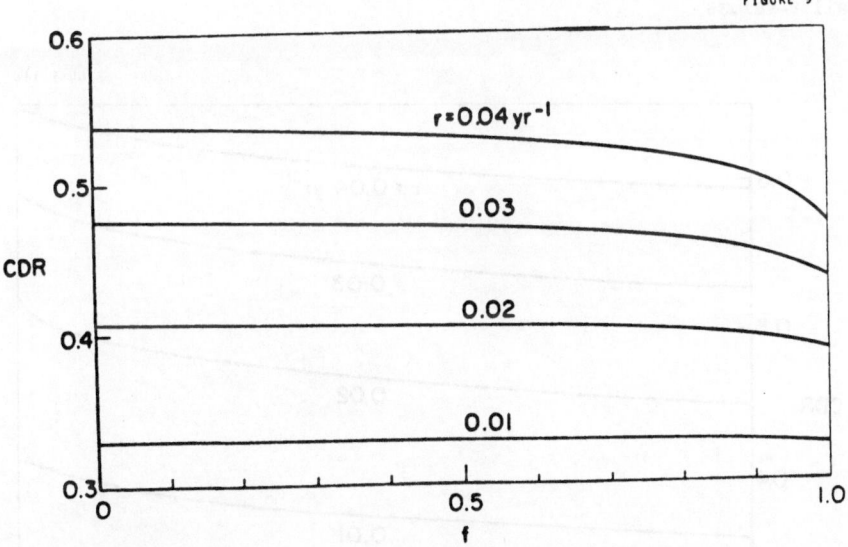

Fig. 9: This figure shows asymptotic values of the "child dependency ratio", CDR, defined as the fraction of the total population who are below the age of 15 years. CDR is shown as a function of f (the fraction of HIV infectees who go on to develop AIDS), for several different values of the growth rate in the AIDS-free population, r, as shown. This figure is for $\wedge = 0.4$ $v = 0.1$, $\mu = 0.02$ (all in yr^{-1}), and $\tau = 15$ yr. The parameter ε has the value $\varepsilon = 0$; that is, all offspring born to infected mothers die (100% mortality from vertical transmission). As discussed in the text, CDR is roughly unaffected by HIV/AIDS for most values of the epidemiological parameters, although there is a noticeable downswing under the combined impact of horizontal and vertical transmission for large values of f in this figure (where $\varepsilon = 0$).

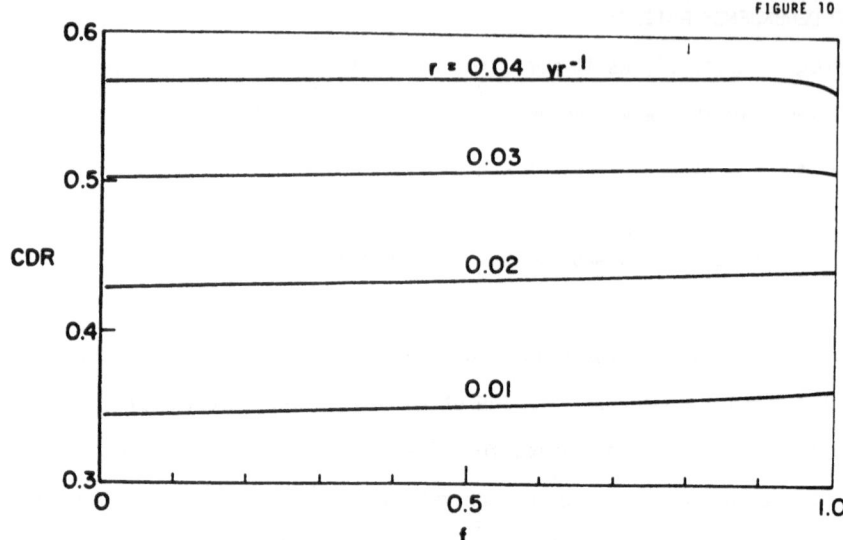

Fig. 10: As for Fig. 9, except here ε = 0.5 (half the offspring born to infected mothers die from AIDS, which currently seems likely to be what happens in practice). All other parameters are as in Fig. 9. In this case, the asymptotic value of CDR is essentially unaffected by HIV/AIDS for all f-values.

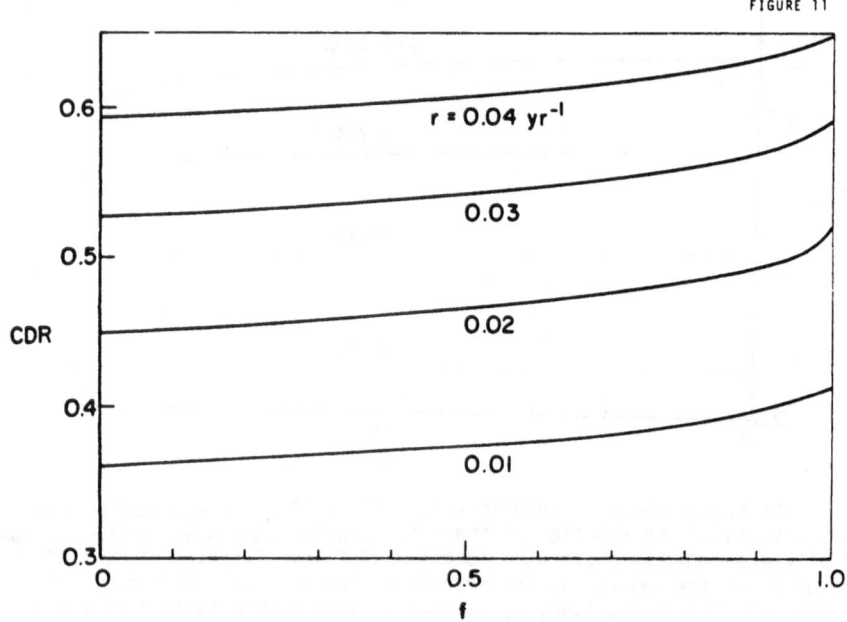

Fig. 11: Again as for Fig. 9, except here ε = 1 (no vertical transmission). In the absence of vertical transmission, it can be seen that, for large f-values, the effects of adult deaths predominate over the depression of overall rates of population growth; the result is that the asymptotic value of CDR rises somewhat above its value in the AIDS-free population, in this limit of ε → 1.

bation time for that fraction who do go on to develop AIDS), $\mu = 0.02 \text{ yr}^{-1}$

(corresponding roughly to a 50 year life expectancy in the absence of AIDS), and $\tau = 15 \text{ yr}$. Figure 12 similarly shows CDR as a function of f for the same set of r-

FIGURE 12

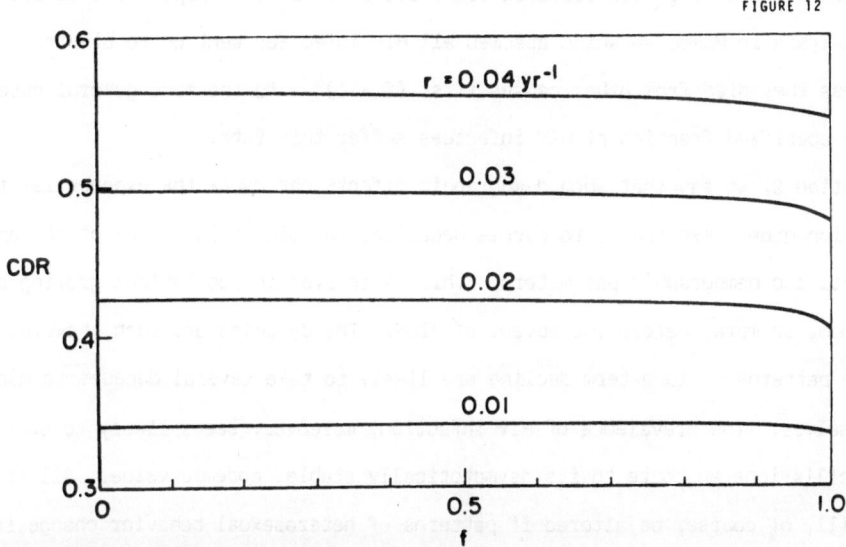

Fig. 12: This figure is to be compared with Fig. 10, except that now the epidemiological parameters $\wedge$ and v have the values $\wedge = 0.2 \text{ yr}^{-1}$ and $v = 0.06 \text{ yr}^{-1}$. All other parameter-values are as in Fig. 10. As in Fig. 10, the asymptotic value of CDR is essentially unaltered by the presence of HIV/AIDS.

values, for a slower-growing epidemic where $\wedge = 0.2 \text{ yr}^{-1}$, $v = 0.06 \text{ yr}^{-1}$, $\epsilon = 0.5$, and the other parameters are as in Figs. 9-11.

The main feature of Figs. 9-12 is that CDR is changed remarkably little by the effects of HIV/AIDS, and is particularly insensitive to the value of f (the fraction of adults killed by horizontally-transmitted AIDS) for a specified value of ϵ (the survival probability of vertically-infected newborns). It appears that the counter-vailing tendencies discussed above essentially cancel each other out; the tendency for adult deaths from AIDS to increase CDR is roughly balanced by the tendency for depressed population growth rates to decrease CDR. This insensitivity of CDR to the exact values of the epidemiological parameters, and especially to f, cannot be arrived at intuitively, but the results shown in Figs. 9-12 suggest to us that such insen-sitivity is likely to remain true in elaborate computations based on more detailed epidemiological and demographic assumptions.

5. CONCLUSION

The demographic impact of heterosexually-transmitted AIDS comes both from adult mortality (resulting from horizontal transmission) and from the effective depression of birth rates (resulting from vertical transmission). In this paper we extend the earlier analysis in MAMcL -- which assumed all HIV infectees went on to die from AIDS, unless they died from other causes first ($f = 1$) -- to the more general case where some specified fraction of HIV infectees suffer this fate.

In Section 2, we saw that such demographic effects can cause the exponential rate of population growth eventually to become negative, for plausible values of the epidemiological and demographic parameters. This is so even in populations growing at 3% per annum, or more, before the advent of AIDS. The dynamics are such, however, that these patterns of long-term decline are likely to take several decades to manifest themselves. The prevalence of HIV infection, moreover, seems likely to exhibit damped oscillations en route to its asymptotically stable, endemic value. All these results will, of course, be altered if patterns of heterosexual behavior change in such a way as to reduce transmission rates.

In Section 3 we explored the effects of HIV/AIDS upon the asymptotic form of age-profiles in total numbers, and in numbers infected. On the one hand, increasing death rates and effectively decreasing birth rates cause the overall rate of population growth to decrease, which means -- other things being equal -- that age-profiles tend to be less steep than is characteristically the case at present in developing countries. This effect, by itself, tends to decrease the ratio of numbers of children to numbers of adults. On the other hand, adult deaths from AIDS obviously tend to increase the ratio of numbers of children to numbers of adults. The results shown in Figs. 7 and 8, and summarized in the "child dependency ratio" (CDR) of Figs. 9-12, suggest that these countervailing effects tend roughly to cancel, leaving long-term CDR-values roughly unchanged, for the parameter-values that are currently thought to pertain in Africa. More generally, however, HIV/AIDS has the capacity to increase or to decrease long-term CDRs, depending on the details of the interplay among the relevant epidemiological and demographic parameters.

This work was supported, in part, by the NSF under grant DMS87-03503 (RMM), and by the Sloan Foundation (RMM).

REFERENCES

Anderson, R.M. and May, R.M. 1986. The invasion, persistence and spread of infectious diseases within animal and plant communities. Phil. Trans. Roy. Soc. B314, 533-570.

Anderson, R.M. Medley, G.F., May, R.M. and Johnson, A.M. 1986. A preliminary study of the transmission dynamics of the human immunodeficiency virus (HIV) the causitive agent of AIDS. IMA J. Math. Appl. Med. Biol., 3, 229-263.

Castillo-Chavez, C., Cooke, K., Huang, W., and Levin, S.A. On the role of long incubation periods in the dynamics of acquired immunodeficiency syndrome (AIDS). Part 1. Single population models. 1989. J. Math. Biology. In press.

Dietz, K. 1988. On the transmission dynamics of HIV. Math. Biosci., 90, 397-414.

Getz, W.M. and Pickering, J. 1983. Epidemiological models: thresholds and population regulation. Amer. Natur., 121, 892-898.

Hyman, J.M. and Stanley, E.A. 1988. A risk-based model for the spread of the AIDS virus. Math. Biosci., 90, 415-473.

May, R.M. 1974. Stability and Complexity in Model Ecosystems. (Second Edition). Princeton University Press; Princeton.

May, R.M. and Anderson, R.M. 1987. Transmission dynamics of HIV infection. Nature, 326, 137-142.

May, R.M., Anderson, R.M. and McLean, A.R. 1988. Possible demographic consequences of HIV/AIDS: I, assuming HIV infection always leads to AIDS. Math. Biosci., 90, 475-505.

Medley, G.F., Anderson, R.M., Cox, D.R. and Billard, L. 1987. Incubation period of AIDS in patients infected via blood transfusion. Nature, 328, 719-721.

APPENDIX A

In an earlier paper (MAMcL), we analyzed the limiting case f = 1, and found inter alia that the asymptotic rate of population change was ρ → -(μ + v) when ε → 0. On the other hand, the analysis in this paper (eq(2.20)) suggests that ρ + μ is always positive, and that ρ → -μ in the limit ε → 0 and f → 1; this analysis, however, assumes γ → ∞ (with this assumption about the upper bound to the age-range for sexual activity and reproduction being implicit in Section 2 and explicit in Section 3). This Appendix gives a more detailed discussion of the various limits, and resolves any apparent inconsistencies.

To begin, we return to Section 3.3 and notice that, for finite γ, eqs(3.26) and (3.25) for J and I, respectively, are replaced by the more general

$$J = \frac{N(\tau)[\lambda(1 - \phi(s)x(\gamma)) - (\rho + \mu + \lambda)\phi(s)y(\gamma)]}{(\rho + \mu + \lambda)(\rho + \mu + v)} \quad , \tag{A.1}$$

$$I = \{N(\tau)[1 - \phi(s)n(\gamma)] - fvJ\} / (\rho + \mu). \tag{A.2}$$

As explained in the main text, these results can be obtained either by integration of the expressions for Y(a) and N(a), or directly from integrating eqs(3.1)-(3.4) over all a (and noting that, asymptotically, ∂F/∂t → ρF). In eqs(A.1) and (A.2), x(γ), y(γ), n(γ) are given by eqs(3.17b), (3.18b), (3.19b) with a = γ, and as before N(τ) = exp[- (ρ + μ)τ]. We have defined

$$\phi(s) = \exp[- (\rho + \mu)s], \tag{A.3}$$

where s is defined as s ≡ γ - τ; that is, φ(s) = G(γ)/N(τ), when μ(a) is a constant. If (ρ + μ) > 0, then in the limit s → ∞ (i.e., γ → ∞) we have φ → 0, and eqs(A.1) and (A.2) reduce to eqs(3.26) and (3.25) (whence we find, from eq(2.20), that indeed ρ + μ > 0 for all f < 1, completing the argument).

The quantities λ and ρ can now be expressed in terms of the basic epidemiological and demographic parameters (ν̂, μ, v, βc, ε, f, τ, γ) by substituting eqs(A.1) and (A.2) for J and I into eqs(3.23) and (3.24).

Before doing this, we make one approximation. The parameters 1/λ and 1/v represent the epidemic's doubling time in its early stages and the average incubation time, respectively. Both times typically are short in comparison with the average age-range for sexual activity and reproduction, $\gamma - \tau$. That is, both λs and vs typically will be significantly larger than unity, which means that $x(\gamma)$ and $y(\gamma)$ will typically be significantly less than unity, as will the difference between $n(\gamma)$ and $(1 - f)$ (see eqs(3.17b)-(3.19b)). Neglecting the terms $x(\gamma)$ and $y(\gamma)$ in comparison with unity, and replacing $n(\gamma)$ by $1 - f$, we obtain from eq(A.2) the result

$$ I = \left\{ N(\tau) \left[1 - (1 - f)\phi(s) \right] - fvJ \right\} / (\rho + \mu). \tag{A.4}$$

J is again given simply by eq(3.26) in the main text.

Substituting eqs(A.4) and (3.26) into eq(3.23) for λ, we obtain, after some algebra,

$$ \lambda = \frac{(\rho + \mu) \left[\beta c - (\rho + \mu + v)(1 - (1 - f)\phi) \right]}{(\rho + \mu + v)(1 - (1 - f)\phi) - fv}. \tag{A.5}$$

Using this result in conjuction with eq(3.24), we obtain, after some algebraic manipulation and rearrangement, the following expression for ρ in terms of basic epidemiological and demographic parameters:

$$ \rho + \mu + v = \left(\frac{\beta c}{\beta c - \theta'} \right) v \left[\epsilon - (1 - f)\phi + \frac{v(1-f)(1-\phi)}{\rho + \mu} \right]. \tag{A.6}$$

Here, as before, $v = \tilde{v} \exp \left[-(\rho + \mu)\tau \right]$ and ϕ is given by eq(A.3) with $s = \gamma - \tau$. The parameter θ' differs from the earlier parameter θ of eq(2.16):

$$ \theta' = fv + (1 - \epsilon)v \left[1 - (1 - f)\phi \right]. \tag{A.7}$$

We can now see explicitly that the order in which we take certain limits can affect the results.

First, if we take the limit $f \to 1$, the second two of the three terms inside the square brackets in eq(A.6) vanish, and also the θ' of eq(A.7) reduces to the θ of eq(2.16). Thus eq(A.6) reduces to the result $\rho + \mu + v = \beta c v \epsilon / (\beta c - \theta)$ obtained in MAMcL. In the further limit when no offspring of infected mothers survive, $\epsilon \to 0$, we have the asymptotic result $\rho \to -\mu - v$.

Second, suppose we begin by explicitly taking the limit $s \to \infty$ and thence (provided $\rho + \mu > 0$) $\phi \to 0$. This gives eq(2.20) of the main text; eq(2.20) implies that $\rho + \mu > 0$, as illustrated for typical parameter-values in Fig. 2. If we now let $\epsilon \to 0$, we find that $\rho \to - \mu$ as $f \to 1$ (cf. Fig. 2).

The resolution of this apparent paradox (which arises most clearly when, as here, we take $\epsilon = 0$) is, of course, that, for $s \to \infty$, ρ falls from $-\mu$ to $-(\mu + v)$ as f increases from $1 - \mathcal{O}(1/s)$ to 1. Thus if we work simply with the limit $f = 1$ (as in MAMcL), we have $\rho \to - (\mu + v) + \mathcal{O}\epsilon$ for small ϵ. But if we work with finite f (but let $s \to \infty$), as in this paper -- vide Fig. 2 -- we have $\rho \to - \mu + \mathcal{O}\epsilon$ as $f \to 1$ (with the ultimate transition to $\rho \to - (\mu + v)$ at $f = 1$ occurring in an infinitesimal window of width $\sim 1/s$ around $f = 1$). This can be seen more explicitly by asking what combination of epidemiological and demographic parameters in eq(A.6) lead to ρ having the value $- \mu$. This combination of parameters can be found by putting $\rho + \mu = 0$ in eq(A.6) to get

$$v[\beta c - fv - f\hat{\nu}(1 - \epsilon)] \neq \beta c\hat{y}[\epsilon + (1 - f)(vs - 1)]. \tag{A.8}$$

This equation can be rearranged to give an expression for f_c, the f-value corresponding to $\rho + \mu = 0$:

$$f_c = 1 - \frac{\beta c[v - \epsilon\hat{\nu}] - v[v + (1 - \epsilon)\hat{y}]}{\beta c\hat{\vartheta}[sv - 1] - v[v + (1 - \epsilon)\hat{\vartheta}]}. \tag{A.9}$$

It is now plain that, for $sv \gg 1$, ρ falls below $-\mu$ only as f approaches very close to unity ($f \to 1 - \mathcal{O}(1/sv)$). As f increases to unity over this window (whose width is of order $1/sv$), ρ falls rapidly to its value of $\rho = - (\mu + v) + \beta c\mathcal{y}\epsilon / (\beta c - \mathbf{\Theta})$ at $f = 1$.

If vs is indeed much greater than unity, there is a kind of "boundary layer" at $f = 1$ in figures such as Fig. 2, within which ρ falls from $-\mu$ to the value obtained by putting $f = 1$ at the outset (see MAMcL). If we were to put $\gamma = 50$ yr rather than $\gamma \to \infty$ in our numerical examples, then with $\tau = 15$ yr and $v = 0.1$ yr^{-1} we have $vs = 3.5$. This is significantly in excess of unity, but is not very large, and the ensuing numerical results show some quantitative differences from those in Figs. 2-12 as $f \to 1$, although the qualitative features are unchanged. It is our view that once we include such details, we might as well embark on fully numerical computations, with realistic expressions for fertility and mortality schedules, $m(a)$ and $\mu(a)$, and so on. Such work is in progress.

Part V.　Fitting Models to Data

FITTING MATHEMATICAL MODELS TO BIOLOGICAL DATA:

A REVIEW OF RECENT DEVELOPMENTS

David Ruppert
School of Operations Research
and Industrial Engineering
Cornell University
Ithaca, New York 14853

Abstract

A common problem is to fit a theoretical model, $y = f(x,\beta) +$ error, relating a measured response, y to a measured vector of predictors, x, and an unknown parameter vector, β. Ordinary least–squares is an appropriate fitting method when the variation of y about the model, i.e., the "error", is normally distributed with a conditional variance, given x, that is constant. In many biological problems, y is nonnegative, right–skewed, and has a conditional variance that is an increasing function of $f(x,\beta)$. In this paper, two methods for fitting models to such data are discussed. The methods are (1) transformation of the response and the model and (2) weighted least–squares. The rationale behind these methods, estimation techniques, and statistical inference (testing and confidence intervals) are discussed.

1. Introduction

This paper reviews recent research on fitting theoretical models to data. Though these methods are applicable outside biology, my own experiences applying these techniques have been primarily with biological data, and this paper will concentrate on that field.

By "theoretical model", I mean a model derived from biological assumptions using mathematical techniques. In contrast, an empirical model takes a convenient, flexible class of functions and fits them to the data. Parameters in theoretical models usually have biological interpretation; e.g., they are migration rates, growth rates, probability of infection, etc. The distinction between theoretical and empirical model will be relevant later when data transformations are discussed.

The theoretical model is assumed to be in the form

$$y_i = f(x_i, \beta), \tag{1.1}$$

where (y_i, x_i), $i = 1,...,n$ are independent observations of a scalar response, y, and a vector, x, of independent variables. The model determines f, so f is a known function, but β is a vector of unknown biological or physical parameters. Equation (1.1) holds exactly only in the absence of error, but in reality there are many sources of random variability. For example, y_i and x_i may be measured with error, β may vary between observations, or y_i may be random with $f(x_i,\beta)$ being only its expectation or, perhaps, median. Also, $f(x_i,\beta)$ may be a simple approximation to the complex relationship between x_i and y_i.

Several statistical problems arise immediately. The foremost is to estimate β well. After that, one may wish to test the model, to predict a future y with a known x, or to control a future y by controlling x. This paper is primarily a review of recent research on parameter estimation, but prediction will also be touched upon.

This review will not attempt to be complete. Indeed, the topic of fitting mathematical models is sufficiently broad to cover a sizeable proportion of the entire statistics literature. Rather, I will concentrate on a related set of statistical methodologies that have been the focus of much of my recent research. Much of this research is discussed much more fully in a book to be published (Carroll and Ruppert 1988), but this paper provides a shorter overview. Also, the later sections of this paper outline recent work not found in the book.

If the conditional distributional of y given x is known, then β can be estimated by maximum likelihood. Since such knowledge is rarely available, biologists traditionally have used nonlinear least–squares where β is estimated by $\hat{\beta}$ that minimizes

$$\sum_{i=1}^{n} [y_i - f(x_i,\hat{\beta})]^2. \tag{1.2}$$

Nonlinear least–squares is often said to be nonparametric since the method can be applied without knowing the probability distribution of the data. However, least–squares is efficient only if the "errors" defined by

$$\epsilon_i = y_i - f(x_i,\beta) \tag{1.3}$$

are normally distributed and homoscedastic (have a constant variance).

In practice, biological data are often right skewed with the conditional variance of y given x an increasing function of the conditional mean. An alternative to least–squares is robust regression, but robust regression estimators are designed for data where $\epsilon_1,...,\epsilon_n$ have a constant variance and are symmetrically distributed, though possibly with tails that are heavier than the normal distribution.

For right–skewed heteroscedastic data, one can postulate that y follows a distribution other than the normal distribution, say the gamma distribution. This is the approach of generalized linear models (McCullagh and Nelder 1983). This technique can be quite useful, especially if the assumed distribution has a "shape parameter" that can be estimated from the data.

In this paper, I will discuss another class of models, transformation and weighting models. These have proved to be effective methods for modeling skewed and heteroscedastic data.

By weighting, I mean estimating the conditional variance of y given x and weighting each observation by the reciprocal of its estimated variance. This technique is sometimes called generalized least–squares or weighted least–squares with estimated weights. Generalized least–squares is suitable for data that are nearly normally distributed but heteroscedastic.

Transformation models assume that a nonlinear function, $h(\cdot)$, of y is normally distributed with a constant variance. The function $h(\cdot)$ is assumed to be in a parametric class, e.g., the power transformation family. Transformation models are suitable for skewed data where the conditional variance of y is a function of its conditional mean.

Transformation/weighting models combine a transformation of y with generalized least–squares, and are flexible enough to model skewed data with the conditional variance of y a function of independent variables.

2. Transformations

The effects of a monotonically increasing transformation $h(\cdot)$ depend upon whether h is concave or convex—we will not be concerned with transformations that are neither.

If h is concave then its first derivative is decreasing. If h is applied to a random variable W, then h contracts the right tail of W while spreading the left tail. As a result $h(W)$ is less right skewed (or more left skewed) than W. Conversely, if h is convex then $h(W)$ is more right skewed than W. These statements are made mathematically precise by van Zwet (1964) in his important study of convex transformations of random variables. Let $\gamma(W)$ be the third–moment skewness coefficient, i.e.,

$$\gamma(W) = E[(W - \mu_W)/\sigma_W]^3, \tag{2.1}$$

where μ_W and σ_W are the expectation and standard deviation of W. Van Zwet proves that if h is convex then $\gamma(W) < \gamma(h(W))$.

Bartlett (1947) shows how transformations affect the relationship between the conditional variance and the conditional mean of y. Suppose that these quantities are related by

$$\text{Var}(y|x) = G[E(y|x)]$$

for some positive function G. Fix x and let $\mu = E(y|x)$. Linearizing h about μ, one has

$$\text{Var}(h(y)|x) = E\{[h(y) - E(h(y)|x)]^2|x\} \tag{2.2}$$

$$\doteq E\{[h(y) - h(\mu)]^2|x\}$$

$$\doteq (\dot{h}(\mu))^2 E\{(y-\mu)^2|x\}$$

$$= (\dot{h}(\mu))^2 G(\mu).$$

Therefore, $\text{Var}(h(y)|x)$ is approximately constant if

$$\dot{h}(\mu) \; \alpha \; \frac{1}{\sqrt{G(\mu)}} .$$ (2.3)

If G is increasing, then h will be concave.

Much biological data is non–negative, right–skewed, and is more variable when the mean is large. It is fortunate that concave transformation reduce both right skewness and this type of heteroscedasticity.

Power transformations are widely used in statistical analysis. They are simple and flexible. The logarithm can be naturally embedded in the power transformation family (Box and Cox 1964). Define

$$y^{(\lambda)} = (y^{\lambda}-1)/\lambda \;\; \text{if} \;\; \lambda \neq 0$$ (2.4)
$$= \log(y) \;\; \text{if} \;\; \lambda = 0.$$

I will call $y^{(\lambda)}$ the modified power transformation. Notice that $y^{(\lambda)}$ is convex when $\lambda > 1$ and concave when $\lambda < 1$. Notice that $y^{(\lambda)}$ is a smooth function of λ even at $\lambda = 0$ and

$$(\partial / \partial y) y^{(\lambda)} = y^{\lambda-1} \;\; \text{for all} \;\; \lambda.$$

3. Transformations and Theoretical Models

When the theoretical model (1.1) is available, a transformation of y alone is disadvantageous since it can destroy the model. However, the model can be preserved by transforming y and $f(x,\beta)$ in the same way. Carroll and Ruppert (1984) propose the following model for fitting theoretical models to data. Let $\{h_{\lambda}(\cdot)\}$ be a family of monotonic transformations, e.g., the modified power transformations (2.4). The transformation parameter λ may be scalar or a vector and varies over some parameter space. Carroll and Ruppert's (1984) "transform–both–sides" (TBS) model is

$$h_{\lambda}(y_i) = h_{\lambda}(f(x_i,\beta)) + \sigma \epsilon_i,$$ (3.1)

where $\epsilon_1,...,\epsilon_n$ are independent, standard normal random variables. By estimating λ, we use the data to choose the transformation to normality and homoscedasticity.

It should be emphasized that the purpose of transformation is to induce the proper error structure for least–squares estimation. Simultaneous transformations of y and $f(x,\beta)$ have been used for other purposes, particularly to linearize the model $f(x,\beta)$. However, the transformation that linearizes may exasperate, rather than reduce, skewness and heteroscedasticity. Box and Hill (1974) give an example where the linearizing transformation induces such marked heteroscedasticity that least–squares estimation after linearization gives unacceptable estimates. Box and Hill (1974) correct the heteroscedasticity by weighted least–squares, and Carroll and Ruppert (1984) analyze the same data by estimating the transformation to homoscedastic errors. Since nonlinear least–squares software is now widely available, transforming models solely to linearize them should now be considered poor statistical practice.

Let $\theta = (\beta, \lambda, \sigma)$. Carroll and Ruppert (1984) estimate of θ by maximum likelihood. The log–likelihood is, up to an additive constant,

$$L(\theta) = -n \log(\sigma) - \sum_{i=1}^{n} \{h_\lambda(y_i) - h_\lambda(f(x_i, \beta))\}^2 / 2\sigma^2$$

$$+ \sum_{i=1}^{n} \log(J_i(\lambda)) \tag{3.2}$$

where $J_i(\lambda)$ is the Jacobian of the transformation from y_i to $h_\lambda(y_i)$. For general transformation families there are several simple methods for maximizing (3.2). The first is to apply the Newton–Raphson procedure or the Fisher method of scoring directly to (3.2). This, of course, requires special software but such software is readily available, say in IMSL or GAUSS. I have experienced nonconvergence difficulties when attempting to minimize (3.2), and I recommend first eliminating σ^2 by maximizing $L(\theta)$ over σ^2. For any fixed β and λ, $L(\theta)$ is maximized in σ^2 by

$$\hat{\sigma}^2(\beta, \lambda) = n^{-1} \sum_{j=1}^{n} \{h_\lambda(y_i) - h_\lambda(f(x_i, \beta))\}^2,$$

and the maximized value of $L(\theta)$ is

$$L_{max}(\beta, \lambda) = -\frac{n}{2} \log(\hat{\sigma}^2(\beta, \lambda)) - \frac{n}{2} + \sum_{i=1}^{2} \log(J_i(\lambda)). \tag{3.3}$$

Note that

$$\sum_{i=1}^{n} \log(J_i(\lambda)) = -\frac{n}{2} \log(1/\dot{J}(\lambda)^2) \tag{3.4}$$

where $\dot{J}(\lambda)$ is the geometric mean of $J_i(\lambda)$, $i = 1,...,n$. Substituting (3.4) into (3.3), we have

$$L_{max}(\beta, \lambda) = -\frac{n}{2} \log\left[\left(\frac{\hat{\sigma}(\beta, \lambda)}{\dot{J}(\lambda)}\right)^2\right] - \frac{n}{2}. \tag{3.5}$$

I have only rarely had difficulties maximizing $L_{max}(\beta, \lambda)$ using MAXLIK on the GAUSS package for PCs, and these rare instances were always with a model that fit poorly or was over–parameterized.

From (3.5) we see that for each fixed λ, $L_{max}(\beta)$ is maximized in β by minimizing $\hat{\sigma}^2(\beta, \lambda)$, that is, by the least–squares estimate of β, say $\hat{\beta}(\lambda)$. Let

$$L_{max}(\lambda) = L_{max}(\hat{\beta}(\lambda), \lambda). \tag{3.6}$$

Computing $L_{max}(\lambda)$ requires only nonlinear least–squares software. If only such software is available, then $L_{max}(\lambda)$ can be computed for all λ on some grid and then $\hat{\lambda}$ can be taken as the value where $L_{max}(\lambda)$ is maximized on this grid. This method is simple but time consuming.

If $h_\lambda(y)$ is $y^{(\lambda)}$, the modified power transformation, then there is a simple method for maximizing $L_{max}(\beta,\lambda)$ simultaneously over β and λ by nonlinear least–squares. By (3.5), the MLE minimizes the sum of squares

$$\left[\frac{\hat{\sigma}(\beta,\lambda)}{\dot{J}(\lambda)}\right]^2 = \sum_{i=1}^n \left\{\frac{[h_\lambda(y_i) - h_\lambda(f(x_i,\beta))]}{\dot{y}^{\lambda-1}}\right\}^2$$

where $\dot{y}$ is the geometric mean of $y_1,...,y_n$; see Carroll and Ruppert (1988, section 4.3) or Giltinan and Ruppert (1987) for details.

The asymptotic covariance matrix, $V_{\hat{\theta}}$, of $\hat{\theta} = (\hat{\beta},\hat{\lambda},\hat{\sigma})$ can be estimated by inverting the Hessian of $-L(\theta)$, also called the observed Fisher matrix. That is,

$$V_{\hat{\theta}} = -\{\nabla^2 L(\theta)\}^{-1}.$$

By a theorem of Patefield (1977), the covariance matrix of $(\hat{\beta},\hat{\lambda})$ is $V_{(\hat{\beta},\hat{\lambda})} = -\{\nabla^2 L_{max}(\beta,\lambda)\}^{-1}$. This is convenient when the Fisher method of scoring is applied to (3.3). Large–sample confidence intervals and tests can be constructed from $V_{\hat{\theta}}$ or $V_{(\hat{\beta},\hat{\lambda})}$ by well–known methods that are described in Carroll and Ruppert (1988, section 4.3). For accurate tests and confidence intervals with moderate to small samples, Efron's (1979, 1982) bootstrap is preferred to inference based on $V_{\hat{\theta}}$.

Model (3.1) can be reexpressed as a model for y rather than $h_\lambda(y)$. Let $g_\lambda(y)$ be the inverse of $h_\lambda(y)$ so that $g_\lambda[h_\lambda(y)] = h_\lambda[g_\lambda(y)] = y$. Then

$$y_i = g_\lambda[h_\lambda(f(x_i,\beta)) + \sigma\epsilon_i]. \tag{3.7}$$

Equation (3.7) models the skewness and heteroscedasticity of y_i as arising from a nonlinear transformation of ϵ_i. It follows from (3.7) that the conditional p–th quantile of y_i, given x_i, is

$$q_p(y_i|x_i) = g_\lambda[h_\lambda(f(x_i,\beta)) + \sigma\Phi^{-1}(1-\alpha)]. \tag{3.8}$$

Since $\Phi^{-1}(1/2) = 0$, (3.8) shows that the conditional median of y_i given x_i is $g_\lambda[h_\lambda(f(x_i,\beta))] = f(x_i,\beta)$. Because its y_i is skewed, the conditional expectation of y_i will not equal $f(x_i,\beta)$ exactly. Model (3.8) combines the theoretical model $f(x_i,\beta)$ for the median of y_i with an empirical model for the error structure. I say empirical model because in most applications, the family $\{h_\lambda\}$ is not derived from biological assumptions and λ and ϵ_i do not have biological interpretations.

$q_p(y_i|x_i)$ can be estimated by substituting estimates of β, λ, and σ into (3.8). By plotting the estimated quartiles (first, third, and median) against x, one can see the heteroscedasticity and skewness of the response. An overlay of the same plot based on the nonlinear regression model will show how far predictions differ between the TBS and the usual regression models. The difference can be substantial; see Carroll and Ruppert (1988, section 4.4) for an example. Estimation of the conditional expectation of y_i is slightly more complicated, but still easy; see Carroll and Ruppert (1988, section 4.4) for a full discussion of inference about y.

Gallagher (1986) has extensively studied the application of TBS to water quality models. He presents the analyses of several real data sets as well as a Monte Carlo study.

Although I have emphasized the transform—both—sides method as a technique for theoretically derived models, Snee (1986) gives a number of interesting examples where transform—both—sides is applied to empirical models.

4. Generalized Least—Squares

If y_i is normally distributed but with a nonconstant variance, then the following model is appropriate:

$$y_i = f(x_i,\beta) + \sigma g(z_i,\beta,\xi)\epsilon_i, \tag{4.1}$$

where $\epsilon_1,...,\epsilon_n$ are independent standard normal random variates, g is a known "variance function", z_i is a vector of independent variables possibly equal to x_i or a subvector of x_i, and ξ is an unknown parameter vector. Examples of $g(z_i,\beta,\xi)$ are

$$g(z_i,\beta,\xi) = z_i^\xi, \tag{4.2}$$

where z_i is a scalar and g does not depend on β, and

$$g(z_i,\beta,\xi) = f(x_i,\beta)^\xi \tag{4.3}$$

where $z_i = x_i$ and the conditional variance of y_i is proportional to a power of the mean. Now let $\theta = (\beta,\xi,\sigma)$. Up to an additive constant, the log—likelihood is

$$L(\theta) = -n \log(\sigma) - \sum_{i=1}^{n} \log g(z_i,\beta,\xi) - \frac{1}{2} \sum_{i=1}^{n} \left[\frac{y_i - f(x_i,\beta)}{\sigma g(z_i,\beta,\xi)} \right]^2. \tag{4.4}$$

$L(\theta)$ can be maximized jointly in β, ξ, and σ. Sheiner and Beal (1985) call the MLE "extended least—squares" because, like least—squares, it is consistent even for data that are not normally distributed. The MLE is not equivalent to generalized least–squares. The GLS estimator has certain robustness properties that the MLE lacks (Carroll and Ruppert 1982), but the MLE is more efficient than GLS when the data are *exactly* normal and the conditional variance of y_i is given *exactly* by $g(z_i,\beta,\xi)$ (Jobson and Fuller 1980).

The GLS can be computed by the following algorithm: (1) Compute the unweighted LS estimate $\hat{\beta}_{(0)}$. (2) Fix β equal to $\hat{\beta}_{(0)}$ in (4.4) and maximize over σ and ξ; call the maximizers $\hat{\sigma}$ and $\hat{\xi}$; (3) Using $\{g(z_i,\hat{\beta}_{(0)},\hat{\xi})^{-2}\}$ as weights, compute the weighted least–squares estimator of β and call it $\hat{\beta}$; (4) Let $\hat{\beta}_{(0)} = \hat{\beta}$ and go back to (2). Iterate until convergence. The details of this algorithm are discussed in Giltinan and Ruppert (1987), where sample SAS programs are given.

The theory of estimating variance functions is given by Davidian and Carroll (1987). Full discussion of the available methods and several examples appear in Carroll and Ruppert (1988, chapter 3).

5. Transformation/Weighting Models

The TBS method was applied to stock–recruitment data in Carroll and Ruppert (1984). Later (Ruppert and Carroll 1985) it was realized that a certain transformation/weighting model contains as special cases several standard methods for fitting the Beverton–Holt stock–recruitment model. Interestingly, this model is the same equation as the Michaelis–Menten model of enzyme kinetics. This served as motivation for a general transformation/weighting model

$$h_\lambda(y_i) = h_\lambda(f(x_i,\beta)) + \sigma g(z_i,\beta,\xi)\epsilon_i, \tag{5.1}$$

where $\epsilon_1,...,\epsilon_n$ are independent standard normal random variables. According to model (5.1), y_i can be transformed to normally distributed errors by $h_\lambda(\cdot)$, but the errors are heteroscedastic with conditional variance $\sigma^2 g^2(z_i,\beta,\xi)$. Model (5.1) is extremely flexible. It separates the mean, or more precisely the median, structure from distributional shape and the variance structure: The conditional median of y_i is given by $f(x_i,\beta)$, distributional shape is specified by $h_\lambda(\cdot)$, and $g^2(z_i,\beta,\xi)$ specifies the variance after transformation.

In the Michaelis–Menten or Beverton–Holt model x is scalar, $\beta = (\beta_1,\beta_2)$, and

$$f(x,\beta) = \frac{1}{\beta_1+\beta_2/x} = \frac{x}{\beta_2+\beta_1 x}. \tag{5.2}$$

Various fitting methods that have appeared in the literature are described in Carroll, Cressie, and Ruppert (1987). Each is a special case of model (5.1) with $h_\lambda(y) = y^{(\lambda)}$, $z_i = x_i$, and

$$h(z_i,\beta,\xi) = x_i^\xi \tag{5.3}$$

for some ξ. The different methods are distinguished simply by different values of λ and ξ.

There has been considerable controversy on how to best fit the Michaelis–Menten model. Some studies (see Currie, 1982) simulate a single, or only a few, error structures. (An error structure is a single value of (λ,ξ).) Not surprising, the method which is the MLE for the simulated error structure shows up as best in the study. What is surprising is that the author concludes that this method is preferred in practice, although no evidence is given that the error structure used in the simulations actually obtains in practice!

Using (5.1)/(5.3), Carroll, Cressie, and Ruppert (1987) show that error structures vary widely, not only between subject matters but even between similar experiments from the same laboratory. Model (5.1)/(5.3) is effective for identifying error structure and results in an efficient estimate of β. Carroll, Cressie, and Ruppert (1987) give an example showing that the TBS model can estimate β with considerably more accuracy than methods currently in use.

The log–likelihood under (5.1) is

$$L(\beta,\lambda,\xi,\sigma) = -n \log \sigma - \sum_{i=1}^{n} \log g(z_i,\beta,\xi) \tag{5.4}$$

$$-\frac{1}{2} \sum_{i=1}^{n} \left[\frac{h_\lambda(y_i) - h_\lambda(f(x_i,\beta))}{\sigma g(z_i,\beta,\xi)} \right]^2 + \sum_{i=1}^{n} \log(J_i(\lambda)).$$

The log-likelihood can be maximized by methods similar to those in sections 3 and 4. Inferential procedures for model (5.1) are discussed in Carroll and Ruppert (1988, chapter 5).

Kettl (1987) has studied model (5.1) with power transformations and power of the mean (POM) weighting, that is, with $x_i = z_i$ and

$$g(x_i,\beta,\xi) = f(x_i,\beta)^\xi. \tag{5.5}$$

Since a power transformation can remove heteroscedasticity precisely of the form (5.5), the TBS model with POM weighting appears redundant at first. However, combining a power transformation with power of the mean weighting has the advantage of allowing the transformation parameter to be chosen primarily to remove skewness. Often the transformation parameter inducing symmetry is substantially different from the one inducing a constant variance. Then power transformation combined with POM weighting can result in a significantly better fit than either alone, particularly for larger data sets; see Rudemo et al. (1987) for an example with 150 observations. Another advantage of a model combining a power transformation with POM weighting is that a context is provided for testing whether power transformation alone fits better than POM weighting alone, or vice versa.

Rudemo et al. (1987) have introduced two models that derive the function $g(z_i,\beta,\xi)$ from a random–effects model for x. Since their two models are similar to each other, we will only describe the first, which uses Berkson's (1950) "controlled variate" model.

Suppose $x = z$ is univariate and that x is the "target value" of the independent vari-
able, while the actual value of the independent variable is $\tilde{x} = x + \sigma_X \delta$. Here δ is a random
variable with mean 0 and variance 1, so that σ_X^2 is the conditional variance of $\tilde{x}$ given x.
An example would be bioassay where x is intended dose and $\tilde{x}$ is actual dose that varies
from x because of measurement errors. Only x, not $\tilde{x}$, is known. Assume further that

$$\sigma_X = \sigma_1 x^{1-\Delta} \tag{5.6}$$

for some σ_1 and Δ, and that the TBS model

$$h_\lambda(y) = h_\lambda(f(\tilde{x},\beta)) + \sigma\epsilon \tag{5.7}$$

holds with the independent variable equal to its actual value, $\tilde{x}$. We, of course, do not know
$\tilde{x}$ and fit the model with the independent variable equal to x. This introduces another
source of error that we will now study. Let $\dot{h}_\lambda(y) = (d/dy)h_\lambda(y)$. Using a Taylor
approximation, we have

$$h_\lambda(y) = h_\lambda(f(x,\beta)) + [\dot{h}_\lambda(f(x,\beta))][\tfrac{\partial}{\partial x} f(x,\beta)]\sigma_X \delta + \sigma\epsilon. \tag{5.8}$$

From (5.6) and (5.8), Rudemo et al. (1987) derive

$$g^2(x,\beta,\Delta) = 1 + Ax^{2-2\Delta}[\dot{h}_\lambda(f(x,\beta)) \tfrac{\partial}{\partial x} f(x,\beta)]^2$$

where $A = (\sigma_1/\sigma)^2$. The parameters in such theoretical variance models will often have
biological interpretations and may allow comparisons between the amounts of variability from
different sources (δ versus ϵ). Such comparisons could lead to important biological insight
or, at least, may be relevant for future experimental designs.

6. Other Developments

Until now, only maximum likelihood estimation of β and of the transformation and
variance parameters λ and ξ has been mentioned. There are several reasons why other
estimators might be used. First, MLEs are notoriously sensitive to bad data or slight
violations of the model assumptions. For example, a single gross error, say due to a
measurement or recording error, can ruin the least–squares estimate of a regression
parameter, and this estimator is, of course, the MLE for normal errors. Moreover, there is no
guarantee that within a given transformation family there exists a transformation to
homoscedastic, normal errors. It might be better to search for a transformation to errors that
are either symmetric or homoscedastic, but not necessarily both.

Carroll and Ruppert (1987) introduce a robust estimator for the TBS model. Their estimator is essentially a weighted maximum likelihood estimator. The weights depend on the data in such a way that observations that are outlying or have an unusually large influence on the MLE are automatically downweighted. Giltinan, Carroll, and Ruppert (1986) introduce similar estimators for weighting models of form (4.1).

Ruppert and Aldershof (1987) extend the transform—to—symmetry estimators of Hinkley (1975) and Taylor (1985) to the transform—both—sides model. Such estimators only attempt to remove skewness, not heteroscedasticity. Ruppert and Aldershof also define estimators that attempt to remove only heteroscedasticity, not skewness. These are defined through a measure of correlation between the squared residuals and the estimated mean responses. Using both these "skewness" and "heteroscedasticity" estimators, they develop a test of the null hypothesis that a transformation exists to both symmetric and homoscedastic errors. Finally, in the case that this null hypothesis is true, they show how to optimally combine the skewness and heteroscedasticity estimators.

Acknowledgment

My research has been supported by NSF grants DMS–8400602 and DMS–8701201.

References

Bartlett, M.S. (1947). The use of transformations. *Biometrics, 3*, 39–52.

Berkson, J. (1950). Are there two regressions? *J. Amer. Statist. Assoc., 45*, 164–180.

Box, G.E.P. and Cox, D.R. (1964). An analysis of transformations. *J. Royal Statist. Soc., B, 26*, 211–246.

Box, G.E.P. and Hill, W.J. (1974). Correcting inhomogeneity of variance with power transformation weighting. *Technometrics, 16*, 359–389.

Carroll, R.J. and Ruppert, D. (1982). A comparison between maximum likelihood and generalized least–squares in a heteroscedastic linear model. *J. Amer. Statist. Assoc., 77*, 878–882.

Carroll, R.J. and Ruppert, D. (1984). Power transformations when fitting theoretical models to data. *J. Amer. Statist. Assoc., 79*, 321–328.

Carroll, R.J. and Ruppert, D. (1987). Diagnostics and robust estimation when transforming the regression model and the response. *Technometrics, 29*, 287–299.

Carroll, R.J. and Ruppert, D. (1988). *Transformation and Weighting in Regression.* Chapman and Hall (to appear).

Carroll, R.J., Cressie, N.A.C., and Ruppert, D. (1987). A transformation/weighting model for estimating Michaelis–Menten parameters. Statistical Laboratory Preprint Series, Preprint #87–20. Iowa State University.

Currie, D.J. (1982). Estimating Michaelis–Menten parameters: bias, variance and experimental design. *Biometrics, 38,* 907–919.

Davidian, M. and Carroll, R.J. (1987). Variance function estimation. *J. Amer. Statist. Assoc., 82,* 1079–1092.

Efron, B. (1979). Bootstrap methods: another look at the jackknife. *Ann. Stat., 7,* 1–26.

Efron, B. (1982). *The Jackknife, the Bootstrap, and Other Resampling Plans.* CBMS–NSF Monograph 39. SIAM, Philadelphia.

Gallagher, D. (1986). *The Application of Data Based Transformations to Parameter Estimation in Water Quality Models.* Ph.D. dissertation. Department of Environmental Sciences and Engineering. School of Public Health. University of North Carolina at Chapel Hill.

Giltinan, D., Carroll, R.J., and Ruppert, D. (1986). Some new methods for weighted regression when there are possible outliers. *Technometrics, 28,* 219–230.

Giltinan, D. and Ruppert, D. (1987). Fitting Heteroscedastic Regression Models to Individual Pharmacokinetic Data Using Standard Statistical Software. Technical Report No. 759. School of Operations Research and Industrial Engineering, Cornell University.

Hinkley, D.V. (1975). On power transformations to symmetry. *Biometrika, 62,* 101–111.

Jobson, F.D. and Fuller, W.A. (1980). Least–squares estimation when the covariance matrix and the parameter vector are functionally related. *J. Amer. Statist. Assoc., 75,* 176–181.

Kettl, E. (1987). Some Applications of the Transform–Both–Sides Regression Model. Ph.D. Dissertation, Dept. of Statistics, Univ. of North Carolina at Chapel Hill.

McCullagh, P. and Nelder, J.A. (1983). *Generalized Linear Models.* Chapman and Hall: New York.

Patefield, W.M. (1977). On the maximized likelihood function. *Sankhya, B,* 39, 92–96.

Rudemo, M., Ruppert, D., and Streibig (1987). Random effects models in nonlinear regression with applications to bioassay. Mimeo Series #1727. Dept. of Statistics, Univ. of North Carolina at Chapel Hill.

Ruppert, D. and Aldershof, B. (1987). Transformations to symmetry and homoscedasticity. Manuscript.

Ruppert, D. and Carroll, R.J. (1985). Data transformations in regression analysis with applications to stock recruitment relationships. In *Resource Management: Lecture Notes in Biomathematics 61,* M. Mangel, editor. Springer Verlag, New York.

Sheiner, L.B. and Beal, S.L. (1985). Pharmacokinetic parameter estimates from several least–squares procedures: superiority of extended least–squares. *J. Pharmacokin. Biopharm. 13,* 185–201.

Snee, R.D. (1986). An alternative approach to fitting models when reexpression of the response is useful. *J. of Quality Technology, 18,* 211–225.

Taylor, J.M.G. (1985). Power transformations to symmetry. *Biometrika, 72,* 145–152.

van Zwet, W.R. (1964). *Convex Transformations of Random Variables.* Mathematisch Centrum, Amsterdam.

INVERSE PROBLEMS FOR DISTRIBUTED SYSTEMS:
STATISTICAL TESTS AND ANOVA

H. T. Banks and B. G. Fitzpatrick
Center for Control Sciences
Division of Applied Mathematics
Brown University
Providence, RI 02912

Abstract: In this note we outline some recent results on the development of a statistical testing methodology for inverse problems involving partial differential equation models. Applications to problems from biology and mechanics are presented. The statistical tests are based on asymptotic distributional results for estimators and residuals in a least squares approach.

1. Introduction

In this note we present a summary of initial efforts in our attempts to develop a rigorous foundation for statistical testing of results for inverse problems involving distributed parameter systems. Our approach is in the spirit of regression analysis or analysis of variance (ANOVA). Before presenting the details of our model comparison statistical methods, we present a brief overview of some of the motivation and related questions involving estimation problems that underlie our efforts.

Typical estimation problems involve finding maximum likelihood or least squares estimators for parameters (possibly space and time dependent), such as $\lambda, \mathcal{V},$ and $\mathcal{D}$, in models described by transport type equations

$$\frac{\partial u}{\partial t} + \frac{\partial}{\partial x}\left(\mathcal{V}u\right) = \frac{\partial}{\partial x}\left(\mathcal{D}\frac{\partial u}{\partial x}\right) + \lambda u$$

with appropriate boundary and initial conditions. These problems arise in diverse biological applications, including transport of labeled substances in brain tissue [BK], insect dispersal in heterogeneous environments [BK],[BKL], climatology [DBW], and bioturbation [BR]. Models for more complex phenomena involve systems with nonlinear and semilinear equations of the form

$$\frac{\partial u_i}{\partial t} + \frac{\partial}{\partial x}\left(\mathcal{V}(t,x,u_i)u_i\right) = \frac{\partial}{\partial x}\left(\mathcal{D}(t,x,u_i)\frac{\partial u_i}{\partial x}\right) + f_i(\lambda,\underline{u})$$

for the population densities $\underline{u} = (u_1, u_2)$ arising in predator–prey interactions involving nonlinear growth and predation and density dependent dispersal [BKM], [O].

In addition to problems involving these parabolic systems, there are important inverse problems in size/age structured population studies [MD],[O], which entail estimation of coefficients in hyperbolic systems of partial differential equations. Often these models involve some type of stochastic growth assumption. For example, one class of inverse problems [BBKW] requires estimation of a random growth rate g, as well as mortality μ and fecundity k, in a stochastic version of the McKendrick/Von Foerster type model with renewal boundary conditions:

$$\frac{\partial u}{\partial t} + \frac{\partial}{\partial x}\Big(g(t,x)u\Big) = -\mu u, \qquad x_0 < x < x_1,$$

$$g(t,x_0)u(t,x_0) = \int_{x_0}^{x_1} k(t,\xi)u(t,\xi)\,d\xi$$

$$g(t,x_1)u(t,x_1) = 0.$$

Under slightly different assumptions (a Markov transition process for evolution in size/age classes), one must estimate the first and second moments g_1 and g_2 of the Markov process in the Fokker–Planck model [B2]

$$\frac{\partial u}{\partial t} + \frac{\partial}{\partial x}\Big(g_1(t,x)u\Big) = \frac{\partial^2}{\partial x^2}\Big(g_2(t,x)u\Big) - \mu u, \qquad x_0 < x < x_1,$$

$$\left[g_1 u - \frac{\partial}{\partial x}(g_2 u)\right]^{x=x_0} = \int_{x_0}^{x_1} k(t,\xi)u(t,\xi)\,d\xi$$

$$\left[g_1 u - \frac{\partial}{\partial x}(g_2 u)\right]^{x=x_1} = 0.$$

The statistical methods we outline below are also quite useful in problems arising in applications other than those of a biological nature. For example, we have made use of these ideas in our studies [BWIC],[BFW] of damping mechanisms in composite material beams. In these investigations the typical model involves the Euler–Bernoulli equation with damping

$$\rho\frac{\partial^2 u}{\partial t^2} + \frac{\partial^2}{\partial x^2}\Big(EI\frac{\partial^2 u}{\partial x^2}\Big) + \Sigma(u) = f, \qquad 0 < x < l.$$

Here the damping Σ contains parameters to be estimated and may be one of several forms:

viscous : $\qquad \Sigma(u) = \gamma\dfrac{\partial u}{\partial t}$

Kelvin – Voigt : $\qquad \Sigma(u) = \dfrac{\partial^2}{\partial x^2}\Big(c_D I\dfrac{\partial^3 u}{\partial x^2 \partial t}\Big)$

time hysteresis : $\qquad \Sigma(u) = -\dfrac{\partial^2}{\partial x^2}\displaystyle\int_{-r}^{0}\dfrac{\alpha e^{\beta s}}{\sqrt{-s}}\dfrac{\partial^2 u}{\partial x^2}(t+s,x)\,ds$

spatial hysteresis : $\Sigma(u) = -\dfrac{\partial}{\partial x}\displaystyle\int_{0}^{l} h(x,\xi)\Big(\dfrac{\partial^2 u}{\partial x \partial t}(t,x) - \dfrac{\partial^2 u}{\partial \xi \partial t}(t,\xi)\Big)\,d\xi.$

Common features of all of these applications include a partial differential equation model and boundary conditions containing parameters that are possibly spatially and time dependent and which must be estimated using observations of the system. Mathematically, we have a parameter dependent dynamical system

$$\dot{u} = \mathcal{A}(t, x, q, u) \tag{1}$$

for the states u in a state space $\mathcal{H}$ and parameters q in a parameter space Q. From observations $\{Y_k\}$ of some function $\{\Gamma(u)_k\}$ of the states (e.g. , we might have $\{u(t_i, x_j)\}$ or $\{u_t(t_i, x_j)\}$, $k = (i, j)$),we wish to estimate the parameters q using a least squares fit to the observations. That is, we seek to solve the problem of minimizing

$$J(q) = \sum_{k=1}^{n} |Y_k - \Gamma(u(q))_k|^2$$

over $q \in Q$, where $u(q)$ is the solution of (1).

In general such problems will involve both an infinite dimensional state space $\mathcal{H}$ and an infinite dimensional parameter space Q, although in many cases of practical interest an a priori parameterization of the elements of Q can lead to problems with finite dimensional parameter space. The difficulties associated with these aspects of the inverse problems have led to a number of studies (see [BKC],[BL],[BCR], [BI] for a sample of some of these) of approximation ideas and computational techniques for solution of the minimization problem. Conceptually, one approximates the state space $\mathcal{H}$ by a family $\mathcal{H}^N$ of finite dimensional state spaces, the parameter space Q by a family of finite dimensional parameter spaces Q^M, and the system (1) by a family of systems

$$\dot{u}^N = \mathcal{A}^N(t, x, q^M, u^N) \tag{2}$$

for approximate states $u^N \in \mathcal{H}^N$ and approximate parameters $q^M \in Q^M$. One then seeks to minimize

$$J^N(q^M) = \sum_{k=1}^{n} |Y_k - \Gamma(u^N(q^M))_k|^2$$

over $q^M \in Q^M$ subject to $u^N(q^M)$ being a solution of (2).

The general question of convergence and stability related to these approximations are important and have been the subject of intensive efforts over the past several years. The question of *convergence of the methods* is as follows: If $\hat{q}^{N,M}$ and $\hat{q}$ denote the solutions of the problems for J^N with (2) and J with (1) as denoted above, then can we guarantee the convergence of the estimators $\hat{q}^{N,M} \rightarrow \hat{q}$, as $N, M \rightarrow \infty$? *Method stability* or continuous dependence of estimators on observations can be posed as follows: If $\{Y^P\}$ is a sequence of observations and $\hat{q}^{N,M}(Y^P)$ the corresponding approximate estimators, does $Y^P \rightarrow Y$ as $P \rightarrow \infty$ guarantee $\hat{q}^{N,M}(Y^P) \rightarrow \hat{q}(Y)$ as $N, M, P \rightarrow \infty$?

The answers to these questions depend, of course, on the approximation properties of $\mathcal{H}^N, Q^M, \mathcal{A}^N$ for $\mathcal{H}, Q, \mathcal{A}$ respectively, on the system being approximated, and on the type of observations (i.e., the mapping Γ above) taken. A rather general theoretical framework which facilitates these analyses has recently been given in [BI]. This framework permits us to establish affirmative answers to the convergence and stability questions for many of the systems from biology and mechanics outlined above, in the cases one uses spline or spectral families in the approximations $\mathcal{H}^N, Q^M$ for $\mathcal{H}, Q$ and a Galerkin–like procedure in approximating the system operator A by A^N. A nonlinear version of the framework is under development, with initial findings reported in [BRR]. The entire theoretical approximation efforts are based on abstract semigroup results, the theory of sesquilinear forms (dissipative and coercive or maximal monotone operators), and compactness properties of Q that are in some sense equivalent to a Tychonov regularization of the inverse problems. For more precise details, the reader should consult [B1], [BI], [BI2], [BRR].

Other important aspects of the inverse problems that are the focus of our attention here include algorithm development and testing, especially in regard to schemes for use on vector and parallel computers. These efforts are particularly important in dealing with nonlinear systems or systems with delays (such as the time hysteresis damping models for beams as mentioned above). We defer any discussions of these issues and turn to the main task at hand: the development of statistical methods for model comparison and validation.

2. Model Comparison and Validation

In many applications of least squares parameter identification, we do not expect to achieve a zero residual (i.e., $J(\widehat{q}) = 0$ in the notation of the previous discussions). Among the reasons for this are that the data may include observation errors, and that the PDE model itself is usually only an approximation of the actual physical or biological process. Given several competing models and their corresponding least squares estimators, it is important to develop quantitative measures which tell us which model fits better. Direct comparison of the residuals is not always easy or useful. We shall outline our investigations of statistical tests from a form of nonlinear regression analysis. Nonlinear analysis will be necessary: even if a PDE model is linear in the unknown parameters, the resulting solution will in general be a nonlinear function of the parameters.

We consider the collection of noisy observations:

$$Y_k = g(x_k) + \varepsilon_k, \quad 1 \le k \le n,$$

where $\{Y_k\}$ is the collection of observations, $g: X \to \mathbf{R}$ is an unknown continuous function, $\{x_k\}$ is a collection of settings at which measurements are made ($x_k \in X \subset \mathbf{R}^m$), and $\{\varepsilon_k\}$ is the noise process. We also have a parameterized function $f(x, q)$ ($\Gamma(u(q))$ in the notation above) to which we wish to fit our observations. In our efforts, the function f arises from a parameter-dependent differential equation which is derived from physical considerations of the system studied, and depending on our measuring instruments, we may have $f = u$,

$f = u_t$, $f = u_{tt}$, or f may be some other function of the solution to our differential equation. To estimate the unknown parameter q, we shall use the least squares cost functional

$$J_n^N(q) = \frac{1}{n} \sum_{k=1}^{n} (Y_k - f^N(x_k, q))^2, \quad q \in Q_{ad}.$$

The superscript N denotes (as above) the approximation that we must make in most applications, for a minimizer of J_n^N over Q_{ad}. We note that g may or may not be representable as $g(x) = f(x, q^{**})$, for some q^{**} (that is, our model may or may not be exactly related to the process that actually generated the observations).

We now list the assumptions we make for the problem, and the results that we have derived. More detailed discussions and proofs may be found in [F],[BF].

(A1) The infinite sequence $\{\varepsilon_k\}$ is composed of independent, identically distributed random variables on a probability space $(\Omega, \mathcal{F}, P)$. Furthermore, $E(\varepsilon_k) = 0$ and $Var(\varepsilon_k) = \sigma^2 < \infty$.

(A2) The function $f: Q \rightarrow C(X)$ is a continuous function. The space Q is a separable topological space, and X is a compact subset of $\mathbf{R}^m$. Also, $g: X \rightarrow \mathbf{R}$ is continuous. The admissible set Q_{ad} is a compact subset of Q. We shall write $f(x, q)$, rather than $f(q)(x)$.

(A3) Our observations correspond to the sequence $\{x_k\}$ in X. There exists a finite measure μ on X such that

$$\frac{1}{n} \sum_{k=1}^{n} h(x_k) \rightarrow \int_X h \, d\mu$$

for each continuous h.

(A4) The functional

$$J^*(q) = \sigma^2 + \int_X (g(x) - f(x, q))^2 \, d\mu(x)$$

has a unique minimizer in Q_{ad} at q^*. Also, the functional

$$J_n^0(q) = \sigma^2 + \frac{1}{n} \sum_{k=1}^{n} (g(x_k) - f(x_k, q))^2$$

has a unique minimizer in Q_{ad} at q_n^0.

(A5) The mapping $f^N: Q \rightarrow C(X)$ is continuous for each N, with $f^N \rightarrow f$, uniformly on each compact subset of Q, as $N \rightarrow \infty$.

Lemma 1: Under (A1)—(A5), $J_n^N \rightarrow J^*$, as $n, N \rightarrow \infty$, with probability one. The convergence is uniform on compact subsets of Q.

Theorem 1: Under (A1)—(A5), $q_n^N \rightarrow q^*$ with probability one, as $N, n \rightarrow \infty$.

This result tells us that, as we make better approximations and take more data, the estimators we get converge to the "true" parameter. Here, "true" parameter means that parameter which provides the best $L^2(X,\mu)$ fit of f to g. It is interesting to note the case of model-generated data; that is, there exists $q^{**} \in Q_{ad}$ such that $g(\cdot) = f(\cdot, q^{**})$. If μ is such that $\int_X (\phi - \psi)^2 \, d\mu = 0 \Rightarrow \phi = \psi$, the uniqueness we require for q^* is trivially verified. We have $q^* = q^{**}$, and $q_n^0 = q^* = q^{**}$, and the data has the form

$$Y_k = f(x_k, q^*) + \varepsilon_k, \quad 1 \le k \le n.$$

Thus, the estimators really do converge to the "true" parameter (that is, the parameter that generated the data).

We now consider the question of hypothesis testing. The test in which we shall be interested is whether or not $Hq^* = h$, where H is a known linear function, and h is a known vector. This general setup includes many applications of interest.

We put $Q_0 = \{q \in Q_{ad} : Hq = h\}$. The test statistic we shall use is $T_n^N = n(J_n^N(\tilde{q}_n^N) - J_n^N(\hat{q}_n^N))$, where $\tilde{q}_n^N$ is a minimizer of J_n^N, subject to the constraint $q \in Q_0$. One justification for the use of this statistic is that it is easily computed from the estimators : in fact, $J_n^N(\tilde{q}_n^N)$ and $J_n^N(\hat{q}_n^N)$ are usually by-products of the optimization routine used to compute $\tilde{q}_n^N$ and $\hat{q}_n^N$. (Recall that the superscript N denotes use of f^N, rather than f, in the cost functional.) Another justification is that the statistic approximates the likelihood ratio statistic in the case of normal errors and model-generated data. In this case, we can write the joint density of $Y_1, \ldots, Y_n$:

$$p(y_1, \ldots, y_n; q) = \left(\frac{1}{2\pi\sigma^2}\right)^{\frac{n}{2}} \exp\left(-\sum_{k=1}^{n} (y_k - f(x_k, q))^2 / (2\sigma^2)\right),$$

Now, the likelihood ratio test, which is uniformly most powerful over a large class of tests (see, for example, [L]), is given by

$$
\begin{aligned}
L_n &= \frac{\sup\{p(Y_1, \cdots, Y_n; q) : q \in Q_0\}}{\sup\{p(Y_1, \cdots, Y_n; q) : q \in Q_{ad}\}} \\
&= \frac{\exp\{-\frac{1}{2\sigma^2} \sum_{k=1}^{n} (Y_k - f(x_k, \tilde{q}_n))^2\}}{\exp\{-\frac{1}{2\sigma^2} \sum_{k=1}^{n} (Y_k - f(x_k, \hat{q}_n))^2\}} \\
&= \exp\{-\frac{n}{2\sigma^2}(J_n(\tilde{q}_n) - J_n(\hat{q}_n))\},
\end{aligned}
$$

where $J_n(q) = \frac{1}{n}\sum_{k=1}^{n}(Y_k - f(x_k, q))^2$ is the cost without approximation.

The likelihood ratio test is performed in the following manner: given $\alpha \in (0,1)$, choose t so that $\sup\{P_q(L_n < t) : q \in Q_0\} \le \alpha$, where P_q denotes the fact that we use the $p(\cdot; q)$ density to compute the event's probability. Note that L_n compares the best the hypothesis has to offer the data with the best that the whole space can do, best meaning maximum likelihood. Since this statistic depends on f, we can not compute it directly. Thus, we use the statistic T_n^N given above. In the following theorems, it will become apparent why we consider the log of the likelihood ratio, rather than the likelihood ratio itself. We now attend to determining a limiting $(n, N \to \infty)$ distribution of T_n^N, which can be used to approximate the probabilities for performing the test.

We add some assumptions which will help us achieve the results of interest.

(A7) The space Q is finite dimensional; i.e., $Q \subset \mathbf{R}^p$, and $q^* \in \text{Int} Q_{ad} \subset \text{Int} Q$.

(A8) The mapping $f: Q \to C(X)$ is a C^2 function.

(A9) As $N \to \infty$, we have $\partial^2 f^N / \partial q^2 \to \partial^2 f / \partial q^2$, uniformly on $X \times Q_{ad}$.

(A10) The matrices $\mathcal{J} = \partial^2 J^*(q^*) / \partial q^2$, and

$$V = 4\sigma^2 \int_X \frac{\partial f(x, q^*)}{\partial q} \frac{\partial f(x, q^*)}{\partial q}^T d\mu(x)$$

are positive definite.

Choose $q_n^{0,N}$ as a minimizer of

$$J_n^{0,N}(q) = \frac{1}{n} \sum_{k=1}^n (g(x_k) - f^N(x_k, q))^2 + \sigma^2.$$

Note that one consequence of (A5) is that $J_n^{0,N} \to J^*$ uniformly on Q_{ad}, as $N, n \to \infty$. Under (A1)—(A5), then, we see that $q_n^{0,N} \to q^*$, as $N, n \to \infty$.

Theorem 2: Under (A1)—(A5),(A7)—(A10), we have

$$\sqrt{n} \frac{\partial J_n^N(q_n^{0,N})}{\partial q} \xrightarrow{D} N(0, V).$$

(Here $\xrightarrow{D}$ denotes convergence in distribution.)

Theorem 3: Assume (A1)—(A5),(A7)—(A10), and that H_0 is true. Then, we have

$$\sqrt{n}(\tilde{q}_n^N - \hat{q}_n^N) \xrightarrow{D} N(0, \mathcal{J}^{-1} V' \mathcal{J}^{-1}), \qquad \text{as } N, n \to \infty,$$

where V' is given by

$$V' = H^T (H \mathcal{J}^{-1} H^T)^{-1} H \mathcal{J}^{-1} V \mathcal{J}^{-1} H^T (H \mathcal{J}^{-1} H^T)^{-1} H.$$

Theorem 4: Under (A1)—(A5),(A7)—(A10), and assuming that H_0 is true, we have that $T_n^N \xrightarrow{D} \frac{1}{2} Z^T \mathcal{J} Z$, where Z has a $N(0, \mathcal{J}^{-1} V' \mathcal{J}^{-1})$ distribution.

Note that the limit distributions depend on g through $\mathcal{J}$, and on σ^2 through V. Some remarks are in order here.

Remark (i): In the case of model–generated data, the above limit theorems become much simpler. We notice that $\mathcal{J}$ is given by

$$\mathcal{J} = 2 \int_X \left[\frac{\partial f(x, q^*)}{\partial q} \frac{\partial f(x, q^*)}{\partial q}^T + \frac{\partial^2 f(x, q^*)}{\partial q^2} (g(x) - f(x, q^*)) \right] d\mu(x).$$

When we have $g(\cdot) = f(\cdot, q^*)$, the first term is just $V/2\sigma^2$, and the second term is 0, yielding $\mathcal{J} = V/2\sigma^2$. This simplification leads to

$$V' = H^T (H\mathcal{J}^{-1}H^T)^{-1} H \cdot 2\sigma^2.$$

We now note that $\mathcal{J}(\mathcal{J}^{-1}V'\mathcal{J}^{-1}) = V'\mathcal{J}^{-1}$, so that

$$\left[\frac{1}{2\sigma^2} \mathcal{J}(\mathcal{J}^{-1}V'\mathcal{J}^{-1}) \right]^2$$

$$= \left(H^T(H\mathcal{J}^{-1}H^T)^{-1}H\mathcal{J}^{-1} \right) \left(H^T(H\mathcal{J}^{-1}H^T)^{-1}H\mathcal{J}^{-1} \right)$$

$$= H^T(H\mathcal{J}^{-1}H^T)H\mathcal{J}^{-1} = \frac{1}{2\sigma^2} V'\mathcal{J}^{-1}.$$

This calculation shows that $\frac{1}{2\sigma^2}\mathcal{J}(\mathcal{J}^{-1}V'\mathcal{J}^{-1})$ is idempotent, with rank r. This tells us that $\frac{1}{2\sigma^2}Z^T\mathcal{J}Z$ has a $\chi^2(r)$ distribution (see [G]). One important aspect of this result is that the limiting distribution, which we shall use to compute $P(T_n^N > t)$, is independent of any unknown parameters. In case σ^2 is unknown, we can use $J_n^N(\widehat{q}_n^N)$ as a consistent estimator, for $J_n^N(\widehat{q}_n^N) \to J^*(q^*) = \sigma^2$. In this case, Slutsky's theorem [B3] yields

$$U_n^N = n \frac{J_n^N(\widehat{q}_n^N) - J_n^N(\widehat{q}_n^N)}{J_n^N(\widehat{q}_n^N)} \xrightarrow{\mathcal{D}} \frac{Z^T\mathcal{J}Z}{2\sigma^2} \sim \chi^2(r).$$

Note that, as a ratio of the reduction in residual to the residual, the statistic U_n^N is of the same form as the statistic used in ANOVA: see [G]. To test the hypothesis, then, we choose a significance level, α. Next, we choose a threshold, t, so that $P(\chi^2(r) > t) = \alpha$. Finally, we compare the statistic $U_n^N = T_n^N/J_n^N(\widehat{q}_n^N)$ to t. If $U_n^N > t$, we reject the hypothesis as false; otherwise, we accept. Note that if we were to perform the test over and over B times, we would see (by the law of large numbers) that

$$\frac{1}{B} \sum_{b=1}^{B} [\# \text{ of } U_n^N\text{'s} > t] \to P(U_n^N > t), \qquad B \to \infty.$$

Remark (ii): If we do not have $g(\cdot) = f(\cdot, q^*)$, computation of the threshold t becomes more difficult, because the limit distribution now depends on the unknown quantities g and q^*. We can, however, replace the matrices V' and $\mathcal{J}$ with estimators $V_n'^N$ and $\mathcal{J}_n^N$, where

$$V_n'^N = 4\sigma^2 \frac{1}{n} \sum_{k=1}^{n} \frac{\partial f^N(x_k, \widehat{q}_n^N)}{\partial q} \frac{\partial f^N(x_k, \widehat{q}_n^N)^T}{\partial q} \to V',$$

and

$$\mathcal{J}_n^N = \frac{\partial^2 J_n^N(\hat{q}_n^N)}{\partial q^2} \to \mathcal{J}.$$

However, if σ^2 is also unknown, it must be estimated as well. One way of doing this is the direct estimation of g.

Suppose that g is known to lie in a set $\mathcal{K}$ which is compact in $C(X)$. Then, the least squares theory outlined above can be applied. Put $\bar{J}(\phi) = \frac{1}{n}\sum_{k=1}^{n}(Y_k - \phi(x_k))^2$, and $\bar{Q}_{ad} = \mathcal{K}$. If the $\{x_k\}$ sequence is chosen so that

$$\int (\phi(x) - \psi(x))^2 \, d\mu(x) = 0 \Rightarrow \phi = \psi,$$

for $\phi, \psi \in \mathcal{K}$, then (A1)—(A4) are satisfied. Furthermore, by using piecewise linear splines (or some other suitable approximation technique), we can find an estimator $\hat{g}_n^M$, which converges to g almost surely. Then, replace g with $\hat{g}_n^M$ in the definition of $V_n'^N$; call the replacement $V_n'^{N,M}$. Now, let $\Phi_n^{M,N}$, Φ denote normal distribution functions with zero mean vector and covariances $V_n'^{N,M}$, V', respectively. Now

$$\int I_{[z^T \mathcal{J}_n^N z > t]}(z) \, d\Phi_n^{N,M}(z) \to \int I_{[z^T \mathcal{J}z > t]}(z) \, d\Phi(z),$$

by dominated convergence. Thus, we can use these approximations to determine the threshold t.

Remark (iii): Another quantity of interest is the power of the test; that is, the ability of the test to reject false null hypotheses. The power function is defined to be

$$1 - \beta(q^*) = P(T_n^N > t | q^* \text{ is the true parameter}).$$

We have included dependence on the true value of the parameter, for if the hypothesis H_0 is false, the distribution of T_n^N may depend on the unknown parameter. This is the case, for example, in most linear statistical models (see [G]). Note that the power will depend on the threshold t set by the chosen significance level, α.

Suppose that the set Q_0 does not contain the true parameter q^*. Since Q_0 is compact, it must be the case that the $\tilde{q}_n^N$ sequence is bounded away from q^*. Now, J^* is minimized uniquely by q^*, so we have that $J_n^N(\tilde{q}_n^N) - J_n^N(\hat{q}_n^N) \geq \eta > 0$, for some η and N, n sufficiently large. Hence, $T_n^N \to \infty$ with probability one, if the hypothesis is false, and the power function goes to 1, for true parameters not in Q_0. This tells us that our test should be very good at rejecting false null hypotheses, if we have enough data and a good approximation to f.

3. Examples

We briefly summarize results for several examples in which we have applied the above theorems to actual data sets and residuals obtained using our least squares techniques.

Example 1. We consider a convection-diffusion model for fluid transport in cat brains. For details see [BK]. We take as our model

$$u_t = \mathcal{D}u_{xx} + \mathcal{V}u_x,$$

with unknown parameter $q = (\mathcal{D}, \mathcal{V})$. To test the hypothesis of no convection, we have $H_0 : \mathcal{V} = 0$, versus the alternative $H_A : \mathcal{V} \neq 0$. We form the statistic

$$U_n^N = n \frac{J_n^N(\tilde{q}_n^N) - J_n^N(\hat{q}_n^N)}{J_n^N(\hat{q}_n^N)},$$

which will have a $\chi^2(1)$ limit distribution.

We apply this test to the data summarized in Table 1 of [BK]. The first data set (K-R 1), which consists of $n = 8$ observations, yields the residuals

$$J_n^N(\tilde{q}_n^N) = 180.17, \qquad J_n^N(\hat{q}_n^N) = 106.15,$$

so that the statistic U_n^N takes the value $U_n^N = 5.579$. Now, the $\chi^2(1)$ distribution has the rejection thresholds 2.71, 3.84, and 6.64, for significance levels .10, .05, and .01, respectively. Thus, we would reject the hypothesis of no convection at levels .10 and .05, and accept at the level .01. The $P-$ value for this statistic is .0182, so any level larger than .0182 would produce rejection; any level smaller, acceptance.

Applying the same test to the data from K-R 7 (again see Table 1 of [BK]), we have $U_n^N = 15.3$, which indicates rejection at all of the above-listed significance levels. In fact, the P-value in this case is 0, to the precision of the χ^2 distribution routine used.

Example 2. We now apply the same model as in the previous example to an insect dispersal problem (see [BKL]). We again test the hypothesis $H_0 : \mathcal{V} = 0$ (no "advection"), versus $H_A : \mathcal{V} \neq 0$. For data sets ST-3 and ST-6, we have $n = 9$ observations, and the statistic calculation yields $U_n^N = 146.45$ for ST-3, and $U_n^N = 221.00$ for ST-6. These are clear-cut cases for rejecting H_0.

We should remark here that the number of observations in both of the above examples is quite low, especially for the application of asymptotic distributions. Furthermore, the residuals shown in Example 1 are large enough to make one consider if the model is correctly specified. The analysis in both examples was performed assuming $g(x) = f(x, q^*)$, and this assumption should be examined more closely.

Example 3. In identification of stiffness and damping parameters for beams, we generally have much more data, in many cases we have $n > 1000$. We include some results for a composite beam model with viscous (γ) and Kelvin-Voigt ($c_D I$) damping mechanisms (see [BWIC]). For this example, we have $n = 1022$ observations. To test $H_0 : \gamma = 0$, we compute the test statistic $U_n^N = 78$. Comparing this to the $\chi^2(1)$ thresholds again yields rejection at all levels. Testing for the Kelvin-Voigt term's importance, we have $H_0 : c_D I = 0$, and the statistic is $U_n^N = 1297$, another clear case of rejection at all levels. From these results we would conclude that both damping mechanisms are important in describing the damping present in these particular data.

Acknowledgements

This research was supported in part under grants NSF MCS 8504316, NASA NAG-1-517, AFOSR-84-0398, and AFOSR-F49620-86-C-0111. Part of this research was carried out while the first author was a visiting scientist at the Institute for Computer Applications in Science and Engineering (ICASE), NASA Langley Research Center, Hampton, VA, which is operated under NASA contract NAS1-18107.

References

[B1] Banks, H. T., On a variational approach to some parameter estimation problems, in *Distributed Parameter Systems*, Sringer Lec. Notes in Control and Info. Sci. 75 (1985),1-23.

[B2] Banks, H. T., Computational techniques for inverse problems in size structured stochastic population models, LCDS-CCS Report 87-41, Division of Applied Mathematics, Brown University, Providence, RI (1987), and Proceedings IFIP Conference on Optimal Control of Systems Governed by Partial Differential Equations, Santiago de Compostela, Spain, July 6-9, 1987, Springer-Verlag, New York.

[BBKW] Banks, H. T., L. W. Botsford, F. Kappel, and C. Wang, Modeling and estimation in size structured population models, LCDS-CCS Report 87-13, Division of Applied Mathematics, Brown University, Providence, RI (1987); Proceedings 2nd Course on Mathematical Ecology, (Trieste, December 8-12, 1986), T. Hallam et. al., editors, World Scientific Publishing Compamy, Singapore (1988),521-541.

[BCK] Banks, H. T., J. M. Crowley, and K. Kunisch, Cubic spline approximation techniques for parameter estimation in distributed systems, IEEE Trans. Automatic Control, $AC-28$ (1983),773-786

[BCR] Banks, H. T., J. M. Crowley, and I. G. Rosen, Methods for identification of material parameters in distributed models for flexible structures, Matematica Aplicada E Computacional 5, no. 2, (1986), 139-168.

[BFW] Banks, H. T., R. Fabiano, and Y. Wang, Estimation of Boltzmann damping coefficients in beam models, LCDS/CCS Technical Report 88-13, Division of Applied Mathematics, Brown University, Providence, RI (1988).

[BF] Banks, H. T., and B. G. Fitzpatrick, Statistical methods for model comparison in parameter estimation problems for distributed parameter systems, J. Math. Bio., to be submitted.

[BI] Banks, H. T., and K. Ito. A unified framework for approximation and inverse problems for distributed parameter systems, Control - Theory and Advanced Technology, 4 (1988),73-90.

[BI2] Banks, H. T., and D. W. Iles, On compactness of admissible parameter sets: Convergence and stability in inverse problems for distributed parameter systems. ICASE report 86-38, NASA Langley Research Center, Hampton, VA, 1986.

[BK] Banks, H. T., and P. Kareiva, Parameter estimation techniques for transport equations with applications to population dispersal and tissue bulk flow models, J. Math. Biol. 17 (1983), 253-272.

[BKL] Banks, H. T., P. Karieva, and P. D. Lamm, Modeling insect dispersal and estimating parameters when mark-release techniques may cause initial disturbances, J. Math. Biol. 22 (1985), 259-277.

[BKM] Banks, H. T., Kareiva, K. Murphy, Parameter estimation techniques for interaction and redistribution models: a predator-prey example, Oecologie, 74 (1987),356-362.

[BL] Banks, H. T., and P. D. Lamm, Estimation of variable coefficients in parabolic distributed systems, IEEE Trans. Auto. Contr. $AC-30$ (1985), 386-398.

[BRR] Banks, H. T., S. Reich, and I. G. Rosen, An approximation theory for the identification of nonlinear distributed parameter systems, LCDS-CCS Report 88-8, Division of Applied Mathematics, Brown University, Providence, RI (1988) and SIAM J. Cont. and Opt., submitted.

[BR] Banks, H. T., and I. G. Rosen, Numerical schemes for the estimation of functional parameters in distributed models for mixing mechanisms in lake and sea sediment cores, Inverse Problems, 3, 1987,1-23.

[BWIC] Banks, H. T., Y. Wang, D. J. Inman, and H. Cudney, Parameter identification techniques for the estimation of damping in flexible structure experiments, *Proc. 20th IEEE Conf. Dec. and Control*, Los Angeles, (1987), 1392-1395.

[B3] Billingsley, P. *Convergence of Probability Measures*, Wiley, New York, New York, 1968.

[DBW] Dexter, F., H. T. Banks, and T. Webb III, Modeling Holocene changes in the location and abundance of Beech populations in eastern North America, Rev. Palaeobotany and Palynology 50 (1987),273-292.

[F] Fitzpatrick, B. G., Statistical methods in parameter identification and model selection, Ph. D. Thesis, Division of Applied Mathematics, Brown University, Providence, RI, 1988.

[G] Graybill, F. *Theory and Application of the Linear Model*, Duxbury, North Scitaute, Massachusetts, 1976.

[L] Lehmann, E. L. *Testing Statistcal Hypotheses*, Wiley, New York, New York, 1986.

[MD] Metz, J. A. J., and O. Diekmann, *The Dynamics of Physiologically Structured Populations*, Springer Lecture Notes in Biomathematics, 68 (1986).

[O] Okubo, A., *Diffusion and Ecological Problems: Mathematical Models*, Springer-Verlag, New York, 1980.

SMALL MODELS ARE BEAUTIFUL: EFFICIENT ESTIMATORS ARE EVEN MORE BEAUTIFUL

D. Ludwig
Departments of Mathematics and Zoology
University of British Columbia

Abstract

There is always a conflict between fidelity to nature and the simplicity of statistical models. This conflict becomes especially acute when data are noisy or uninformative, as is often the case for problems in resource management. In such situations, the most effective model must sacrifice fidelity in order to obtain efficient estimators. This point is illustrated for a simple linear model, which is solved completely. An application from fisheries management also illustrates the virtues of simple models. It also shows how equally simple models can differ sub stantially in their estimation efficiency.

1. Introduction

The operations of aggregation and simplification are essential in forming an effective basis for action. Although we all make such choices on a regular basis, there is no theory which is both specific enough to be useful in particular cases and broad enough to be a useful general tool. Some statistical theories are well enough developed to aid intuition. Box and Jenkins(1976) and their followers have provided a procedure to guide the choice and validation of time series models. Similarly, Belsley, Kuh and Welsch(1980) have considered the diagnosis and therapy of regression models. These theories do not go far enough to address the problems of resources and environment which we would like to consider.

In the next section, I shall simplify some aspects of the approach of Belsley, Kuh and Welsch in order to obtain a complete solution for a very simple problem in regression. The solution reveals the important distinction between prediction losses and efficiency losses in evaluating model performance.

In the third section, I present an example from fisheries management to show how the same ideas can be applied in a much more complicated situation. An additional principle emerges: even very simple and accurate models can be less effective than other models chosen for their statistical properties.

This research was supported by the National Science and Engineering Research Council of Canada, under Grant Number A-9239

2. Linear Regression

The basic model of linear regression may be cast in the form

$$Y = X\beta + \epsilon. \tag{2.1}$$

where β is a vector of p parameters. X is an n by p design matrix. $X\beta$ is the true vector of output from the input X. and ϵ is the vector of errors in the observations. We shall assume that the components of ϵ are independent and normally distributed with mean zero and variance σ_ϵ^2. If $X_{ij} = x_i^{j-1}$. then we have a polynomial regression

$$Y_i = \sum_{j=1}^{n} \beta_j x_i^{j-1} + \epsilon_i. \tag{2.2}$$

The regression estimate must be chosen with two concerns in mind:

1. We would like to include as much as possible in the model in order to obtain as accurate a prediction as possible.
2. We wish to have estimates which are insensitive to the errors in the observations.

In statistical jargon, we wish to have efficient estimates. These two objectives are in conflict with each other, as we shall see below. A similar phenomenon occurs if we attempt to receive a weak signal from a fading radio station as we drive away from it. We can turn up the volume in order to produce a greater output from the weak signal, but then we also increase the amount of noise or static which we receive. The situation can be made more tolerable by filtering the output signal, to mask the most offensive noise. But then we are also losing part of the information in the signal. As we increase the size of our regression model (and sometimes also the sophistication of our approach) we are in effect turning up the gain on our input. In order to obtain tolerable results, it is often necessary to filter the output by restricting our model or simplifying it in certain ways.

In order to simplify the analysis. we shall assume that any linear combinations of old parameters are acceptable as new parameters. i. e. the dimension of the problem may be reduced in any way which suits our purposes. Such changes make a great mathematical simplification, but they might well be unacceptable in certain practical or legal situations. where estimates of certain types are mandated.

The Singular Value Decomposition

The Singular Value Decomposition (SVD) consists in finding an orthogonal matrix V such that the product XV has orthogonal columns. In fact, we may write

$$XV = U\Lambda, \tag{2.3}$$

where U has orthogonal columns of unit length, and Λ is a p by p diagonal matrix. The elements of Λ are the singular values of X. A convenient alternate form for (2.3) is

$$X = U\Lambda V^T. \tag{2.4}$$

The calculation of V, U and Λ can be reduced to a spectral decomposition of $X^T X$, since it follows from (2.4) that

$$X^T X = V\Lambda^2 V^T. \tag{2.5}$$

The columns of V are the eigenvectors of $X^T X$, and the elements of Λ^2 are the eigenvalues of $X^T X$. The Singular Value Decomposition is ordinarily computed directly, without forming $X^T X$. Although the SVD is an old idea, its usefulness for statistical theory has not been appreciated until recently. See Golub(1968), and Belsley, Kuh and Welsch(1980).

Now we may define a new set of parameters by the equation

$$\alpha = V^T\beta \tag{2.6}$$

In view of (2.4), the original system (2.1) becomes

$$Y = U\Lambda\alpha + \epsilon. \tag{2.7}$$

The great advantage of the system (2.7) is that the columns of U are orthogonal unit vectors. This makes the system of equations defining the least-squares estimator easy to solve, and the solution $\hat{\alpha}$ inherits desirable statistical properties, as we shall see below.

Least Squares Solution

In order to obtain the least squares solution of the problem, project Y onto the space spanned by the columns of U, which we denote by $U_i, i = 1, \ldots, p$. Thus

$$Y_{LS} = \sum_{i=1}^{p} U_i U_i^T Y. \tag{2.8}$$

The corresponding form for $\hat{\alpha}$ is

$$\hat{\alpha}_i = \frac{1}{\lambda_i} U_i^T Y. \tag{2.9}$$

If the elements of ϵ are independent and normally distributed (as was assumed above) then each of the components $\hat{\alpha}_i$ is normally distributed, and they are independent of one another, because of the orthogonality of U. Therefore the variance of $\hat{\alpha}_i$ is given by

$$\sigma_{\alpha_i}^2 = \frac{1}{\lambda_i^2}\sigma_\epsilon^2. \tag{2.10}$$

A more general possibility is to choose a subset of the possible regression variables: let I be a subset of the integers from 1 to p. Then we may set

$$Y_I = \sum_{i \in I} U_i U_i^T Y. \tag{2.11}$$

Each term which is included in the sum (2.11) will improve Y_I as an approximation to Y, but will result in an increase in the variance of the combined estimates, due to the corresponding term in (2.10). How can we decide whether or not to include i in I? A simple answer is to apply a significance test to the corresponding estimate $\hat{\alpha}_i$. However, such a procedure takes no account of the effect of such decisions upon the ultimate goal of the modeling exercise.

Expected Loss function

We may attempt to measure the effect of an incorrect estimate $\hat{\beta}$ by introducing a quadratic loss function: let

$$Q(\hat{\beta}) = [L(\hat{\beta} - \beta)]^2. \tag{2.12}$$

where L is a linear function. We may express Q in terms of α by setting

$$\ell_i = LV_i. \tag{2.13}$$

where V_i is the i-th column of the matrix V. Then,

$$L(\hat{\beta} - \beta) = \sum_{i \in I} \ell_i(\hat{\alpha}_i - \alpha_i) - \sum_{i \notin I} \ell_i \alpha_i. \tag{2.14}$$

From (2.7) we see that

$$\alpha_i = \frac{1}{\lambda_i}U_i^T(Y - \epsilon). \tag{2.15}$$

and

$$\hat{\alpha}_i - \alpha_i = \frac{1}{\lambda_i}U_i^T \epsilon. \tag{2.16}$$

If (2.15) and (2.16) are substituted into (2.14), and the result is employed in (2.12), we obtain

$$Q(\hat{\alpha}) = \left[\sum_{i \in I} \frac{1}{\lambda_i}\ell_i U_i^T \epsilon - \sum_{i \notin I} \ell_i \alpha_i\right]^2. \tag{2.17}$$

Finally, we take the expectation of Q over the distribution of ϵ. Again using the orthogonality of U_i and the normality and independence of $U_i^T \epsilon$, we obtain the expected loss

$$\mathcal{E}[Q(\hat{\alpha})] = \left(\sum_{i \notin I} \ell_i \alpha_i \right)^2 + \sum_{i \in I} (\frac{\ell_i}{\lambda_i})^2 \sigma_\epsilon^2. \tag{2.18}$$

Interpretation of the expected loss

The fundamental formula (2.18) reveals that the expected loss is a sum of two terms. The first term (the prediction loss) can be reduced by increasing the size of I, or even eliminated by choosing the least squares solution. Of course, if our model is considered as a restriction of a sequence of larger and larger models, then the prediction error relative to the larger models will increase. The second term (the efficiency loss) is independent of any larger models in which the present model is embedded. The efficiency loss depends upon the magnitude of the errors in Y, i. e. σ_ϵ^2. It also depends upon the singular values of the design matrix. An optimal estimation scheme must make a tradeoff between the prediction loss and the efficiency loss. Since σ_ϵ^2 ordinarily cannot be controlled, an optimal choice of estimation scheme must concentrate upon the prediction loss and the singular values λ_i. If the singular values decrease rapidly beyond a certain point, then the choice of estimation scheme is easy: keep the parameters which correspond to large singular values, and discard those which correspond to smaller ones. In the case of parameters with large estimates, the corresponding prediction loss asssociated with dropping the parameter might force retention of the corresponding parameter. This point and related concepts are treated at some length in Draper and Smith(1981).

The question now arises: how is the behavior of the singular values related to the structure of the design matrix X? If X has only two columns and they are nearly collinear, then the second singular value of X will be close to zero, since hardly anything is left of the second column of X after the component in the direction of the first column is removed. Likewise, if X has three columns and they nearly lie in a plane, then the third singular value will be small, and so forth for higher dimensions: small singular values correspond to near dependencies among the columns of X. Thus we can use the SVD as a diagnostic tool for regression and parameter fitting. Further details are given in Belsley, Kuh and Welsch(1980).

In order to get a feel for the behavior of singular values, I have computed them for two design matrices. The first corresponds to the polynomial regression (2.2). Here x consists of 30 equally spaced values between 0 and 1, and the columns of X are formed by taking j between 0 and 4. The resulting singular values are (6.9, 2.6, .62, .10, .011). Not only do the singular values decrease, but the ratio of successive singular values increases sharply from 2.7 to 9.3 as i increases. Thus the deterioration accelerates as i increases. It should be remembered that the efficiency loss increases with the squares of these ratios. Clearly the efficiency loss will be extreme unless the number of terms in the regression is held to 2 or 3, or unless the errors in Y are exceptionally small.

The second design matrix was constructed by modifying the last four columns of X to be $\sin(j\pi x_i)$, $j = 1,\ldots,4$. In this case, the singular values are (6.6, 3.8, 3.8, 3.8, 1.2). This is not very surprising, since the functions $\sin(j\pi x)$ are orthogonal when integrated over the unit interval. Thus the columns of the second design matrix are nearly orthogonal when a discrete sum is taken instead of the integral. In the second case, one could perhaps use all 5 parameters in an explanatory model. This example illustrates how important a proper design is if parameters are to be estimated with accuracy.

Real data sets tend to be more similar to the first example than the second. Well designed experiments are often extremely difficult or expensive or impossible to carry out. Observational data almost never provide the kind of systematic contrasts which produce healthy numbers of singular values substantially different from zero. Under these circumstances, it is tempting to try including everything known about the problem in the parameter estimation, in the hope of reducing the prediction loss to an acceptable level. But the preceding analysis shows that such a procedure can produce hopelessly inefficient parameter estimates. To make matters worse, the final decision-makers involved may not be aware of the lack of reliability of the estimates.

The Moral

We conclude that in case of poor design or uninformative data, only minimal models are likely to produce acceptable expected losses. The dominant term in the total loss is not likely to be the prediction loss, but rather the efficiency loss. It follows that the most effective model is unlikely to be very realistic, but rather to have robust statistical properties.

3. Fisheries Management

The preceding points are illustrated in recent work on parameter estimation for fisheries managment. The basic data consist of a sequence of fishing efforts E_t, $t = 1,\ldots,n$, and a sequence of catches C_t, $t = 1,\ldots,n$. The objective is to choose a sequence of catches which maximizes some measure of long-term yield. This problem is much more complicated than might seem at first sight, since the stock dynamics are unknown and the relationship between stock size and effort is also unknown. A comprehensive discussion of this problem is given in Walters(1986). For our present purposes, we shall narrow the objective to one of choosing a (constant) value of effort which, if maintained indefinitely, will maximize the sum of discounted harvests. Thus we must arrive at an estimate $\hat{E}$, and study the distribution of the associated value function.

Surplus Production Model

The following models are based upon the assumption that there is a stock with a biomass B_t, and that the catch is related to the stock biomass by

$$C_t = B_t(1 - e^{-qe_t}),\tag{3.1}$$

where the effective effort e_t is related to the observed effort E_t by

$$e_t = E_t e^{v_t}.\tag{3.2}$$

The sequence $\{v_t\}$ is assumed to independently and normally distributed with mean 0 and variance σ_v^2. The stock remaining after harvest is denoted by S_t:

$$S_t = B_t - C_t.\tag{3.3}$$

We shall assume that the following year's biomass is related to the stock size by

$$B_{t+1} = F(S_t)e^{w_t}.\tag{3.4}$$

The function F is unknown, as is the parameter q and the sequence $\{v_t\}$, and the sequence $\{w_t\}$, which is assumed to be independently and normally distributed. The methods which are to be tested may make assumptions about the form of F which are quite different from those used to generate the catch and effort data.

Stock and Recruitment

The Ricker model makes the assumption that

$$F(S) = Se^{\alpha - \beta S}.\tag{3.5a}$$

Another possibility is the Power Model

$$F(S) = e^\alpha S^\beta.\tag{3.5b}$$

Still another possibility is the general model of Schnute(1985):

$$F(S) = e^\alpha S(1 - \beta\gamma S)^{1/\gamma}.\tag{3.5c}$$

Finally we shall consider a generalized Ricker model

$$F(S) = Se^{\alpha - \beta S^p}.\tag{3.5d}$$

In all of these models the parameters α and β are to be estimated. Additional parameters such as γ or p are specified as part of the model. In addition to these relatively simple surplus production models, there are more complicated age-structured models proposed by Deriso(1980) and Schnute(1985), which will not be described here.

Calculation of the Estimated Optimal Effort and its Value

In each case, the parameters q, α and β are to be estimated from the data on catch and effort. Once these parameters are estimated, the optimal value of the constant effort $\hat{E}$ can be obtained in two stages: first an optimal stock size $\hat{S}$ is obtained by maximizing $\delta \bar{F}(s) - S$ as a function of S, where δ is a discount factor (usually set equal to .9). Then the corresponding effort $\hat{E}$ is obtained from the condition that a constant effort of magnitude $\hat{E}$ produce a stock size of $\hat{S}$. In tests of the methods with simulated data, the preceding estimates of $\hat{E}$ can be compared with the "exact" values from the model and parameters used in generating the data. Thus we can compute the "exact" optimal effort E^* and the value of using such an effort, which we denote by V^*. The latter value can be compared with the long term return from using the effort $\hat{E}$, which we denote by $\hat{V}$. The ratio $\hat{V}/V^*$ is used as our measure of loss for the estimation scheme. Further details are given in Ludwig, Walters and Cooke(1987).

Implications of the Linear Theory

The theory of Section 2 is not applicable to the present case, because the parameters appear nonlinearly in the Surplus Production Model and in the value (loss) function described above. However, the ideas of that section may be applied in a heuristic way. For instance, in the cases of the Ricker or Power models, the parameters α and β appear linearly in the regression of $\log(B_{t+1}/S_t)$, if q is assumed to be fixed. The columns of the corresponding design matrix X consist of a column of ones (corresponding to α), and a column consisting of S_t (corresponding to β), in the case of the Ricker model. The Singular Value Decomposition is immediate in this case: the columns of W consist of ones and $S_t - \bar{S}$. The singular values are thus

$$\lambda_1 = \sqrt{n},$$

$$\lambda_2 = \sqrt{\sum_{t=1}^{n}(S_t - \bar{S})^2}.$$

It follows that both parameters can be well determined only if the variance of the stock size is substantial. This simple result can also be obtained by a variety of methods. It raises a profound difficulty, since the objective of management is often stated to be sustained yields, with small variation. Such an objective prevents the effective determination of the stock-recruitment parameters.

Similar remarks can be made about the determination of the parameter q. It appears only through being multiplied by the efforts E_t, just as the parameter β appears only in association with S_t. We may conclude that q will be well determined only if E_t varies substantially in the data. Moreover, the determination of all three parameters will be impossible unless E_t

and S_t vary independently. Thus it is clear that one is likely to be able to determine all of the parameters in the model only if special actions are taken to ensure adequate contrast in effort and stock size, and if the effort and stock size sequences are not too highly correlated.

Such phenomena were observed by Ludwig and Hilborn(1983) in the course of their attempts to find strategies to obtain informative sequences of efforts. They found that the uncertainty in $\hat{\beta}$ could be approximated from the linear regression which assumes that q is known. If $\hat{\beta}$ is poorly determined, their most successful strategy would relax fishing effort for a while, in order to generate substantial variation in the stock size. On the other hand, if q was poorly determined (as judged by the sensitivity of the likelihood function to changes in q) then a period of very high efforts was indicated. If all of the parameters appeared to be adequately determined, then the estimated optimal effort $\hat{E}$ was chosen.

Complicated versus Simple Models

A noteworthy feature of the simulations of Ludwig and Hilborn was that the data were generated with an age-structured model, but the estimation used only the Ricker model. The Ricker model seemed to be perfectly adequate for management, in spite of its poor correspondence with the "true" model. This phenomenon was observed in more systematic tests in Ludwig and Walters(1985), where the age-structured model of Deriso was compared with the Ricker model. Other more recent work in progress shows that the age-structured model of Schnute(1985) does not perform as well as the Ricker model, even when the data are generated using the more complicated model. The relationship between the data and the parameters is less direct in the more complicated models: it is necessary to use data from three or more years at once in order to obtain each residual, whereas it is only necessary to use data from successive years in the Ricker model. This complication in the estimation schemes might be the cause of their poorer performance.

Correct versus Incorrect Models

One feature of the discussion of Section 2 that is certainly borne out by our experience is that the prediction loss (due to an incorrect model) is less important than the efficiency loss (due to poorly determined parameters). In a sense, all of our example were cooked up to show this feature: it will be present wherever information is scarce. Information will generally be scarce for a managed situation because management restricts the behavior of the system in certain ways, and thereby prevents us from obtaining information about its behavior if those restrictions were not present.

This raises a final question: can models be chosen to minimize the efficiency loss? We have carried out some experiments to explore this issue. Instead of merely comparing complicated and simple models, we can compare across the collection of simple surplus production models (3.5a-d). The results of Ludwig, Walters and Cooke(1987) show that the Ricker model is superior to the Power model, even when data are generated using the Power model. Similarly, the Ricker model is superior to all of the models in the class of Schnute's general model (3.5c), except for itself. But within the class of generalized Ricker models, the model with $p=2$ is superior to the Ricker model, and the model with $p=3$ is superior to the model with $p=2$, and so forth. But there is no model which is superior to all of the others. The reason for this behavior is that efficiency errors decrease as p increases, but the prediction loss increases as p moves away from the "correct" p^*. Therefore any model can be beaten by a model with a higher p than p^*, but not if p is too much greater than p^*.

4. Is this a Special or a General Phenomenon?

The results of Section 2 are certainly very special. The only justification for such a simple treatment is the clarity with which the expected loss can be derived and interpreted. In more complicated situations, no such simple demonstration is feasible. However, the distinction between prediction error and efficiency error is just as important in more complicated cases. In fact, the more complicated the underlying situation, the more likely it is that the information available will be inadequate. Therefore we may expect the efficiency error to be even more dominant in the total loss function than it is in simple cases. However, since parameter estimates themselves are difficult to determine in complicated situations, it is often thought to be not worth the trouble to obtain estimates of the efficiency of the estimators. These examples emphasize that we ignore efficiency losses at our peril.

References

Belsley, D. A., E. Kuh, R. E. Welsch. 1980. Regression Diagnostics: Identifying Influential Data and Sources of Colliniearity. J. Wiley: New York.

Box, G. E. P., G. M. Jenkins. 1976. Time Series Analysis Forecasting and Control. Holden-Day: oakland California.

Deriso, R.B. 1980. Harvesting strategies and parameter estimation for an age-structured model. Can. J. Fish Aquat. Sci. 37:268-282.

Draper, N. R., H. Smith. 1981. Applied Regression Analysis, second edition. J. Wiley: New York.

Golub, G. 1968. Least Squares, singular values and matrix decompositions. Aplikace Mathematiky 13:44-51.

Ludwig, D., R. Hilborn. 1983. Adaptive probing strategies for age-structured fish stocks. Can. J. Fish. Aquat. Sci. 40:559-569.

Ludwig, D., C. Walters. 1985. Are age structured models appropriate for catch- effort data? Can. J. Fish. Aquat. Sci. 42:1066-1072.

Ludwig, D., C. Walters, J. Cooke. 1987. Comparison of two models and two methods for catch and effort data. Submitted to Natural Resource Modeling.

Schnute, J. 1985. A general theory for analysis of catch and effort data. Can. J. Fish. Aquat. Sci. 42:419-429.

Walters, C. 1986. Adaptive Management of Renewable Resources. Macmillan: New York.

Part VI.   Dynamic Properties of Population Models

INFERRING THE CAUSES OF POPULATION FLUCTUATIONS

(or, chaos meets data in population ecology)

Stephen Ellner

Biomathematics Program, Department of Statistics
North Carolina State University
Raleigh, NC 27695-8203, USA

Abstract

Many natural populations fluctuate in abundance, in ways that appear random to the eye and to some standard time-series analyses. This paper reviews the methods, and results, of recent attempts to determine if the fluctuations are chaotic rather than random. Early "indirect" methods (fit a model to the data, and see if the model is chaotic) found no evidence of chaos, but the results are sensitive to the choice of model and parameter-estimation methods. "Direct" methods (based on reconstructing trajectories in time-delay coordinates) appear to reveal "strange" (i.e., fractal) attractors that are a hallmark of chaotic dynamics. Reconstruction is model-free, but offers no way of objectively evaluating one's visual impression of the attractor. Fractal dimension calculations, using a new maximum-likelihood method for short time-series (Ellner 1988), can be used to test for fractal structure (non-integer dimension) if there's enough data. For some measles incidence data, dimension calculations suggest an attractor with dimension between 2 and 3, supporting the visual impression of a strange attractor underlying the fluctuations.

1. Introduction

The logistic difference equation $x_{t+1} = rx_t(1-x_t)$ with $r=3.7$ produces erratic, chaotically fluctuating values of x_t (Figure 1a). Contemplating this time-series, Solomon (1979) asked, "How could the trajectory of this completely determined process ever be distinguished from the sample path of a stochastic one?" The implication is that there is no way of analyzing the sequence of x_t values, to tell if the fluctuations are chaotic or random.

Fortunately, Solomon's question has a simple answer, shown in Figure 1b. Graphing x_{t+1} vs. x_t, the relationship is seen to be perfectly deterministic, though highly nonlinear. At the time, however, Solomon's opinion was just about *every* ecologist's opinion: chaotic population dynamics are "indistinguishable from stochastic growth" (Thomas et al. 1980; for similar opinions see also May & Oster 1976).

Figure 1. (Top) The chaotic time-series x_t produced by the logistic difference equation with $r = 3.7$, $x_0 = 0.5$; (Bottom) The time-series is seen to be deterministic, by plotting the values as ordered pairs (x_t, x_{t+1}).

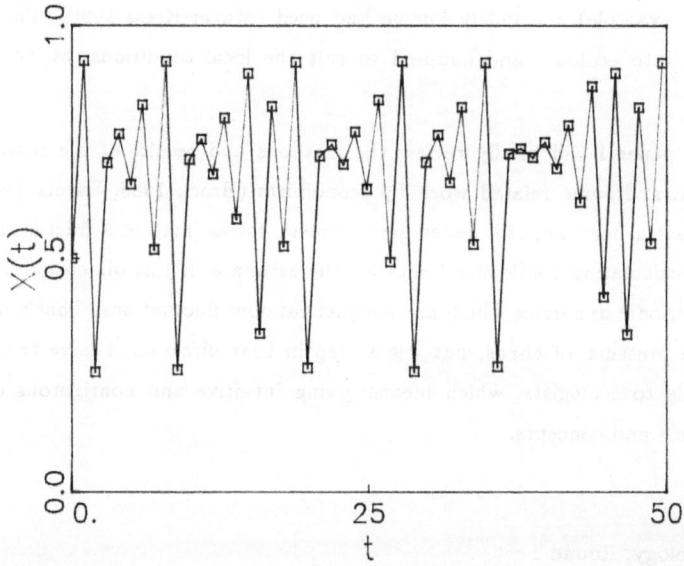

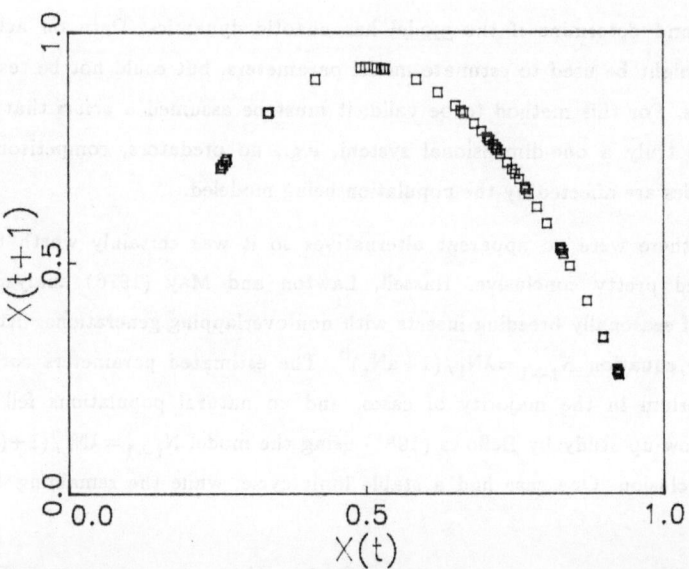

In the ecological literature the idea still persists that "for most practical purposes, and certainly for ecologists, the chaotic regime may be regarded as one in which simple and deterministic relationships...give apparently random dynamics" (May 1986). In physics, however, methods for identifying low-dimensional chaos in experimental data (of which Fig. 1 is the simplest example) are widely known and used (Mayer-Kress 1986). The methods have been imported into ecology, and adapted to suit the local conditions, by Schaffer, Kot and coworkers.

In this paper I will briefly review the methods and results of the search for chaos in ecological data, and some related work by economists (Brock 1986, Sayers 1987). I will also present a statistical method, still under development, for estimating a fractal dimension of the dynamics and calculating confidence limits for the estimate. If this dimension is >1 but finite, it indicates aperiodic dynamics which are not just random fluctuations. That's not the same as testing for the presence of chaos, but it's a step in that direction. I have tried to make this paper accessible to ecologists, which means giving intuitive and nonrigorous explanations of some basic terms and concepts.

2. Chaos in ecology, Round 1

When ecologists began to wonder if any real populations behaved chaotically, the only available method was indirect: model the population dynamics using a difference equation like the logistic, and determine if the model has chaotic dynamics. Data on actual population fluctuations might be used to estimate model parameters, but could not be tested directly for signs of chaos. For this method to be valid, it must be assumed a priori that the population dynamics are truly a one-dimensional system, e.g., no predators, competitors, or resources whose dynamics are affected by the population being modeled.

Still, there were no apparent alternatives so it was certainly worth trying, and the results seemed pretty conclusive. Hassell, Lawton and May (1976) analyzed data on 28 populations of seasonally breeding insects with non-overlapping generations, fitting the data to the difference equation $N_{t+1} = \lambda N_t / (1+aN_t)^b$. The estimated parameters corresponded to a stable equilibrium in the majority of cases, and no natural populations fell in the chaotic region. A follow-up study by Bellows (1981) using the model $N_{t+1} = \lambda N_t / (1+(aN_t)^b)$ reached the same conclusion. One case had a stable limit cycle, while the remaining 13 had a stable equilibrium.

So chaos was not found in nature; what about the laboratory? Experiments on laboratory populations avoid the problems of unmodeled interactions and an uncontrolled environment. Again, the results were quite uniform. 20 of 25 lines of *Drosophila melanogaster* raised in the laboratory had stable equilibria for the estimated model parameters (Mueller & Ayala 1981), and in no case could the null hypothesis of a stable equilibrium be rejected. 58 laboratory populations of *Drosophila* from 27 species were all found to have a globally stable equilibrium (Thomas et al. 1980). Additional data on 10 of those species under a different food regime were fit to the "θ-logistic" model $N_{t+1} = N_t + rN_t(1 - (N_t/K)^\theta)$, and none of the species had parameters in the chaotic region (Phillipi et al. 1987).

In Hassell et al.'s analysis, Nicholson's laboratory blowflies were the only example of parameters in the chaotic region, and they were chaotic also in Brillinger et al.'s (1980) more detailed model. But Nisbet & Gurney (1982) concluded, after their fastidious analysis, that the blowfly oscillations were limit cycles. Thus it was entirely reasonable for Nisbet & Gurney (1982), after surveying all the available evidence, to express the belief that "deterministic stability is the rule rather than the exception, at least with insect populations".

3. Reconstruction.

The breakthrough, which made it possible to analyze data directly for evidence of chaos, was the method of "reconstruction" in time-delay coordinates (Takens 1981). The method is simplest for the case of a single-variable time-series. So suppose that the data are one-dimensional, in continuous or discrete time: $x(t)$, $0 \le t \le T$ or $x(t_i)$, $i = 1,2,...,N$. It is assumed that x is one variable out of many, in a deterministic differentiable dynamical system (e.g., a system of ordinary or partial differential equations). From the values of x we create the M-dimensional time-series

$$X(t) = (\; x(t), x(t+\tau), x(t+2\tau), \ldots x(t+(M-1)\tau)\;).$$

with $\tau > 0$ arbitrary (continuous time) or an arbitrary multiple of the time between measurements (discrete time). M is called the "embedding dimension". For example, if our data are $x(1)$, $x(2)$, $x(3)$,..., and we take M=3 and $\tau=2$, then

$$X(1) = (\; x(1), x(3), x(5)\;)$$
$$X(2) = (\; x(2), x(4), x(6)\;)$$
$$X(3) = (\; x(3), x(5), x(7)\;),$$

and so on. $X(t)$ is then a sequence of points in 3-dimensional space. The same method is used if x is k-dimensional, and the dimension of X is then kM.

Takens (1981) proved that for M sufficiently large, the attractor for $X(t)$ is qualitatively the same as the full system's attractor. Informally, the attractor is where the system winds up if you let it run long enough (assuming such a place exists). Reconstruction lets us see the attractor, if we are content with <u>qualitative</u> information about the <u>attractor</u> (e.g., "is the attractor a periodic orbit?"; technically, the map taking the full system to X is generically an embedding of the attractor into Euclidean M-space if M is large enough). Quantitative reconstruction is impossible: we can't determine the actual values of the unmeasured state variables. Also, the dynamics off the attractor might not be faithfully reconstructed, in general.

How large M must be depends on the attractor dimension, which is generally unknown. In practice M is increased until $X(t)$ looks well-behaved (smooth trajectories without crossings) and/or the estimated dimension of the reconstructed attractor stops increasing. For mathematical details and rigor, and a review of reconstruction as applied to physical and chemical systems, see the review by Eckmann & Ruelle (1985).

4. Chaos in Ecology, Round 2

This round belongs to Schaffer & Kot, who picked up reconstruction and ran with it, attracting some followers in the process. A small sample of their results is shown in Figure 2 (while Schaffer & Kot have also examined several ecological models, I will discuss here only their analyses of data). In each case the original time-series had dimension 1, and the embedding dimension is 3; the original discrete-time data have been smoothed in some cases, and interpolated.

Schaffer & Kot (1986) describe these as "some apparent examples of real-world chaos". All of them had been interpreted before as periodic, with superimposed "noise" due to environmental fluctuations (in *Thrips* the periodicity was seasonal with period 1 year; the others had apparent multi-year periodicities). The reconstructed trajectories appear to trace out an attractor of dimension 2 (or higher, for measles), rather than a 1-dimensional periodic orbit. For measles, the case is strengthened by applying the same reconstruction to the output of an SEIR model with seasonal variation in the disease-transmission rate, and noting the striking similarity to the reconstructed data (Figure 2e). (SEIR models are standard epidemiological models in which the population is classed as Susceptible, Exposed, Infectious, or Removed [dead or immune]; see e.g. Aron & Schwartz 1984). Schaffer (1985) gives the details of the measles reconstruction, and lists 10 population data-sets that are similarly suggestive of chaos (including Nicholson's blowflies).

While the trajectories in Figure 2 do look like chaos, visual impressions may be misleading, and may not convince a sceptic who prefers other sorts of models. Bulmer (1974) observed that the 19th-Century Canadian lynx cycle (one of Schaffer & Kot's (1986) "apparent examples") had a fairly regular period but an irregular amplitude, which were "obvious in data on the lynx cycle, and in the correlogram and periodogram calculated from the data". To account for these he proposed the model

Figure 2. Examples of population fluctuations reconstructed in time-delay coordinates by Schaffer & Kot. (a) the lynx cycle in Canada: (b) outbreaks of the insect *Thrips imaginis*; (c) a microtine rodent cycle; (d) measles incidence in Baltimore, 1900-1927; (e) an SEIR model with chaotic dynamics due to seasonal variation in the transmission rate; note the similarity to (d). Reproduced from Schaffer & Kot (1986).

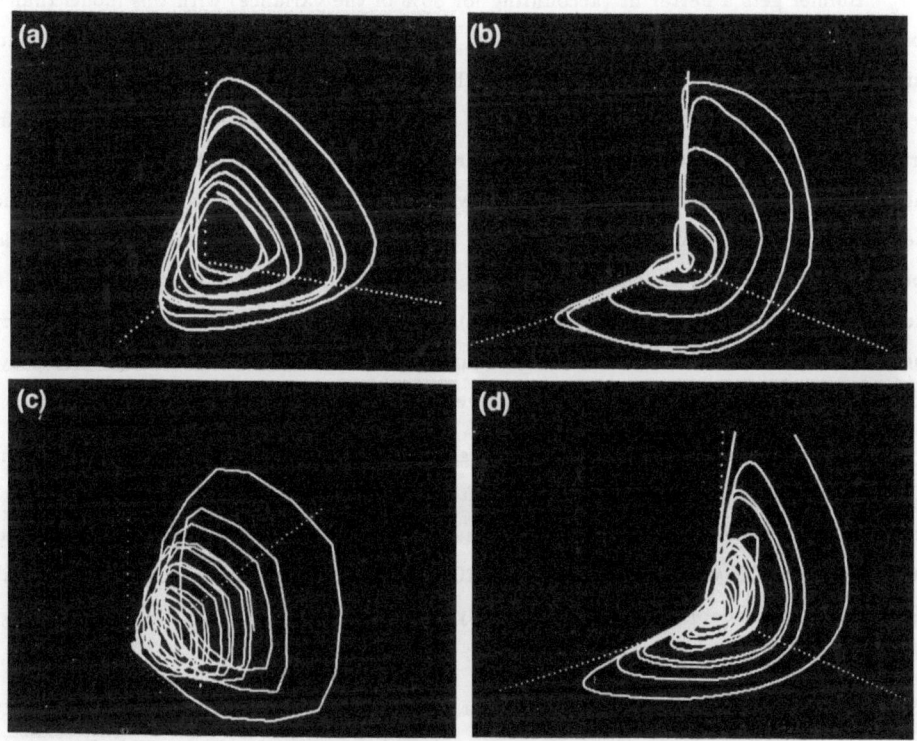

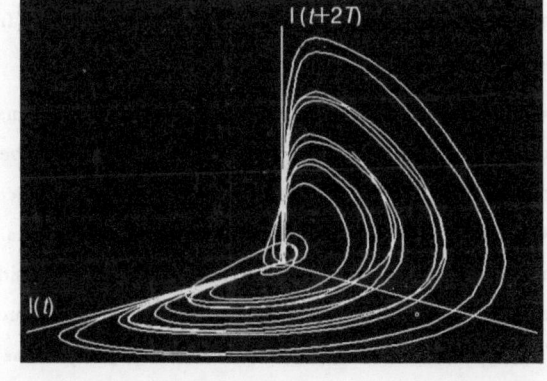

$$x_t = \mu + \alpha \sin w(t-\phi) + \beta x_{t-1} + e_t$$

where $x_t = \log_{10}$(number of lynx trapped in year t), μ is the mean of x, and the e's are random errors. The sin term accounts for the regular periodicity, while the autoregressive term makes the amplitude fluctuate.

Bulmer gets a better fit (accounting for 86% of the variance) with this model than with models suggested previously by Bartlett and Moran, and judged the residuals to be random. This was taken as confirming the validity of the underlying model, and of the significance level for the periodic component ($\alpha \neq 0$ at P<.001) calculated on the basis of the model. Significant periodic cycles with period near 10 years were found in a total of 10 Canadian species, and Bulmer (1974) suggested (without direct evidence) that the common period was due to interactions (documented or plausible) with the snowshoe hare. Bulmer (1976) developed stochastic models in which the hare follows a noisy limit cycle (conjectured to result from interactions between hare and its food-plants), and the lynx cycle is driven by the hare cycle. The power spectrum of the lynx cycle was found to have the qualitative form predicted by the models, and there were even a few quantitative agreements.

However, Bulmer "eyeballing" his residuals and declaring them random is no more rigorous than "eyeballing" the same data after reconstruction, and declaring that they look chaotic. Indeed, Finerty's (1980, Fig. 7) reanalysis found evidence of a single dominant cycle with period near 10 years in only 4 of the 10 species for which this was claimed by Bulmer (1974). Their methods of analysis differed only in some apparently minor details, but that was enough to change the results.

Analysis of the infectious disease data raises similar problems. Anderson et al. (1984) performed spectral analyses of measles, mumps, and pertussis incidence. This is a standard statistical method, in which a time-series is expressed as a sum of periodic components (sines & cosines). The "spectral density" is a curve showing the relative contribution to the sum by each of the periodic components, as a function of frequency (frequency=1/period, so e.g. a frequency of .5 means a period of 2).

Anderson et al. found consistent peaks in the spectral density at a period of 1 year, and also at multiples of one year: 2 years for measles, 3 years for pertussis and mumps. Figure 3 shows an example of spectral analysis applied to measles incidence. Even in a short time-series, the spectral density has 2 clear (and highly significant) peaks, so the fluctuations are not purely random. The nonzero spectral density in frequencies outside the peaks can be viewed as random error (presumably due to random environmental fluctuations) in the time-series framework. This leads to a description of the disease dynamics as a noisy periodic orbit, whose period is 2 or 3 years. The multi-year periods observed in the data are roughly in accord with the period of damped oscillations in simple compartment-models for the diseases, and can be

Figure 3. Conventional spectral analysis of measles incidence in Baltimore.
Reproduced from Kot et al. (1988). (top) The time-series. (bottom) The spectral density has
two clear peaks corresponding to periods of 1 and 2 years.

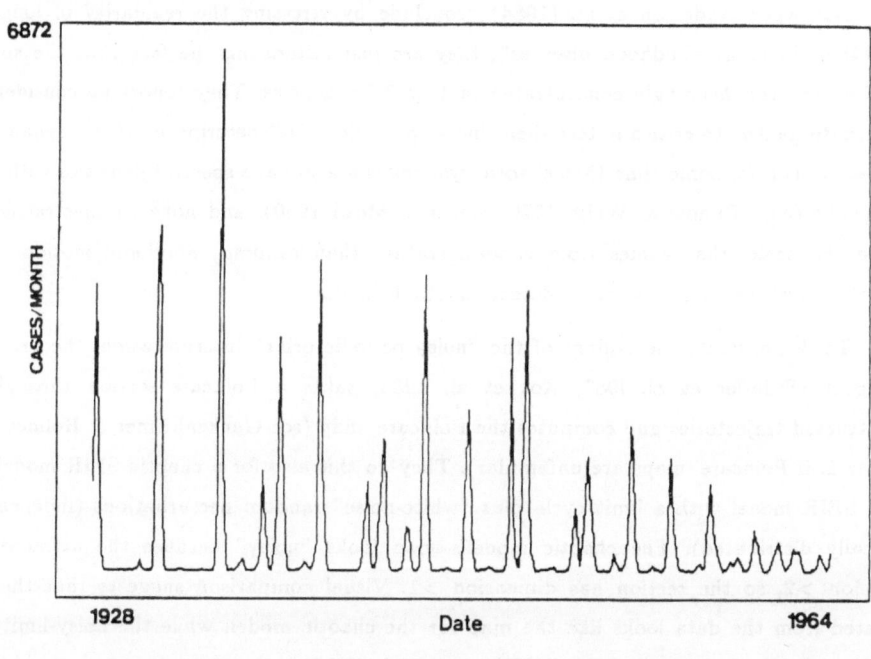

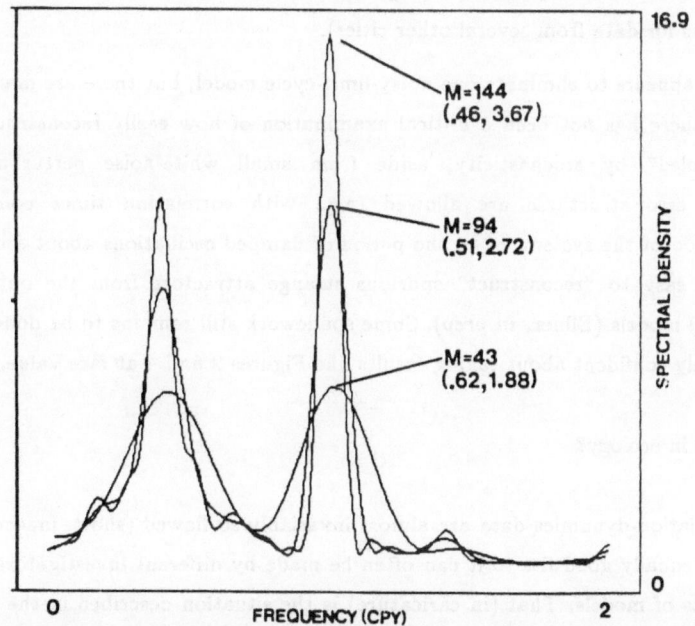

interpreted as modes repeatedly kicked into action by the environmental fluctuations (Anderson et al. 1984, May 1986).

But when Anderson et al. (1984) "conclude by stressing the regularity of long-term fluctuations in many childhood diseases", they are just reiterating the fact that the spectral densities are overwhelmingly concentrated in 1 or 2 frequencies. They report no consideration of alternate models to critically test their "noisy periodic orbit" description of the dynamics. It has been known for some time that chaotic systems can also have spectral densities with a few clear peaks (e.g., Bunow & Weiss 1979, Smith & Mead 1980), and nonzero spectral density outside the peaks that comes from chaotic (rather than random) aperiodic motion. Thus, spectral analysis by itself can fail to detect chaotic behavior.

To demonstrate the failings of the "noisy periodic orbit" interpretation, the pro-chaos contingent (Schaffer et al. 1987, Kot et al. 1988) takes a Poincare'-section through the reconstructed trajectories and computes the Poincare' map (see Guckenheimer & Holmes 1983, Chapter 1, if Poincare' maps are unfamiliar). They do this also for a chaotic SEIR model, and for an SEIR model with a limit cycle plus "white-noise" random perturbations (independent, identically distributed). The chaotic model's map looks "noisy" because the attractor has dimension >2, so the section has dimension >1. Visual comparison suggests that the map estimated from the data looks like the map for the chaotic model, while the noisy-limit-cycle model has a "map" that's just mush (Figure 4; Schaffer (*pers. comm.*) has recently obtained similar results for data from several other cities).

This appears to eliminate one noisy-limit-cycle model, but there are many more. To my knowledge, there has not been a critical examination of how easily reconstruction techniques can be "fooled" by stochasticity, aside from small white-noise perturbations. If more complicated error-structures are allowed (e.g., with correlation times comparable to an intrinsic period of the system, or to the period of damped oscillations about a limit cycle), it is distressingly easy to "reconstruct" spurious strange attractors from the output of periodic (non-chaotic) models (Ellner, *in prep*). Some spadework still remains to be done, before we can feel completely confident about taking results like Figures 2 and 4 at face value.

5. Invariants in ecology?

Population-dynamics data are almost invariably so flawed (short, inaccurate, etc.) that more or less equally good fits to it can often be made by different investigators predisposed to different sorts of models. That (in caricature) is the situation described in the last section, for chaotic vs. stochastic explanations of population fluctuations. "Goodness of fit" criteria are often useful for model-choice, but they make less sense if completely different qualitative descriptions of the data are being compared. Each modeler can pay some more attention to

Figure 4. Poincare "maps". (top left) chaotic SEIR model for measles; (top right) SEIR model with a limit cycle, and small white-noise random perturbations; (bottom) measles incidence in Baltimore 1928-1963. Reproduced from Kot et al. (1988).

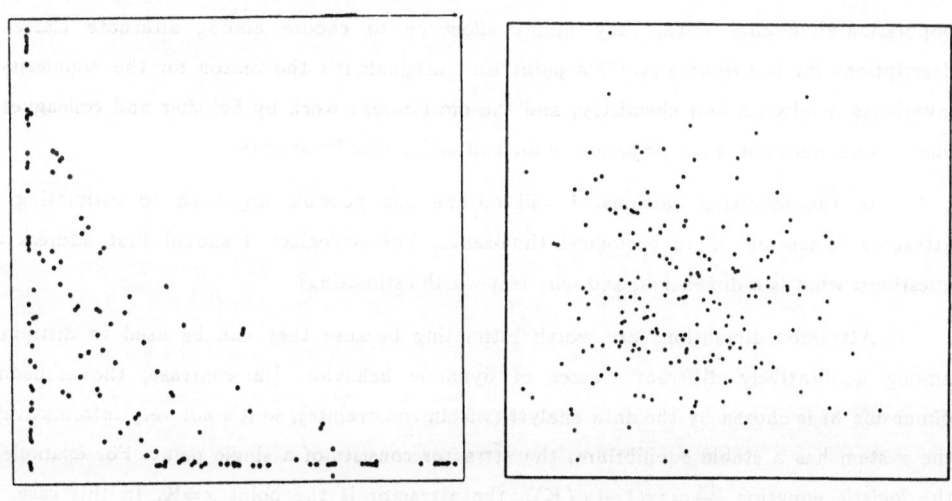

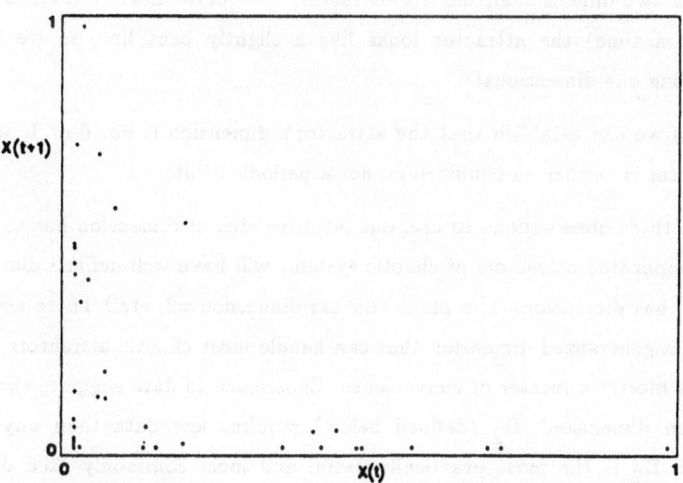

details, and improve the fit.

Dynamical systems theory has identified some numerical quantities, called "invariants", which characterize qualitative aspects of dynamics (e.g., is it chaotic?). The most widely used invariants in the physical sciences are various "dimensions", the "entropy", and the "Lyapunov exponents" (Mayer-Kress 1986), all of which can be estimated directly from the reconstructed dynamics of the full system. If these quantities could be estimated from population-abundance data, they might allow us to choose among alternate classes of descriptions for the dynamics. This point isn't original: it's the reason for the popularity of invariants in physics and chemistry, and the most recent work by Schaffer and colleagues on disease dynamics has involved serious efforts at estimating invariants.

In the following sections, I will outline one possible approach to estimating the attractor dimension of an ecological time-series. For ecologists, I should first address two questions: what is a dimension, and why is it worth estimating?

Attractor dimensions are worth estimating because they can be used to distinguish among qualitatively different classes of dynamic behavior. [In contrast, the embedding dimension M is chosen by the data analyst (within constraints), so it's not very informative]. If the system has a stable equilibrium, the attractor consists of a single point. For example, in the logistic equation $\frac{dx}{dt} = rx(1 - (x / K))$, the attractor is the point x=K. In this case, the system is moving in one dimension, but the attractor is a zero-dimensional subset of the state-space. In a two-species predator/prey model with a limit cycle, the system is moving in the plane, which is two-dimensional, but its attractor is a curve. Locally (i.e., looking at it one small piece at a time) the attractor looks like a slightly bent line, so we can regard the attractor as being one-dimensional.

Thus, if we can establish that the attractor's dimension is not 0 or 1, we can conclude that the attractor is neither an equilibrium, nor a periodic orbit.

To put these observations to use, our intuitive idea of dimension has to be generalized, so that the complicated attractors of chaotic systems will have well-defined dimensions (and so that a line still has dimension=1, a plane still has dimension=2, etc.) There are many sensible ways to define a generalized dimension that can handle most chaotic attractors, and the choice among them is mostly a matter of convenience. Experience to date suggests that estimation of the "correlation dimension" D_2 (defined below) requires less data than any of the known alternatives, so D_2 is the most practically useful and most commonly used dimension. (The notation D_2 refers to an infinite family of generalized dimensions $\{D_q, q \geq 0\}$ introduced by Renyi (1970)).

D_2 has the useful property that if the data are a series of independent continuous random variables (e.g., normally distributed), then D_2 equals the embedding dimension M;

more precisely, the true D_2 is $+\infty$, and the calculated values of D_2 increase without limit as M is increased (Ben-Mizrachi et al. 1984, Holzfuss & Mayer-Kress 1986). Thus, D_2 can in principle distinguish between deterministic and random fluctuations. If the estimated D_2 is > 1, but has an asymptotic maximum value as M is increased, then the dynamics are aperiodic but nonrandom. Integer attractor dimensions > 1 (e.g., an attractor that looks like the surface of a smooth donut) can be produced by linear systems with several periods whose ratio is irrational. But if D_2 is > 1 and not an integer, then the dynamics (although not necessarily chaotic) could not be produced by a linear system (Guckenheimer & Holmes 1983).

For ecological applications, the problem is to estimate the dimension and other invariants from short, noisy data sets. A physicist can feel that a series of 500 values is very "short" (Abraham et al. 1986). An ecologist or economist (Brock 1986) usually has less data, with larger measurement errors and an uncontrolled, time-varying environment. Applied to such data, sometimes the physicists' methods work and sometimes they don't. When they do work, it may not be clear how the results should be interpreted; for example, would an estimated dimension of 1.25 suggest that you have a periodic orbit (since $1.25 \approx 1$), or that you don't (since $1.25 > 1$)? Unlike the physicist, we need reliable confidence intervals to go along with our estimates of invariants.

6. Estimating D_2

To define D_2, let δ be the distance between 2 random points chosen independently according to the invariant measure on the attractor (the invariant measure is the probability distribution that describes the relative amounts of time the system spends in different parts of the attractor). Let F be the cumulative distribution function of δ, i.e. $F(r) = P(\delta \leq r)$. If there is a number d such that $F(r) \sim r^d$ as $r \downarrow 0$, then $D_2 = d$; otherwise D_2 is undefined. F is an example of a "correlation integral" in physics terminology, hence the name "correlation dimension" for D_2.

Grassberger & Procaccia (1983) first applied D_2 to dynamical systems, and suggested estimating it from a time-series $x_1, x_2, \ldots x_N$ by computing

$$\hat{F}(r) = \frac{\#\{ \text{ pairs } x_i, x_j \text{ with } i \neq j, \| x_i - x_j \| \leq r \}}{N(N-1)}$$

(actually, assuming N would always be very large they used N^2 rather than N(N-1) in the denominator). Plotting $\log \hat{F}(r)$ vs. $\log r$, D_2 is the slope at small values of r. This implicitly assumes that $\log F(r) = a + D_2 \log r$ for $r \approx 0$, which is not always true, but experience so far suggests that the linear approximation is adequate in practice for experimental data analysis.

Eckmann and Ruelle (1985) note that this procedure is "not entirely justified

Figure 5. Grassberger-Procaccia plots (log $\hat{F}$ vs. log r); N=length of the time-series. (top left) Henon map with a=1.4, b=0.3, N=100; (top right) logistic map at the transition to chaos (r=3.5699456), N=500; (bottom right) a chaotic SEIR model, N=2400; (bottom left) measles incidence in Baltimore, 1928-1963, N=432. The solid line shows the "scaling region" used for the measles dimension estimates. The embedding dimensions are 2,1,4, and 4 respectively. For the Henon and logistic, the graphs show the average over 100 calculations of $\hat{F}$ with different, randomly chosen initial values of x_0.

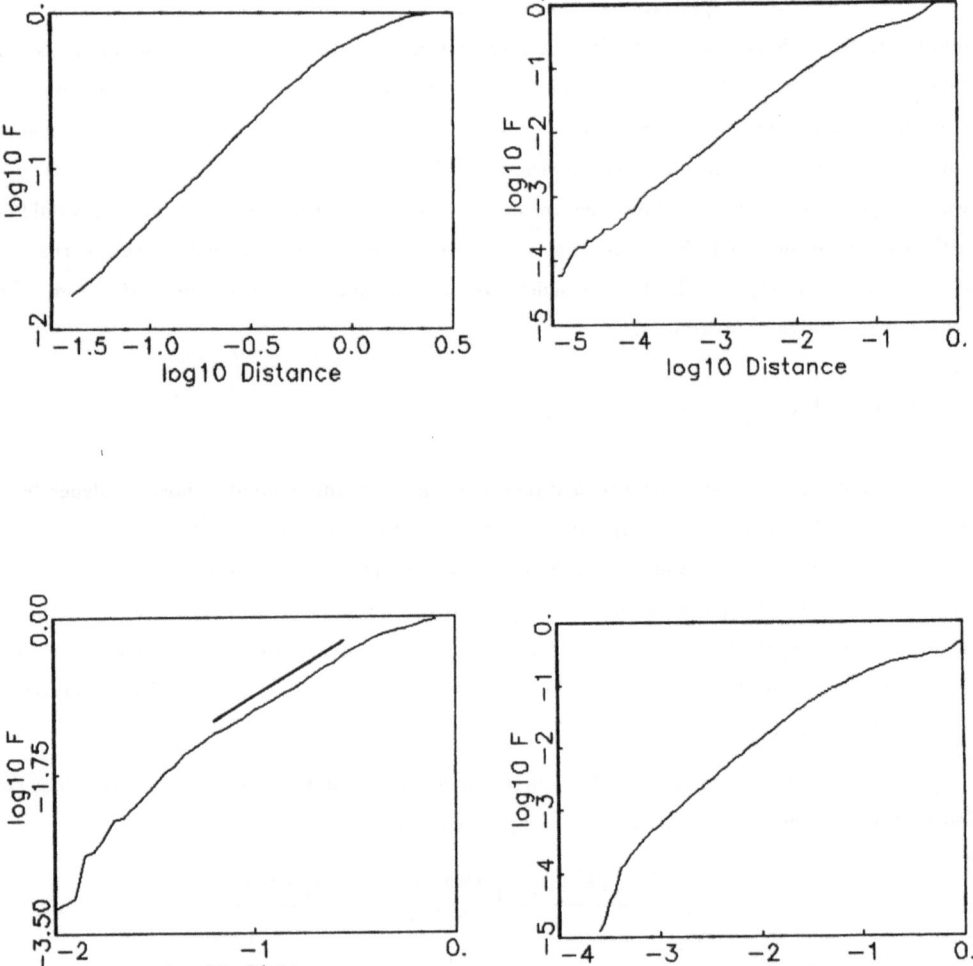

mathematically, but nevertheless quite sound". However, they were summarizing physicists' experience with experimental data and computer simulations of dynamical systems. With small, noisy time-series there are problems. At "large" values of r, approaching the diameter of the attractor, the slope of log $\hat{F}$ decreases to 0. At "small" values of r there are distortions due to noise (e.g., measurement error or roundoff error) and due to finite-sample effects (e.g., $\hat{F}=0$ for r< smallest interpoint distance in the time-series). The linear approximation for log $\hat{F}$ therefore holds only on a finite "scaling region" $\gamma_0 \leq r \leq \gamma_1$. In practice, the scaling region shrinks and eventually may vanish as the embedding dimension is increased, making it difficult to determine if the embedding dimension is large enough for successful reconstruction.

Error estimation is also a problem. The usual approach is to calculate log $\hat{F}$ at several values in the scaling region, and do a standard linear regression. The slope of the regression line is the estimate of D_2, and the standard error of the slope is the reported error estimate. However, the basic linear regression assumption of independent, normally distributed errors isn't satisfied in this case, so the supposed "standard error of the slope" is invalid. As Caswell & Yorke (1986) note (regarding estimation of a different dimension), error estimates obtained by standard linear regression can be orders of magnitude smaller than the actual difference between the estimate and the true dimension. The "reference point" procedure suggested by Holzfuss & Mayer-Kress (1986) also ignores the actual error structure of an empirical cumulative distribution function, and uses an inappropriate weighted-least-squares method.

The actual distribution of $\hat{F}$ could be used as the basis of valid error-estimation for the Grassberger-Procaccia procedure, and this approach is being investigated (Sayers 1987). Takens (1985) suggested using maximum likelihood rather than least-squares estimation, and gave asymptotic confidence intervals for the case of $\gamma_0=0$ and a large number of observations. I have extended the method to allow $\gamma_0>0$ and a small number of observations, and it seems to work well if the scaling region is well-defined. Interpoint distances $\delta > \gamma_1$ are discarded; for $\delta < \gamma_1$ define

$$\rho = \min \{ T, \ln(\gamma_1/\delta) \}, \text{ where } T = \ln(\gamma_1/\gamma_0).$$

If $\ln F(r) = a + D_2 \ln r$ for $\gamma_0 < r < \gamma_1$, then the distribution of ρ is easily shown to be a right-censored exponential, with smooth density $f(\rho)=D_2 \exp(-\rho D_2)$ on $0<\rho<T$, and a point-mass at $\rho=T$ holding the probability from the truncated tail.

Parameter estimation for this distribution is a problem with a long history in reliability theory, since it corresponds to mean-lifetime estimation with a constant failure rate and fixed experiment duration T. Given values $\rho_1, \rho_2, \ldots \rho_n$, the maximum likelihood estimator of D_2 is

$$\hat{D} = R/\sum_{i=1}^{n} \rho_i, \text{ where } R = \#\{ \rho_i < T \}. \tag{1}$$

(Deemer & Votaw 1955). Approximate confidence intervals are most easily constructed using the asymptotic distribution of the relative error as the sample size increases: Gaussian with mean 0 and variance $[n(1-(\gamma_0/\gamma_1)^{D_2})]^{-1}$ (Deemer & Votaw 1955). Using $\hat{D}$ to estimate D_2, the approximate $100(1-\alpha)\%$ confidence interval for D_2 is

$$\hat{D} \pm Z_{\alpha/2}\hat{D}/\sqrt{n\,(1-(\gamma_0/\gamma_1)^{\hat{D}})} \qquad (2)$$

where Z_α is the probability-α critical value of the standard normal distribution.

To leading order in $1/n$ as $n\to\infty$, (2) is the same as the asymptotic confidence intervals previously suggested in the literature (Bartlett 1953; Deemer & Votaw 1955; Bartholomew 1963). It is the simplest (too simple to have been published?) since it doesn't require iterative solution of nonlinear equations, and it was also the most accurate in Monte Carlo simulations with $20-100$ distances and $.5\leq D_2\leq 3$. Barlow et al. (1968) give an implicit formula for exact confidence intervals, but for >35 values (2) was much more accurate than numerical evaluation of the "exact" intervals (using double-precision IMSL). While (2) is only exact as $n\to\infty$, it is acceptably accurate for $n\geq 50$: in Monte Carlo simulations with pseudorandom values having right-censored exponential distributions, nominal 95% confidence intervals had coverage frequencies .945 to .952 (2-sided) and .938 to .969 (1-sided), with standard errors $\leq$ .005. For the procedure described below, this means that there must be 100 or more values in the time-series.

7. Does it work?

In the intended application, the data are a sequence of points $x_1, x_2, \ldots, x_N$. The apparent scaling region was identified by calculating local slopes at 100-125 values of r (evenly spaced on the logarithmic scale) using linear regression of log $\hat{F}$ on log r through 5 adjacent values of r. Since the x's are presumably measured with some error, there are at most $N/2$ independent distances (i.e., $N/2$ independent draws of a ρ) that can be obtained. Therefore n $= N/2$ was used in equation (1), by choosing at random n interpoint distances $\leq \gamma_1$. To reduce the effects of sampling variability on the calculated $\hat{D}$, this procedure was repeated m times (m = 10 to 25) and the average $\hat{D}$ was used. Having done that it is tempting to divide the confidence interval width by $\sqrt{m}$, but that doesn't work (in theory and in simulations) because the replicates of $\hat{D}$ are not independent. For now, I set conservative confidence intervals by using n=N/2 in equation (2).

Table 1 summarizes the results on two time-honored test cases: the "critical" logistic

Table 1. Confidence intervals and estimation of D_2 for time-series produced by chaotic dynamical systems: the Henon map with a=1.4, b=0.3, and the "critical" logistic map at the transition to chaos (r=3.5699456). N = length of the time-series. In all cases n = N/2 was used in equation (2) to compute nominal 95% confidence intervals. Coverage frequencies and other statistics were obtained by 2000 Monte-Carlo simulations with randomly chosen initial conditions and 250 iterations prior to generating the time-series used to estimate the dimension. ( )= standard error of estimate. "Power" is the frequency of 2-sided confidence intervals that contained no integer values. Embedding dimensions M=2 for Henon, M=1 for logistic.

	Henon N=500	Henon N=100	Logistic N=500	Logistic N=100
Coverage frequency by 95% confidence intervals				
2-sided bounds	1.0 (0)	.998 (.001)	.997 (.001)	.998 (.001)
Unbounded above	1.0 (0)	1.0 (0)	.964 (.004)	.918 (.006)
Unbounded below	.999 (.001)	.997 (.001)	1.0 (0)	1.0 (0)
Relative error of $\hat{D}_2$ over the simulations				
mean	.001 (.001)	.03 (.003)	.085 (.002)	.224 (.001)
standard deviation	.03	.081	.026	.058
Power	.984 (.003)	.044 (.005)	1.0 (0)	1.0 (0)

Table 2. Estimates of D_2 for disease incidence. N=length of the time-series. Values in parentheses are the 95% confidence interval from equation (2) with n=N/2.

Embedding Dimension	SEIR model N=2400	Baltimore measles 1928-1963, N=432	Milwaukee Chickenpox 1926-1965, N=480
3	2.16 (2.09, 2.24)	2.17 (1.98, 2.36)	3.55 (3.19, 3.92)
4	2.24 (2.15, 2.32)	2.50 (2.27, 2.72)	4.03 (3.60, 4.47)
5	2.23 (2.13, 2.35)	2.38 (2.15, 2.61)	4.50 (3.98, 5.02)
6	2.23 (2.12, 2.34)	2.57 (2.32, 2.82)	4.31 (3.83, 4.78)

map at the transition to chaos, and the Henon map. The estimates of D_2 are fairly good and the confidence intervals are generally conservative, as expected. However these are easy targets: there is no noise aside from roundoff error, the minimal embedding dimension is known, and the scaling regions in $\hat{F}$ are fairly unambiguous (Figure 5; the small wiggles in the logistic $\hat{F}$ are real, an example in which the linear approximation is not exact).

A less tame and more realistic test-case was provided by W.M. Schaffer, 2400 monthly values of the number of infectives from a chaotic SEIR model (Figure 6). Because the model output (like the measles data) has so much seasonal variability, the best results are obtained by treating each month separately (this amounts to looking at 12 Poincare' sections through the attractor; Schaffer et al. 1987). For each embedding dimension the 12 monthly time-series were rescaled to make the mean or the maximum of the interpoint distances $=1$ (whichever gave the more well-defined scaling region). The intramonth distances were then combined into a single distance distribution from which $\hat{F}$, $\hat{D}$, etc. were computed. The dimension of the full attractor is obtained by adding 1 to the section dimension. This method is valid in theory: the true dimensions of the attractor sections are identical (since they are diffeomorphic images of

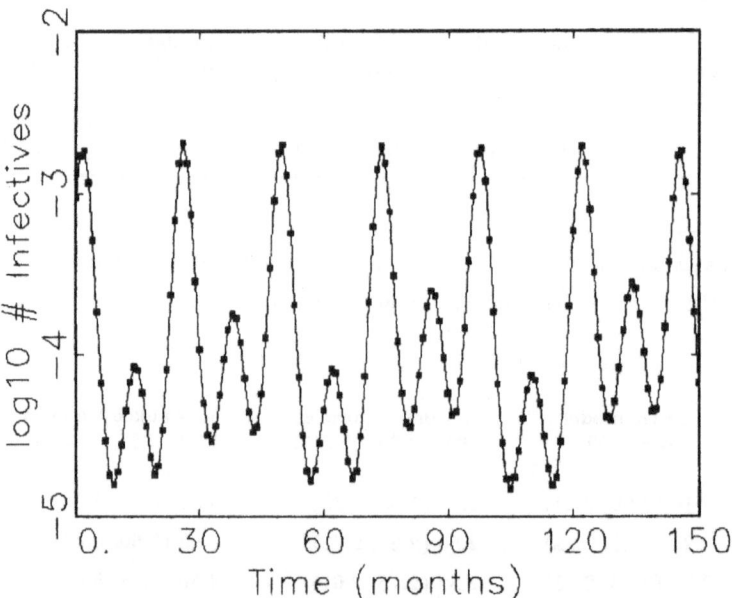

Figure 6. A typical segment of the output from the chaotic SEIR model of Figure 5(c) and Table 2. The complete "data set" consists of 2400 monthly values for the number of infectives.

each other), are 1 less than the attractor dimension, and are not affected by a linear rescaling. A reasonably good scaling region (Figure 5) is obtained, with consistent results as the embedding dimension is increased (Table 2); the confidence intervals are wider in higher embedding dimensions because (γ_0/γ_1) is larger. The analytic bound $D_2 \leq D_1 = 2.39$ is consistent with the Table 2 results, and confirms that the qualitative nature of the attractor has been identified correctly (i.e., fractal with $2 < D < 3$; D_1 was computed by the Kaplan−Yorke formula from the Lyapunov exponents for the model, W.M. Schaffer, *pers. comm.*)

The same procedure was applied to measles incidence in Baltimore, 1928-1963, and to chickenpox in Milwaukee, 1926-1965 (Table 2; data provided by W.M. Schaffer). Identifying the scaling region is rather subjective for these data-sets, so that the values obtained are certainly not definitive. Nonetheless, there is an appreciable difference between the values for measles and those for chickenpox, which suggests that the measles and chickenpox dynamics are qualitatively different. Chickenpox seems to be higher-dimensional, or more strongly affected by random perturbations. For a periodic orbit with white-noise perturbations in embedding dimension M, we should theoretically get an apparent dimension of 1, or M, or both 1 and M in two different scaling regions (Ben-Mizrachi et al. 1984, Brock 1986). In practice, the dimension estimates may continue to increase as M is increased, but with the estimated D_2 remaining less than M (Holzfuss and Mayer-Kress 1986). The chickenpox D_2's are consistent with this scenario (a low-dimensional attractor plus "noise"), but a higher-dimensional attractor can't be ruled out. [Other lines of evidence (Schaffer et al. 1987), and estimates of D_2 using an improvement on the method presented here (Ellner, 1988) tend to favor the noisy-periodic-orbit interpretation for chickenpox.] For the measles data a low embedding dimension appears to be adequate, and the values of D_2 indicate a "strange" attractor whose dimension is stricty between 2 and 3.

8. Conclusions

It should be apparent that much work remains to be done on estimating D_2. The results are good for perfectly accurate data and a clean scaling region, even with realistically short time-series for ecological applications. However real-world data require *ad hoc* pre-treatments to extract a scaling region, with poorly defined limits and some legitimate doubt still as to whether it's real or imaginary. A good deal of subjective choice on these matters preceded the calculations on disease incidences. For now, the best I can say for highly "spiky" data is that the estimates are fairly robust ($\pm 10-20\%$) over several different sensible ways of combining the 12 monthly data-sets (e.g., cf. Table 2 with the estimates of Schaffer et al.

1987). A different method of identifying the scaling region, which uses sliding windows from the order statistics of the interpoint distances (rather than $\hat{F}$) as the basis for maximum-likelihood estimation of the "local slope", can give much better results with noise-free but very short data-sets (Ellner 1988); how well it does with noisy data remains to be seen.

The problem of identifying the scaling region is not unique to D_2. For example, in the more familiar application of fractal dimensions to spatial data (e.g., the dimension of a coastline or leaf surface) the fractal structure is absent at lengths larger than the region being studied, and at molecular length-scales. However, for comparing models with data or comparing different populations, all of $\hat{F}$ may be relevant, not just the apparent scaling regions. Large differences in the $\hat{F}$'s could indicate qualitative differences in the dynamics, even if it isn't feasible to estimate the dimensions accurately. Econometricians have begun testing the adequacy of linear time-series models, using methods based on Brock's (1986) "Residual Test Theorem": if a finite-order autoregressive model

$$x_t = \sum_{j=1}^{m} a_j x_{t-j} + \epsilon_t$$

is fit to the data, the residuals $\{\epsilon_t\}$ will generically have the same invariants as the original data $\{x_t\}$. If the model correctly describes the data, the residuals should be Gaussian white noise. This is tested by comparing the $\hat{F}$'s of (a) the residuals, (b) random shufflings of the residuals, and (c) pseudorandom Gaussian sequences with the same mean and variance as the residuals. While statistical theory for this method has not yet been developed, the numerical results can give striking evidence for inadequacy of an autoregressive model (Sayers 1987).

At present, it seems that dimensions can be estimated from quite short data-sets, but that dealing with noisy data remains problematic. Research is currently quite active on how one can best extract a chaotic "signal" from "noisy" interference (Abraham et al. 1986; Albano et al. 1986, 1987, *pers. comm.*; Broomhead & King 1986). That work essentially stops at the point where γ_0 and γ_1 have been estimated (and uses the standard point-estimate of D_2), so it is complementary to research on error-estimation methods. Dimension estimation consequently poses some well-defined open statistical problems, whose solutions would be of immediate interest to biologists, economists, and even some physicists whose experimental systems can't be kept stationary for long (Albano et al 1986).

Acknowledgements: I thank Mark Kot, Leslie Real, and H.R. van der Vaart for comments on the manuscript, Doug Nychka for hallway-and-coffeepot consulting, and Bill Schaffer for posing the problem and providing data. Research supported in part by NIH−BRSG Grant RR7071.

REFERENCES

Abraham, N.B., A.M. Albano, B. Das, G. de Guzman, S. Yong, R.S. Goggia, G.P. Puccione, and J.R. Tredicce (1986) Calculating the dimension of attractors from small data sets. Physics Letters 114A, 217-221.

Albano, A.M., N.B. Abraham, G.C. de Guzman, M.F.H. Tarroja, D.K. Bandy, R.S. Goggia, P.E. Rapp, I.D. Zimmerman, and T.R. Bashore (1986). Lasers and brains: complex systems with low-dimensional attractors. pp 231-240 in G. Mayer-Kress ed., *Dimensions and Entropies in Chaotic Systems: Quantification of Complex Behavior.* Springer-Verlag, Berlin-Heidelberg-New York-Tokyo.

Albano, A.M., A.I. Mees, G.C. de Guzman, and P.E. Rapp (1987) Data requirements for reliable estimation of correlation dimensions. pp. 207-220 in: H. Degn, A.V. Holden, and L.F. Olsen eds., *Chaos in Biological Systems.* Plenum Publishing Corp., New York.

Anderson, R.M., B.T. Grenfell, and R.M. May (1984). Oscillatory fluctuations in the incidence of infectious disease and the impact of vaccination: time series analysis. J. Hyg., Camb. 93, 587-608.

Aron, J.L. and I.B. Schwartz. (1984). Seasonality and period-doubling bifurcations in an epidemic model. J. Theor. Biol. 110, 665-679.

Barlow, R.E., A. Madansky, F. Proschan, & E.M. Scheuer. (1968). Statistical estimation for the "burn-in" process. Technometrics 10, 51-62.

Bartholomew, D.J. (1963). The sampling distribution of an estimate arising in life-testing. Technometrics 5, 361-374.

Bartlett, M.S. (1953). On the statistical estimation of mean life-times. Philosophical Magazine Ser. 7 vol. 44, 249-262.

Bellows, T.S. (1981). The descriptive properties of some models for density-dependence. J. Animal Ecology 50, 139-156.

Ben-Mizrachi, A., I. Procaccia, & P. Grassberger (1984). Characterizaton of experimental (noisy) strange attractors. Phys. Rev. 29A, 975-977.

Brillinger, D.R., J. Guckenheimer, P. Guttorp, and G.F. Oster (1980). Empirical modeling of population time-series data: the case of age and density dependent vital rates. In G.F. Oster ed., *Some Mathematical Questions in Biology* 13. American Mathematical Society, Providence RI.

Brock, W.A. (1986). Distinguishing random and deterministic systems: abridged version. J. Economic Theory 40, 168-195.

Broomhead, D.S. and G.P. King (1986). Extracting qualitative dynamics from experimental data. Physica 20D, 217-236.

Bulmer, M.G. (1974). A statistical analysis of the 10-year cycle in Canada. J. Anim. Ecol. 43, 701-718.

Bulmer, M.G. (1976). The theory of predator-prey oscillations. Theor. Pop. Biol. 9, 137-150.

Bunow, B. & G.H. Weiss (1979). How chaotic is chaos? Chaotic and other "noisy" dynamics in the frequency domain. Math. Biosciences 47, 221-237.

Caswell, W.E. & J.A. Yorke (1986). Invisible errors in dimension calculations: geometric and systematic effects. pp. 123-136 in: G. Mayer-Kress ed., *Dimensions and Entropies in Chaotic Systems: Quantification of Complex Behavior.*Springer-Verlag, Berlin-Heidelberg-New York Tokyo.

Deemer, W.L. and D.F. Votaw, Jr. (1955). Estimation of parameters of truncated or censored exponential distributions. Ann. Math. Stat. 26, 498-504.

Ellner, S. (1988). Estimating attractor dimensions from limited data: a new method, with error estimates. Physics Letters 133A, 128-133.

Eckmann, J.-P., and D. Ruelle (1985). Ergodic theory of chaos and strange attractors. Rev. Mod. Phys. 57, 617-656, 1115 (addendum).

Finerty, J.P. (1980). *The Population Ecology of Cycles in Small Mammals.* Yale University Press, New Haven and London.

Grassberger, P. & I. Procaccia (1983). Characterization of strange attractors. Phys. Rev. Lett. 50, 346-349.

Guckenheimer, J. & P. Holmes (1983). *Nonlinear Oscillations, Dynamical Systems, and Bifurcations of Vector Fields.* Springer Verlag, New York-Berlin-Heidelberg-Tokyo.

Hassell, M.P., J.H. Lawton, and R.M. May (1976). Patterns of dynamical behavior in single-species populations. J. Anim. Ecol. 45, 471-486.

Holzfuss, J. & G. Mayer-Kress (1986). An approach to error-estimation in the application of dimension algorithms. pp. 114-122 in: G. Mayer-Kress ed., *Dimensions and Entropies in Chaotic Systems: Quantification of Complex Behavior.* Springer-Verlag, Berlin-Heidelberg-New York-Tokyo.

Kot, M., W.M. Schaffer, G.L. Truty, D.J. Graser, and L.F. Olsen (1988). Changing criteria for imposing order. Ecological Modelling 43, 75-100.

May, R.M. (1986). When two and two do not make four: nonlinear phenomena in ecology. Proc. R. Soc. Lond. B228, 241-266.

May, R.M. and G. F. Oster (1976). Bifurcations and dynamic complexity in simple ecological models. Amer. Natur. 110, 573-599.

Mayer-Kress, G., ed. (1986). *Dimensions and Entropies in Chaotic Systems: Quantification of Complex Behavior.* Springer Series in Synergetics v.32, Springer-Verlag, Berlin-Heidelberg-New York-Tokyo.

Mueller, L.D., and F.J. Ayala (1981). Dynamics of single-species population growth: stability or chaos? Ecology 62, 1148-1154.

Nisbet, R.M. and W.S.C. Gurney (1982). Modelling Fluctuating Populations. John Wiley & Sons, New York.

Phillippi, T.E., M.P. Carpenter, T.J. Case, and M.E. Gilpin (1987). Drosophila population dynamics: chaos and extinction. Ecology 68, 154-159.

Renyi, A. (1970). *Probability Theory.* Elsevier North-Holland, Amsterdam.

Sayers, C. (1987) Diagnostic tests for nonlinearity in time-series data: an application to the work-stoppage series (manuscript).

Schaffer, W.M. (1985). Can nonlinear dynamics elucidate mechanisms in ecology and epidemiology? IMA J. Math. Applied in Medicine & Biology 2, 221-252.

Schaffer, W.M. and M. Kot (1986). Chaos in ecological systems: the coals that Newcastle forgot. Trends in Ecology & Evolution 1, 58-63.

Schaffer, W.M., L.F. Olsen, G.L. Truty, S.L. Fulmer, and D.J. Graser (1987). Periodic and chaotic dynamics in childhood infections. To appear in H. Haken ed., *From Biological to Chemical Organization*, Springer-Verlag, Heidelberg.

Smith, R.H. & R. Mead (1980). The dynamics of discrete-time stochastic models of population growth. J. Theor. Biol. 86, 607-627.

Solomon, D. (1979) On a paradigm for mathematical modeling. In G.P. Patil & M. Rosenzweig eds., *Contemporary Quantitative Ecology and Related Ecometrics*, International Co-operative Publishing House, Fairland, MD.

Takens, F. (1981). Detecting strange attractors in turbulence. Lecture Notes in Math. 898, 366-381. Springer-Verlag, Heidelberg.

Takens, F. (1985). On the numerical determination of the dimension of an attractor. Lectures Notes in Math. 1125, 99-106. Springer-Verlag, Heidelberg.

Thomas, W.R., M.J. Pomerantz, and M.E. Gilpin (1980). Chaos, asymmetric growth, and group selection for dynamical stability. Ecology 61, 1312-1320.

STOCHASTIC GROWTH MODELS:
RECENT RESULTS AND OPEN PROBLEMS

Richard Durrett
Dept. of Math., White Hall
Cornell U., Ithaca, NY 14853

Abstract. In this paper we will describe some recent work in interacting particle systems and state some open problems. We will discuss three models: Richardson's model, an epidemic model and the contact process.

1. Richardson's model. In this model the state at time t is $\xi_t \subset Z^2$. (The model can be defined in any dimension but here we will stick to the two dimensional case since it is the one which is most important for applications.) We think of the sites in ξ_t as being occupied by an object that we call a "particle" and that you should think of as being a plant or an immobile animal (e.g. barnacle or mussel). The set of occupied sites evolves according to very simple rules:

If $x \in \xi_t$ then $x \in \xi_{t+s}$ for all $s \geq 0$.
If $x \notin \xi_t$ then $P(x \in \xi_{t+s} | \xi_t) = s | \#$ of occupied neighbors$| + o(s)$.

Here y is a neighbor of x if $\|x-y\| = 1$ where $\|x-y\| = |x_1-y_1| + |x_2-y_2|$, and $o(s)$ means that the missing terms when divided by s go to 0 as $s \to 0$. In the terms of Markov chain theory, particles are born at a rate equal to the number of occupied neighbors.

In Richardson's model ξ_t always fills up all of Z^2 so attention focuses on how ξ_t grows. The next result (due to Richardson (1973) says the set of occupied sites has a limiting shape:

Theorem 1. There is a convex set D so that for any $\epsilon > 0$
$$t(1-\epsilon)D \cap Z^2 \subset \xi_t \subset t(1+\epsilon)D \quad \text{for all t sufficiently large.}$$

To simplify the statement we have omitted the qualifying phrase "with probability one" here and will continue to do this below. Loosely speaking ξ_t looks like $tD \cap Z^2$ when n is large. In one dimension $D = [-1,1]$ but we don't know much about D when d = 2, except for the trivial observations that it has the same symmetry as Z^2, and it $\supset \{x : \|x\| \leq 1\}$. This state of affairs exists because the result is proved by using the subadditive ergodic theorem to show that if τ_k $= \inf \{t : (k,0,..0) \in \xi_t\}$ then
$$\text{as } k \to \infty \quad \tau_k/k \to \mu \equiv \inf E(\tau_j/j) \tag{1}$$
where $E(\tau_j/j)$ stands for the expected value of the random variable in parentheses. The

expression for the limiting constant is mathematically nice because the infimum exists, is nonnegative and finite. It does not lend itself well to computation (except of course for upper bounds). This situation is typical of the results we will discuss in this paper. We have results which describe the qualitative features of the model but very little quantitaive information.

(1) gives the strong law of large numbers for the first passage times. An interesting unsolved problem related to this is to prove the corresponding central limit theorem, i.e. find constants c_k so that

$$(\tau_k - \mu k)/c_k \to \text{a normal distribution} \tag{2}$$

In d=1 the result holds with $c_k = k^{.5}$. In d > 1 smaller norming constants may be needed but there is little consensus about what to guess and rigorous results on this question are almost nonexistent. We cannot even show that when $c_k = k^{.999}$ the limit in (2) is 0. For this result and other information about first passage percolation, a family of models which generalizes the one considered above, see Kesten (1986).

2. An Epidemic Model.

The next process can be used to model the spread of a disease or a forest fire. The state of the process at time t is a function $\xi_t : Z^2 \to \{1,i,0\}$. The states 1, i, and 0 have the following meanings in the two interpretations:

	fire	epidemic
1	tree	healthy
i	on fire	infected
0	burnt	immune

We have chosen measles for the disease because in that case once you have had that disease you cannot have it again. With a forest fire in mind, the dynamics of the model can be described as follows:

If $\xi_t(x) = 0$ then $\xi_{t+s}(x) = 0$ for all $s \geq 0$.
When a tree catches fire it burns for an amount of time with distribution F.
If $\xi_t(x) = 1$ then $P(\xi_{t+s}(x) = i \,|\, \xi_t) = \lambda s |\# \text{ of neighbors on fire}| + o(s)$.

In what follows we will be interested in what happens when the initial state is given by:
$\xi_0(0) = i$ and $\xi_0(x) = 1$ for $x \neq 0$.
Let $I_t = \{x : \xi_t(x) = i\}$ be the set of infected individuals at time t (here and for the rest of the paper we will use the epidemic interpretation), and let $\Omega_\infty = \{I_t \neq \phi \text{ for all } n\}$ be the event that the infection does not die out. The probability of Ω_∞ is a nondecreasing function of λ, so we define $\lambda_c = \inf\{\lambda : P_\lambda(\Omega_\infty) > 0\}$. The first thing we have to do is prove that $\lambda_c < \infty$. To do this we let

$$p(\lambda) = \int_0^\infty (1 - e^{-\lambda t}) \, dF(t)$$

be the probability that x tries to infect a neighbor y while it is infected. To get bound on λ_c we use a result of Kuulasmaa (1982) to conclude

$$.5 \leq p(\lambda_c) \leq .5927 \tag{3}$$

the left and right hand sides being the critical probabilities for bond and site percolation in two dimensions.

In the bond percolation model there is an unoriented edge (called a bond) connecting each pair of neighbors. Bonds are open with probability p and closed with probability 1–p and the state of different bonds are independent. In this model attention focuses on $\Omega_\infty = \{$ the origin belongs to an infinite connected set of open bonds $\}$, and it is known that $p_c = \inf \{$ p: $P_p(\Omega_\infty) > 0 \} = 1/2$. The site percolation model is similar but (i) in the definitions above bond is replaced by site which refers to a point in Z^2 and (ii) the critical value is only known through simulations. (For more about percolation see Kesten (1982).)

To make the connection between the epidemic model and percolation we draw an arrow from x to y if x tries to infect y while it is infected. These arrows are present with probability $p(\lambda)$. Arrows emanating from a given site are dependent, but Kuulasmaa's theorem allows us to compare the model with one in which (a) the arrows are independent and (b) either all the bonds are open or all are closed. Case (a) occurs when F is constant and (b) occurs when F has mass (1–p) at 0 and mass p at ∞. Once we prove that (a) is equivalent to bond percolation (this is NOT obvious) and (b) is equivalent to site percolation (this is easy), (3) results.

In the last paragraph we argued that set of sites whichever become infected is the same as in a percolation model. Using percolation techniques and ideas from Cox and Durrett (1981) we can prove a "shape theorem" for the model under consideration:

Theorem 2. Let $J_t = \{x : \xi_t(x) = 0\}$ be the set of sites which are immune at time t. There is a convex set D so that for all $\epsilon > 0$

$$t(1-\epsilon)D \cap J_\infty \subset J_t \subset t(1+\epsilon)D \quad \text{for all t sufficiently large.}$$

A proof is given in Cox and Durrett (1989). Notice that in contrast to Theorem 1, J_t does not contain $t(1-\epsilon)D \cap Z^2$, just $t(1-\epsilon)D \cap J_\infty$, but this is the best we can hope for since $J_\infty =$ the set of sites which will ever become infected.

3. The Contact Process. In this model the state at time t is $\xi_t \subset Z^2$ and we think of the points in ξ_t as occupied. The system evolves according to the following rules:

If $x \in \xi_t$ $P(x \notin \xi_{t+s} | \xi_t) = s + o(s)$.
If $x \notin \xi_t$ $P(x \in \xi_{t+s} | \xi_t) = \lambda s |\# \text{ of occupied neighbors}|/4 + o(s)$.

We divide by 4 in the last definition so that the birth rate from an isolated particle is λ. The contact process can be thought of as a branching process in which there can only be one particle per site so if we let $\lambda_c = \inf \{\lambda : P(\xi_t^0 \neq \phi \text{ for all } t) > 0\}$ then $\lambda_c \geq 1 =$ the critical value for a branching process. (See Athreya and Ney (1972) for information about these processes.) As in the case of the epidemic, we do not know what λ_c is, but numerically $\lambda_c = 1.65$.

For the developments which follow, it will be useful to describe how the contact process is implemented in a computer. To do this when $\lambda \geq 1$ (the interesting case), we pick a site at random from the finite set of sites under consideration, and call it x. If x is vacant we pick one of its neighbors y at random and make x occupied if y is. If x is occupied, we pick a number at random from (0,1) and kill the particle if (only if) the number chosen is less than $1/\lambda$. We repeat the proceedure to simulate the system. If there are m sites, then $t\lambda m$ cycles correspond roughly to t units of time.

In the last two sections we have seen that starting from a finite set the system expands linearly and has an asymptotic shape. This is true again here, but we will have to introduce a few concepts to state the limit result. We begin by considering what happens in the process ξ_t^1 starting from all sites occupied, i.e. $\xi_0^1 = Z^d$. Now Z^d is the largest state (in the partial order $\supset$), and the model has the property that if $A \supset B$, and we use the computer implementation to contstruct two versions of the process ξ_t^a and ξ_t^b starting from A and B respectively then we will have $\xi_t^a \supset \xi_t^b$ for all t. If we let $A = Z^d$ and $B = \xi_s^1$ in the last observation, then we see that ξ_t^1 is larger than ξ_{t+s}^1 in the sense that the two random sets can be constructed on the same space with $\xi_t^1 \supset \xi_{t+s}^1$.

Once one understands the last paragraph a simple argument shows that as $t \to \infty$ ξ_t^1 decreases to a limit we call ξ_∞^1, where the convergence occurs in the sense that $P(\xi_t^1 \cap C \neq \phi)$ decreases to $P(\xi_\infty^1 \cap C \neq \phi)$ for all finite sets C. It follows from Markov chain theory that ξ_∞^1 is an equilibrium distribution for the process, i.e. if the initial state has this distribution, then this will be the distribution at all $t \geq 0$. If $\lambda < \lambda_c$ then ξ_∞^1 is not interesting — it is ϕ with probability 1 — but if $\lambda > \lambda_c$ it is a nontrivial equilibrium distribution. The reader will note that $\lambda = \lambda_c$ has been left out in the last statement. Presumably this value falls under the first case (recall that critical branching processes die out), but this is a very difficult open problem.

With the equilibrium distribution introduced, we are now in a position to describe the limiting behavior starting from a finite set. Suppose we use the computer implementation to run two versions of the process, one starting with only 0 occupied, and the other starting with all of Z^d occupied, and we call the two resulting processes ξ_t^0 and ξ_t^1. The "shape theorem" in this setting is:

Theorem 3. If $\lambda > \lambda_c^*$ then there is a convex set D so that when $\xi_t^0 \neq \phi$ for all t, then for any $\epsilon > 0$ we have

$$t(1-\epsilon)D \cap \xi_t^0 \subset \xi_t^1 \subset t(1+\epsilon)D \text{ for all t sufficiently large.}$$

The statement of the result is made contorted by the fact that ξ_t^0 may become ϕ, in which case it stays ϕ for all time. The theorem tells us that when this does not occur ξ_t^0 look roughly like $\xi_\infty^1 \cap tD$. In words, it is a linearly growing "blob in equilibrium", or more poetically, it is an "expanding gray disc". The disc is called gray because the equilibrium state has correlations which are exponentially decaying, and hence if we look at the configuration of occupied (white) and vacant (black) cells from a distance all we will see is the average value (a shade of gray).

The second thing which must be explained is : What is λ_c^*? Let λ_c^n be the critical value for the contact process defined on $Z \times \{-n,...n\}$. As $n \uparrow \infty$, $\lambda_c^n \downarrow$ a limit we call λ_c^*. $\lambda_c^* \geq \lambda_c$ and presumably $=$ holds, but we can only prove the result for $\lambda > \lambda_c^*$. Durrett and Griffeath (1982) proved the result above only for $\lambda > \lambda_c^0 =$ the critical value for the contact process on Z. It took five years before Durrett and Schonmann (1987) could improve the result to the conclusion given above. It is an important and difficult open problem to prove that $\lambda_c^* = \lambda_c$ or to find another method of proof which allows us to prove the result for all $\lambda > \lambda_c$.

For more on interacting particle systems, see Liggett (1985). Durrett (1985), Tautu (1986), and the volume containing Durrett and Schonmann (1987) will fill you in on more recent developments. Of course, I like the treatment in Durrett (1988).

REFERENCES

K.B. Athreya (1972) Branching Processes, Springer Verlag, New York.

J.T. Cox and R. Durrett (1981) Some limit theorems for percolation processes with necessary and sufficient conditions. Ann. Prob. 9, 583–603.

J.T.Cox and R. Durrett (1989) Limit theorems for the spread of epidemics and forest fires. Stoch. Proc. Appl. 30, 171–191.

R. Durrett (1984) Oriented percolation in two dimensions, Ann. Prob. 12, 999–1040.

R. Durrett (1985) Particle Systems, Random Media, and Large Deviations. AMS contemporary Mathematics Vol. 41., American Math. Society, Providence, RI.

R. Durrett (1988) Lecture Notes On Particle Systems and Percolation, Wadsworth Publishing Co., Monterey, CA

R. Durrett and D. Griffeath (1982) Contact processes in several dimensions. Z. für Wahr. 59, 535–552.

R. Durrett and R.H. Schonmann (1987) Stochastic growth models, in Percolation Theory and the Ergodic Theory of Interacting Particle Systems, ed. by H. Kesten. Vol. 8 in the IMA Volumes in Math and its Applications, Springer Verlag, New York.

H. Kesten (1982) Percolation Theory for Mathematicians. Birkhauser, Boston.

H. Kesten (1986) Aspects of first passage percolation, in Lecture Notes in Math, Vol. 1180, Sringer Verlag, New York.

K. Kuulasmaa (1982) The spatial general epidemic and locally dependent random graphs. J. Appl. Prob. 19, 745–758.

T. Liggett (1985) Interacting Particle Systems. Springer Verlag, New York.

D. Richardson (1973) Random growth in a tesselation. Proc. Camb. Phil. Soc. 74, 515–528.

P. Tautu (1986) Stochastic Spatial Processes. Lecture Notes in Math, Vol. 1212, Springer Verlag, New York.

USE DIFFERENTIAL GEOMETRY WITH THE SECRET INGREDIENT: GRADIENTS!

Ethan Akin
Mathematics Department
The City College
137 Street and Convent Avenue
New York, NY 10031

Amongst the programs for various applications, I would like to insert a sales pitch. The mathematical device I will describe has already served me well and I hope to convince you of its general usefulness.

Our story begins with Fisher's selection differential equations:

$$\frac{dp_i}{dt} = p_i(m_{ip} - m_{pp})$$

(1)
$$m_{ip} = \Sigma_j \, p_j m_{ij}$$

$$m_{pp} = \Sigma_i \, p_i m_{ip} = \Sigma_{i,j} \, p_i p_j m_{ij}.$$

Here the index i varies over I, the finite set of gametic genotypes. p_i is the relative frequence of type i in the gene pool so that p is a vector in the unit simplex Δ in R^I, i.e. $\Delta = \{x \in R^I : x_i \geq 0 \; \forall \, i \; \& \; \Sigma_i x_i = 1\}$. m_{ij} is the fitness (= growth rate) of zygote type ij. This is assumed constant and since ij is an unordered pair the fitness matrix (m_{ij}) is symmetric.

For this system Fisher proved what he termed the Fundamental Theorem of Natural Selection: define the mean fitness function $\bar{m}$ on Δ by $\bar{m}(p) = m_{pp}$ and compute on a solution path of (1) that:

(2)
$$\frac{d\bar{m}}{dt} = 2 \, \Sigma_i \, p_i (m_{ip} - m_{pp})^2.$$

This is nonnegative, vanishing only at an equilibrium, a rest point, for the system. Thus, $\bar{m}$ is a Lyapunov function for (1). It is increasing on all nonequilibrium solution paths.

Later, Kimura proved, [7], a maximum principle for the system. Regarding the right hand side of (1) as the components of a vector field X^m on Δ, he stated that X^m always points in the direction of greatest increase for the function $\bar{m}$. This last phrase says to the mathematician that X^m is the gradient vector field of $\bar{m}$, or at least some positive multiple thereof. But when you compute the gradient of $\bar{m}$ you get the wrong formula. The resulting vector does not point in the X^m direction.

The paradox is resolved by going back to Kimura's proof. There you are reminded that the concept of gradient depends not only upon the function but upon the metric used to measure the length of the vectors. To evaluate the rate of increase for each direction you need to pick out which vector on the ray has unit length.

The length of a vector is measured by using an inner product, a Euclidean metric, on the vector space. Given X and Y in $\mathbf{R}^I$ the usual dot product is

$$(3) \qquad _o(X,Y) = \Sigma_i\, X_i Y_i$$

and the length of X, denoted $_o\|X\|$, is defined to be

$$(4) \qquad _o\|X\| = \sqrt{_o(X,X)}$$

motivated by the Pythagorean Theorem.

Kimura used a cagy definition of unit vector (= "unit variance") different from the usual one and that is what made his proof work.

Shahshahani and Conley, [8], observed that these results can be interpreted using the differential geometry device of a Riemannian metric. This is an inner product which varies from point to point. Thus, the length of a vector is taken to depend upon its point of attachment. For example, $X^m(p)$, whose components appear in (1), is regarded as attached to the point p. Given a path function $x(t)$ the velocity vector dx/dt at t is assumed attached to the position vector $x(t)$.

The Shahshahani metric is defined on the open set $\mathbf{R}^I_+ = \{x \in \mathbf{R}^I: x_i > 0\; \forall i\}$. If X and Y are vectors of $\mathbf{R}^I$ attached to $x \in \mathbf{R}^I_+$, then

$$(3') \qquad (X,Y)_x = \Sigma_i\, x_i^{-1} X_i Y_i$$

and the length of X with respect to the Shahshahani metric is defined by:

(4') $\|X\|_x = \sqrt{(X,X)}_x.$

To see the purpose of this novelty let us look at a dynamical system on $\mathbb{R}^I_+$

(5) $\dfrac{dx}{dt}i = X_i = x_i\xi_i$

with X_i and ξ_i the absolute and relative growth rates, respectively, of x_i. They are functions of the position x.

Now what does it mean for the vector function X to be the gradient of a real potential function U on $\mathbb{R}^I_+$? Recall that the directional derivative of U in the Y direction is given by the formula:

(6) $d_x U(Y) = \Sigma_i \dfrac{\partial U}{\partial x_i} Y_i$

with the partials evaluated at x.

This is linear in Y and the gradient vector at x, $\mathrm{grad}_x U$, is the unique vector such that

(7) $d_x U(Y) = {}_o(\mathrm{grad}_x U, Y)$

for all Y in $\mathbb{R}^I$. By comparing (6) and (3) we get the usual formula for the components of the gradient:

(8) $(\mathrm{grad}_x U)_i = \dfrac{\partial U}{\partial x_i}$ for all i,

at every point x of $\mathbb{R}^I_+$.

For the Shahshahani gradient at x, $\nabla_x U$, we change the inner product in (7):

(7') $d_x U(Y) = (\nabla_x U, Y)_x.$

Using (6) and (3') we get

(8') $(\nabla_x U)_i = x_i \dfrac{\partial U}{\partial x_i}$ for all i.

So we can compare the conditions that the vector field X be a gradient for the two geometries:

$$X = \text{grad } U \qquad\qquad X = \nabla U$$

(9)

$$X_i = \frac{\partial U}{\partial x_i} \qquad\qquad \xi_i = \frac{\partial U}{\partial x_i}$$

In other words, (5) is a Euclidean gradient system if the absolute rates X_i are the partial derivatives of the potential function U, while it is the Shahshahani gradient if the relative rates are the partials of U. Hence, the central claim of this subject: Where the natural focus of a problem is upon absolute growth rates then the usual Euclidean geometry is appropriate. Where, instead, we focus upon relative growth rates, it is the Shashahani geometry which applies. If you reflect upon the ubiquitous use of relative growth rates to define models in biology and economics, you will observe that this assertion, while vague, is quite a strong claim.

To apply this to the selection equations we note that for a system on the simplex Δ or its interior $\overset{\circ}{\Delta}$ $(= \Delta \cap \mathbb{R}_+^I)$ conditions (7) or (7') are only required to hold for vectors Y in the subspace $\mathbb{R}_0^I = \{Y \in \mathbb{R}^I : \Sigma_i \, Y_i = 0\}$ parallel to Δ. Furthermore, the restricted gradient vectors, denoted $\overline{\text{grad}}_p U$ and $\overline{\nabla}_p U$, must themselves lie in $\mathbb{R}_0^I$. Thus, we obtain these latter gradients by projecting perpendicularly onto $\mathbb{R}_0^I$:

$$(\overline{\text{grad}}_p U)_i = \frac{\partial U}{\partial x_i} - \frac{1}{n} \Sigma_j \frac{\partial U}{\partial x_j}$$

(10)

$$(\overline{\nabla}_p U)_i = p_i \left(\frac{\partial U}{\partial x_i} - \Sigma_j \, p_j \frac{\partial U}{\partial x_j}\right),$$

where n is the number of types in I.

From (10) it is easy to check that the selection dynamical system of (1) is exactly the Shahshahani gradient $\overline{\nabla}(\frac{1}{2}\overline{m})$ where $\overline{m}$ is the mean fitness function. The Shahshahani geometry turns up because the gametic fitnesses $m_i (= m_{ip}$ at p) are relative growth rates.

Other, unrelated, systems in biology and economics have turned out to be Shahshahani gradients in the same way (see [6] and [4]). The detailed geometry of the Shahshahani metric is described in [1] and, in a more leisurely way, in [5].

The gradient property, when it holds, is very useful. For example, the sum of gradient fields is the gradient of the sum. As a contrast with this we will consider recombination.

For multilocus models the selection equations, (1), must be corrected because crossing over in the ij zygote may produce gametes of type neither i nor j. To account for recombination we add to $X^m = \bar{\nabla}(\frac{1}{2}\bar{m})$ a new vector field R on Δ. In the two-locus-two-allele case R was first described by Kimura.

R admits a very nice Lyapunov function, a normalized version of entropy, but it is not a gradient. Thus, there arose the problem of finding some Lyapunov function for the sum $X^m + R$. The question was how should mean fitness be adjusted to get some generalized fitness function which the sum of selection and recombination acts to maximize? The answer is that no such function exists in general. Cycles, i.e. nonconstant periodic solutions, may occur. This is incompatible with a function satisfying the Lyapunov property of increasing on nonconstant solutions.

The original theorem, in [1], is quite general. If R is any vector field which is not a Shahshahani gradient then within the family of vector fields $\{X^m + R\}$ Hopf bifurcations occur and so there are cycles for some selection matrix (m_{ij}). Construction of examples and analysis of the types of cycles required a detailed study of the two-locus-two-allele model ([2] and [3]). Here the region in the parameter space where Hopf bifurcations occur is explicitly described. The explicit description allows a normal form computation which shows that the cycles can be limit cycles, i.e. structually stable, attracting cycles can occur in these models.

Once you stop expecting convergence to equilibrium, cycles are the first thing you look for. They turned up so easily that I would now expect more complicated recurrence to be hidden away in multilocus models. I conjecture that the parametrization procedures used to describe the Hopf bifurcation variety can be used to build other normal forms and so to discover strange attractors. As yet, however, their occurrence in multilocus models remains an open problem.

References

[1]   Akin, E. (1979) The Geometry of Population Genetics. Lect.
Notes in Biomath. vol. 31, Springer , New York.

[2]   _____ (1983) Hopf Bifurcation in the Two Locus Genetic
Model. Mem. of the AMS vol. 284.

[3]   _____ (1987) Cycling in Simple Genetic Systems II: The
Symmetric Case. in Dynamical Systems, Proc. Sopron, Hungary (ed.
Kurzhanski and Sigmund). Lect. Notes in Econ. and Math. Systems
vol. 287: 139-154, Springer, New York.

[4]   _____ (1987) Competitive Growth of Firms in an Industry.
in Economic Evolution and Structural Adjustment, Proc. Berkeley,
Calif. (ed. Batten, Casti and Johansson). Lect. Notes in Econ.
and Math. Systems vol. 293: 86-115, Springer, New York.

[5]   _____ (to appear) The Differential Geometry of Population Genetics and Evolutionary Games.

[6]   _____ and Lacker, M. (1984), Ovulation Control: The
Right Number or Nothing. J. Math. Biol. 20:113-132.

[7]   Kimura, M. (1958) On the Change of Population Fitness by
Natural Selection. Heredity 12:145-167.

[8]   Shahshahani, S. (1979) A New Mathematical Framework for
the Study of Linkage and Selection. Mem. of the AMS vol. 211.

OBSTACLES TO MODELLING LARGE DYNAMICAL SYSTEMS

John Guckenheimer
Mathematics Department
Cornell University
Ithaca, NY 14853

Abstract

This paper describes some of the difficulties in bringing modern dynamical systems theory to bear upon the simulation of dynamical phenomena in complex models of natural processes.

1. Introduction

Mankind is making a significant impact upon the global environment. We read daily about predictions of the future effects of increasing carbon dioxide and chlorine in the atmosphere and of acid rain in our lakes and forests. Predicting the course of these phenomena is a large, complicated problem. Computer models that seek to simulate the dynamics of these processes are one avenue to such predictions. This lecture addresses the technology involved in studying the dynamics of large simulation models. The problems that face us are challenging and, if the alarmists are correct, life threatening. Faced with predictions of dire consequences for our current behavior, we want to make the best possible judgments about our future behavior. The analysis of large simulation models is one part of this process. This lecture assesses this analysis from the perspective of dynamical systems theory.

Twenty years ago, Smale (1967) published a paper that outlined a comprehensive theory of differentiable dynamical systems based primarily on geometric and topological methods. Ten years ago, Feigenbaum (1978) discovered the presence of universal metric properties in the bifurcation of dynamical systems. These two events are historical markers in the development of a successful theory for understanding the long term behavior of dynamical systems, even when those dynamics are complicated and chaotic. There have been numerous applications of this theory to problems in a wide range of disciplines. How can we broaden the sphere within which sound application of dynamical systems theory can be made in the context of large simulation models? That is the question I consider here. I shall draw heavily upon my own personal experience and refer to several problems in which I have been interested during the past fifteen years.

Many of the developments in dynamical systems theory have drawn substantial insight from computer studies. Guckenheimer et al. (1977) is an early study of chaotic dynamics in a population model. The visualization of trajectories within the phase space of a dynamical system has been an important tool for the theory. Consequently, our understanding of dynamical behavior that is not easily represented in one, two or three dimensions is much more limited than our understanding of these low dimensional situations. There are other reasons, however, that understanding higher dimensional systems has been a complicated task from a practical point of view. With the hope that tools can be developed that allow one to better understand large dynamical models of chemical, biological, engineering, environmental or economic systems, I would like to bring forward issues that I see as immediate obstacles in this process.

Let me begin with a specific example. Ten years ago, "chaos" was a new discovery among population biologists. David Auslander, George Oster and I were interested in determining whether chaotic models could be systematically developed and applied to the analysis of fluctuating populations. Laboratory systems living in a constant environment seemed to be the most suitable examples to study. We decided to pick the most extensive population data we could find from a laboratory ecosystem as a test case. The data we chose was the work of A. J. Nicholson, an Australian entomologist who spent some twenty years studying blowflies in the laboratory. Some of Nicholson's populations underwent large fluctuations with more or less well defined periodicity. The published data (Nicholson 1957) included population numbers from a single population that was maintained for a period of approximately two years and counted every day or two. We set out to develop a detailed deterministic, age-structured model for this population in an effort to see whether a chaotic model could reproduce the irregularities of the fluctuations in these population cycles. The model itself did little more than update the population size based upon its current age structure and assumed density dependent birth and death rates of each age class in the population.

After four years, two graduate student careers and many generations of laboratory blowflies, we reached a halfhearted conclusion. Our general conclusion was that the models worked quite well, both producing chaotic behavior reminiscent of the laboratory population and giving reasonably good quantitative predictions of the population numbers for a couple of cycles in the population after the vital parameters were "tuned" to the data. Charlie Wu, one of the graduate students, spent well over a year reproducing Nicholson's experiment and also trying to measure the vital parameters of birth and death rates for the model we had formulated (Wu 1976). Nonetheless, the model had a lot of parameters (on the order of twenty-five) and the process of fitting parameters to the data was ad-hoc. The conclusions from the model were sensitive to small details involving the biology of the blowflies. For example, the pattern of egg laying among healthy blowflies was to lay eggs in batches of about one hundred at intervals of three to four days. At the population level, this left the question whether the birth rates of adult flies in the model should correspond to the staggered egg laying of an individual or be chosen as a population average. Not all individuals laid their eggs at exactly the same ages, so the average reproductive rate of the population was a much smoother function of age than of an individual's reproductive rate. Staggered egg laying behavior in the model seemed to lead more easily to chaotic population fluctuations.

Eventually we tired of the project, partly because we did not see how to rationally proceed in matching the data and model more effectively. One of the questions that we addressed occasionally during this project is one that I will return to today: was it feasible to reasonably determine whether a simpler model than the ones we had studied at length might produce the same dynamical behavior? It was not clear how many age classes the model should have, whether discrete or continuous age classes produced a better correspondence between the laboratory and model, etc. The questions that we addressed in this population study seem to me to be typical of modelling and simulation endeavors when there is thoroughly limited access to some components of the model. The difficulties that eventually led us to tire of our investigations are common ones in scientific studies and need further attention from dynamicists. Too often, extensive effort devoted to the development of a model for some dynamical process is not followed by the development of an intuitive understanding of the behavior derived from the model. The reasons why this happens are easy to understand, but the issues are frequently overlooked.

Abstractly, the mathematical problem that is being solved is the following: given a system of n differential or difference equations with k parameters, one would like to describe a comprehensive picture of the trajectories that evolve from arbitrary initial conditions and arbitrary parameters. Computers are good at simulating the evolution of individual trajectories, but it is a much harder task to develop a phase portrait that shows how the evolution of different initial conditions are related. It is still harder to describe the bifurcations that occur as the qualitative structure of these phase portraits changes. An essential problem here is one of size. As the dimension of the phase space and the number of parameters in the model grow, the computational resources needed for a comprehensive study grow rapidly. Moreover, the style of computing that has led to a good understanding of low dimensional systems has been highly interactive. No one has taught a computer how to assemble a phase portrait for a general dynamical system from its numerical computations, so human judgment still plays a large role in guiding the computations. Thus more extensive computation is not only a matter of decreasing step sizes and letting the computer work harder as is true to some extent when solving PDE's. There is more information that must be assembled and interpreted by people. As an example, investigating the behavior of a system with ten parameters for two different values of each parameter yields over a thousand different systems to explore. Since determining a single phase portrait is already a task that can be complicated, it should be clear that the strategy of naively enlarging what has been done for low dimensional systems dependent on one or two parameters to larger systems with many parameters is an overwhelming task. Even in this age of advanced scientific workstations and gigaflop supercomputers, the computational task is not readily surmountable. Thus, before formulating a highly detailed model of a large complex system, like the International Biological Program studies twenty years ago (Blair 1973), I suggest that some thought be given to the question of what can be reasonably done with the model once formulated. It also seems prudent that there should be further development of the technology involved in numerical investigations of large dynamical systems.

To proceed further with these technical developments, we can look for problems that seem to lie on the border between the small systems that we know how to explore readily and large models that appear hopeless. One of my favorite examples of such a problem lies in the area of chemical kinetics. There are chemical reactors that are known to produce persistent oscillations or chaotic behavior in a well stirred, spatially homogeneous environment. The most extensively studied system of this kind is the "Belousov-Zhabotinsky" reaction in a stirred reactor (Argoul et al. 1987). Naively, the kinetics of such a reactor should be described by the laws of mass action for solution chemistry. Determining all of the intermediate species in the reaction and decomposing the reaction into elementary steps should yield a system of differential equations that gives a comprehensive model for the reactor. Many years ago now, Field and Noyes (1974) produced such a scheme for the Belousov-Zhabotinsky reaction, but no one has yet produced numerical solutions of this reaction scheme, or any other, that produces quantitative agreement with the experimental observations. Indeed, it seems to me scandalous that there is (to my knowledge) no oscillating or chaotic chemical reactor for which a quantitatively accurate model has been produced. Despite this, I suspect that most chemists would assert that the basic chemistry contained in the law of mass action is an adequate basis for understanding these reactors. Assuming that they are correct, this seems to me a natural area for the development of good intermediate sized dynamical models of a scientific problem. With stamina, one should even be able to carry out experimentally the determination of the rate constants that are the model parameters for these systems.

Having expressed a rather pessimistic view of what can be done with large simulation models, I want to turn to some of the things that dynamical systems theory has done and discuss strategies that might be successful in dealing with larger problems than the ones that have been conquered in the past. These strategies involve (1) trying to determine whether substantially smaller models might give rise to the same type of dynamics found in a large model, and (2) trying to locate the complicated dynamical behavior in the model by more intelligent means other than a blind search of the parameter space. The first task uses the concept of "fractal dimension" and techniques for measuring the fractal dimension of attractors. The second task uses the theory of multiparameter bifurcation theory.

2. Dimensional Analysis

I have borrowed here the term dimensional analysis from a different context to refer to the study of the size of attractors. If A is an attractor in a phase space M, the goal is to make some type of quantitative assessment of whether it might be possible to find a dynamical system of with a substantial smaller phase space N that has an attractor B with properties that are qualitatively similar to those of A. There are two issues that one can focus upon in this endeavor. One is to measure the size of the attractor itself in terms of some type of dimension, and the second is to try to determine a smaller, invariant submanifold of the phase space in which the attractor lies. The second approach is much more readily feasible when working with numerical simulations, so that properties of the system off the attractor are much more easily examined than when observing an experimental or natural system. With a minimum of technical detail, here is an outline of the two procedures.

The definition of dimension that seems most suitable for application to the study of attractors involves the concept of the Hausdorff dimension of a measure (Young 1983). The underlying idea is that the volume of d-dimensional sets should scale as the dth power of their radius. Thus the length of a line segment is its diameter, the area of a disk is proportional to the square of its diameter, the volume of a sphere is proportional to the cube of its diameter, etc. Given a way of measuring volumes in a set A, one can look for scaling behavior of subsets of small behavior. Let $B(r,x)$ denote the points of A within distance r of x, and let $V(r,x)$ be the volume of $B(r,x)$ with respect to some measure. The "pointwise dimension of A at x" is the limit of $\log(V(r,x))/\log(r)$, provided that the limit exists. If the limit does exist and is independent of x (for almost all x with respect to the measure), then one calls this the Hausdorff or pointwise dimension of the measure (Farmer et al. 1983). There are a number of other closely related concepts of dimension and dimension-like quantities that have been examined for attractors, but it seems to me that the pointwise dimension is the best match between mathematical elegance and usefulness for estimation.

To implement an estimation of attractor dimension, one must begin with data that represents points in a phase space. With observational data, one seldom has the means to record enough measurements to provide a complete representation. Indeed, the most common situation is that one records a single quantity at periodic time intervals that give a projection of the position of the trajectory in phase space onto a single coordinate axis in the phase space. For example, in a population one counts total population numbers, or in a chemical reactor, one monitors a single concentration. For "generic" measurements on the phase space, it is possible to reconstruct a faithful representation of the attractor in the phase space by using the method of delays suggested originally by Ruelle. From the time series $x(t)$, one makes a multiple time series $X(t) = (x(t), x(t+d), ..., x(t+kd))$. Here d is the delay between samples that are used in constructing the vector and $(k+1)$ is the embedding dimension. There is a lore concerning the values of d and k that work most effectively for the dimension calculations that follow (Mayer-Kress 1986). The optimal situation is one in which it is not necessary to use the method of delays because enough measurements were made originally to separate distinct points on the attractor; i.e., the map from the attractor A into the space of measurements is injective.

Assume now that one does have a representation of the attractor in terms of a k-dimensional space of measurements. The next step in estimating the dimension of the attractor is to define a measure that seems appropriate for the attractor. The ansatz that is used here is to assume that the attractor is ergodic and that the observed trajectory is typical. With these assumptions, the ergodic theorem implies that the proportion of time that a trajectory spends in a given subset of the attractor is proportional to the measure (volume or probability) of the subset. Thus the volume of a set of the form $B(r,x)$ is approximately the proportion of points in the observed trajectory whose distance from x is smaller than r. By computing the distances between all of the observed points and x and then sorting these distances, one easily obtains an approximation of the function $V(r,x)$. From this, an estimate of the pointwise dimension at x is obtained. Statistical fluctuations can be reduced substantially by averaging over different choices of x.

The estimates that come from this procedure tend to be crude for a number of reasons (Guckenheimer 1984). Some of these reasons are based on statistical considerations, some are based in the lack of smoothness of the function $V(r,x)$ for a fractal attractor, and some are based in geometric considerations. The dominant effect, however, is that a good estimate of the dimension of an attractor depends upon the ability to give a good approximation to the logarithm of $V(r,x)$ for small values of r. In turn, the computational procedure for estimating $V(r,x)$ requires that its value be significantly larger than $1/N$ with N the number of points in the computed or observed trajectory. As the dimension d grows, the length of trajectory required to estimate $V(r,x)$ for fixed r increases exponentially with d. Unless the dimension of the attractor is quite small, close returns to previously visited points require times that are prohibitive in most studies. A reasonable geometric picture of the small scale structure in high dimensional attractors seems virtually impossible to obtain. It is unclear whether the type of geometric analysis associated with dynamical systems theory has much to contribute to the study of systems with high dimensional attractors. We can say that the dimension of the attractor is large and we can attempt to compute a variety of statistical averages, but the distinction between the system as a dynamical system with a chaotic attractor and a stochastic system becomes blurred if the evolution of the system from nearby initial conditions cannot be observed.

A vivid example of these types of considerations is provided by the problem of weather forecasting. Presumably, to a high degree of accuracy, the weather can be described by the evolution of the solutions of the fluid equations describing the atmosphere. Thinking of the global atmosphere as a dynamical system is a useful guide, but it does not seem to be a practical way of making predictions. If there were a low dimensional attractor for the atmospheric equations, that one could plausibly use "analogs" as a basis for weather prediction. If today's weather pattern resembles the pattern seen on October 17, 1453, then tomorrow's should resemble the pattern seen on October 18, 1453. Attempts to find analogs in past weather patterns have been made with little success (Lorenz 1969). Let us see why these were almost certainly doomed to failure. Ignoring the complications of seasonality, we have at best some 10,000 daily observations of the atmosphere to draw upon in searching for analogs. This provides a very meager database in which to look for close analogs in a multidimensional space. In a four dimensional unit cube, one can place 10,000 points so that the closest distance between any pair is 1/10. Since the dimension of the atmospheric attractor is almost certainly larger than four and since simulations of global atmospheric models indicate that there is a strong sensitive dependence on initial conditions, poor analogs are all we expect to find and they are not likely to form a good basis for prediction.

The second approach to dimensional analysis is suitable for numerical computations in which one can investigate the trajectories of initial conditions that start very close to one another. Within a strange attractor, there are unstable directions along which trajectories separate at an exponential rate. The rates of exponential divergence or convergence of trajectories are called Lyapunov exponents. Suppose now that an invariant, attracting submanifold is sought in the phase space of a system. This invariant manifold should contain the attractor being studied, but it will be considerably larger so that it can encompass the fractal structure of the attractor. The theoretical basis for the theory of inertial manifolds is that one should look for these invariant manifolds by searching for uniform volume contraction. The sum of the k largest Lyapunov exponents gives the rate at which most k dimensional volumes in the phase space expand or contract. If k is chosen to be the minimal integer so that the sum of the k largest exponents is negative, then there is hope of finding a contracting, inertial manifold. Suitable versions of this strategy have been employed to construct inertial manifolds for several infinite dimensional dynamical systems described by partial differential equations (Foias et al. 1988). Application of the theory produces the assurance that there is a finite dimensional subsystem which contains the attractors of the full infinite dimensional system. Note, however, that the dimension of the subsystems obtained from the theory are likely to be large, substantially larger than is suitable for practical modelling in many cases.

This pessimistic view of the world of high dimensional attractors is tempered by the fact that the technique we have described for estimating dimension does provide a way of determining whether a system resembles one with a low dimensional attractor. This can be done easily and effectively with about as much effort as is required to compute a power spectrum using a fast fourier transform algorithm. It does happen occasionally that interesting systems are found that have low dimensional strange attractors. The Belousov-Zhabotinsky reactor appears to be such a case. In the regime in which its dynamics are chaotic, the observed features of this dynamical behavior closely resembles dynamics that are found in families of three dimensional vector fields. This suggests that some smaller model than the Field-Noyes equations ought to give a reasonably good picture of the dynamics, but as I discussed earlier, a comprehensive understanding of this problem has been elusive.

3. Multiparameter Bifurcations

A second area in which dynamical systems theory might help in modelling large systems is in the study of bifurcations in multiparameter families of equations. Bifurcations are qualitative changes in the dynamical behavior of a system that occur as a parameter is varied. There is a classification of the most common bifurcations into several types that indicate the changes occurring in a system (Guckenheimer and Holmes 1983, Arnold 1983). In systems that depend upon more than one parameter, the locus S of parameter values at which bifurcation occurs gives a stability diagram for the parameter space. Generally, S can be a complicated set. From a knowledge of S and the types of bifurcations occurring on each part of S, it is possible to follow the changes in the dynamics of the system as the parameters vary. The question that is relevant for applications is the extent to which one can find S.

At this juncture, an idea from Thom's Catastrophe Theory (Thom 1975) can help. The idea is that of an organizing center. Think of S as a stratified set built from sheets of codimension one that intersect along subsets of codimension two that meet other such subsets at points of codimension three, etc. Thom's idea is that a local analysis of model problems ("normal forms") should reveal the way in which strata of low codimension meet at a parameter value of high codimension. To the extent that points of high codimension can be located and the analysis of the associated normal forms can be performed, it gives one the means of finding a rich variety of dynamical behavior within the system.

This program of investigating and locating bifurcations of high codimension has been a very fruitful technique for a growing number of applications. Some types of bifurcations are determined by conditions that do not require integration of the vector field but are determined by solving equations that are formulated in terms of the Taylor expansion of the vector field at a point. For example, requiring that a vector field have an equilibrium point whose linearization has a zero eigenvalue of multiplicity k gives a bifurcation of codimension at least k. When k is two, a rigorous analysis shows that the typical situation has Hopf bifurcations giving rise to periodic orbits and homoclinic bifurcations in which periodic orbits disappear while their period becomes unbounded as a saddle point is incorporated into the orbit. If k is larger than two, then numerical studies of the typical situation show that chaotic dynamics can be found for parameter values close to those at which the codimension three bifurcation occurs.

The technology for locating bifurcations of high codimension in applications is not in a totally satisfactory state. Numerical algorithms frequently make the implicit assumption that one is dealing with problems that are generic or nondegenerate in some sense. When these conditions fail, then the methods either fail to work well or they fail entirely. For example, Newton's method does not work well when faced with the task of locating a multiple zero of a function. This disease of numerical methods can cause difficulty when they are applied to bifurcation problems indiscriminately. Bifurcations occur precisely where some type of nondegeneracy condition breaks down, so one must be careful in the choice of computational algorithms. Consider, for example, the problem of finding saddle-node or limit point bifurcations for a vector field $X(x)$. These bifurcations are located at points where $X(x) = 0$ and $DX(x)$ has an eigenvalue zero. Thus Newton's method applied to X cannot be expected to locate the bifurcation points readily. To apply Netwon's method robustly, it is more sensible to look for solutions of the extended system of equations $X(x) = 0$ and $\det(DX(x)) = 0$ in the product of the phase space with one of the parameter directions. Studying this extended system requires that one must be able to compute the Jacobian derivative of X, and to apply Newton's method one needs its derivative.

As the above example illustrates, the direct computation of bifurcations requires that one be able to compute and work with the derivatives of a dynamical system. As the codimension of bifurcations increases, one is likely to need higher and higher derivatives. For high dimensional systems, the expressions of these grow rapidly with the order of the derivative. Root finding algorithms decrease in their capacity to solve systems of equations as the size of the system grows. Thus, the prospect of carrying through the calculation of where organizing centers occur in a system is problematic. Nontheless, it seems to me that effort in this direction is a sensible way to determine the richness of the dynamical behavior that one initially expects to find in a dynamical system with many parameters. The software package AUTO (Doedel 1986) provides a step in this direction.

There are two cautionary notes concerning this process with which I end. The first is that the problem of reconstructing the dynamics associated with normal forms of high codimension is itself a substantial problem. Moreover, the number of undetermined coefficients in a normal form that may influence the dynamics grows with the codimension of the problem. Consequently, the dynamical analysis of normal forms of high codimension quickly begins to look as bad as the large, many parameter problems that we started with. In most problems, though, I would expect that a knowledge of the existence of codimension two and three bifurcations will quickly lead us to the intricacies of the dynamical behavior in the models.

The second cautionary note is that there is a tension between qualitative and quantitative aspects of bifurcation diagrams. Studies of multiparameter bifurcation in a number of systems yield aspects of the behavior that occur in tiny regions of the parameter space. Phenomena that are an essential aspect of a coherent qualitative picture of the dynamical behavior of a system can be ephemeral objects to find computationally. In these cases, it seems that the only way that such phenomena are likely to be seen in the computations is through a systematic search for them that is based upon a theoretical understanding of what must occur. The practical significance of this paradoxical situation is not clear to me, and it seems to me that we need more experience with simulation models before it will be.

Acknowledgments: This research was partially supported by the National Science Foundation, the Air Force Office of Scientific Research and the Army Research Office through the Mathematical Sciences Institute at Cornell University.

References

1. Argoul, F., Arneodo, A., Richetti, P., Roux, J. C. and Swinney, H. L. (1987) Chemical chaos: from hints to confirmation, Accts. in Chemical Research.
2. Arnold, V. I. (1983) Geometrical methods in the theory of ordinary differential equations, Springer-Verlag, New York.
3. Blair, W. F. (1973) Big Biology: The US/IBP program, Dowden, Hutchinson and Ross.
4. Dangelmayr, G. and Guckenhiemer, J. (1987) On a four parameter family of vector fields, Arch. Rat. Mech. Anal. 97, 321–352.
5. Doedel, E. (1986) AUTO: software for continuation and bifurcation problems in ordinary differential equations, Calif. Inst. Tech, Pasadena, CA.
6. Farmer, J. D., Ott, E. and Yorke, J. (1983) The dimension of chaotic attractors, Physica 7D, 153–180.
7. Feigenbaum, M. J. (1978) Quantitative universality for a class of nonlinear transformations, J. Statistical Physics, 19, 25–52.
8. Field, R. J. and Noyes, R. M. (1974) Oscillations in chemical systems IV. Limit cycle behavior in a model of a real reaction, J. Chem. Phys., 60, 1877–1884.
9. Foias, C., Sell, G. and Temam, R. (1988) Inertial manifolds for nonlinear evolutionary equations. J. Diff. Eq.,73, 309–352.
10. Guckenheimer, J. (1984) Dimension estimates for attractors, Contemp. Math. 28, 357–367.

11. Guckenheimer, J. and Holmes, P. (1983) Nonlinear oscilllations, dynamical systems and bifurcations of vector fields, Springer-Verlag, New York.
12. Guckenheimer, J., Oster, G. and Ipaktchi, A. (1977) Dynamics of deterministic population models, J. Math. Biology, 4, 101–147.
13. Lorenz, E. (1969) Atmospheric predictability as revealed by naturally occurring analogues, J. Atmos. Sci., 26, 636–646.
14. Mayer-Kress, G. (ed.) (1986) Dimensions and entropies in chaotic systems, Springer-Verlag, New York.
15. Nicholson, A. J. (1957) The self-adjustment of populations to change, Cold Spring Harbor Symp. Quant. Biology 22, 153–173.
16. Smale, S. (1967) Differentiable dynamical systems, Bull. Am. Math. Soc. 73, 747–817.
17. Thom, R. (1975) Structural stability and morphogenesis, Benjamin.
18. Wu, Y. C. (1976) An experimental and theoretical study of population cycles of the blowfly, Phaenicia sericata (Calliphorides), in a laboratory ecosystem. Ph.D. dissertation, University of California, Berkeley.
19. Young, L. S. (1983) Entropy, lyapunov exponents, and Hausdorff dimension in differentiable dynamical systems, IEEE Trans. Circuits and Systems, 30, 599–606.

ERRATA

LECTURE NOTES IN BIOMATHEMATICS 81
C. Castillo-Chavez, S.A. Levin, C.A. Shoemaker (Eds.),
Mathematical Approaches to Problems in Resource Management
and Epidemiology, ISBN 3-540-51820-7

1. Due to a technical error, pages 46 and 47 of the above
volume were interchanged.

2. Further Figs. 1-4 of the contribution by J.M. Conrad
were omitted. Please insert this sheet after p. 66.

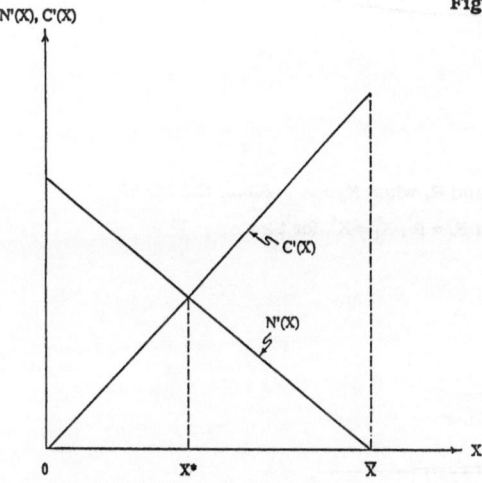

Figure 1. A Graph of Net Marginal Benefits, **N'(X)**, and Net Marginal Costs, **C'(X)**.

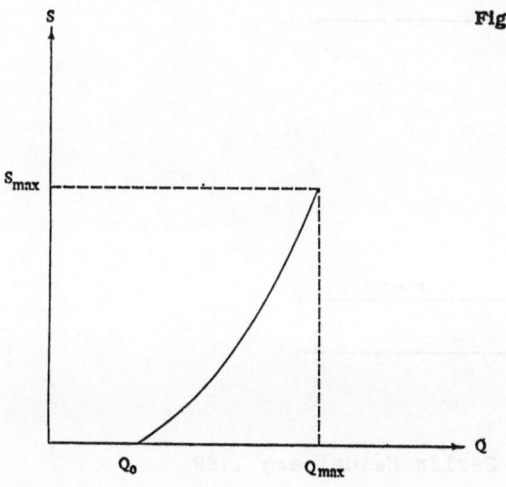

Figure 2. The Transformation Function $\phi(Q,S) \equiv 0$ for Commodity Q and Residual S.

Figure 3. The Econosphere or Spaceship Earth, where materials balance ultimately requires **A = B + C, C = D** and thus **A = B + D** (the mass of material inputs equals the sum of residuals).

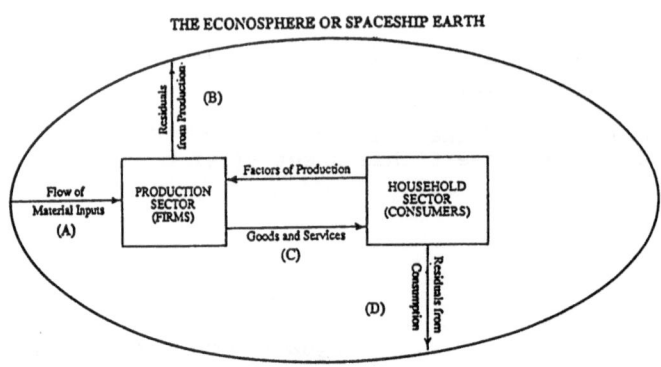

THE ECONOSPHERE OR SPACESHIP EARTH

Figure 4. Approach Paths for X_t and R_t when $X_0 > X^*$. Along the MRAP $\dot{R}_t = 0$ for $0 \leq t \leq \hat{t}$ then $\dot{R}_t = R^*$, $\dot{X}_t = X^*$, for $t \geq \hat{t}$.

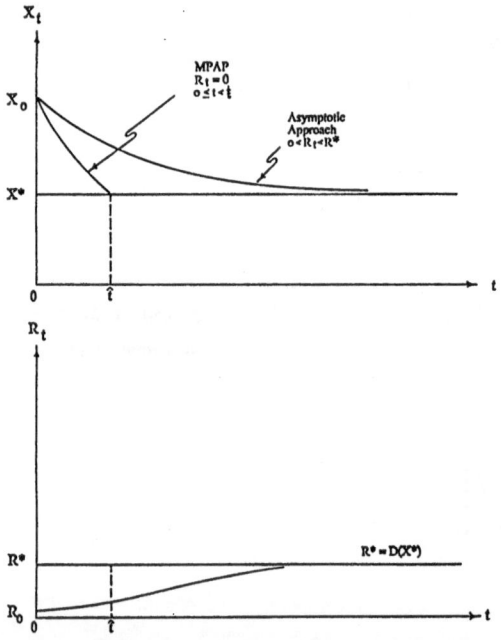

Journal of Mathematical Biology

ISSN 0303-6812   Title No. 285

For mathematicians and biologists working in a wide variety of fields – genetics, demography, ecology, neurobiology, epidemiology, morphogenesis, cell biology – the **Journal of Mathematical Biology** publishes:

● papers in which mathematics is used for a better understanding of biological phenomena
● mathematical papers inspired by biological research, and
● papers which yield new experimental data bearing on mathematical models.

A selection of papers from recent issues:

C. Castillo-Chavez, H. W. Hethcote, V. Andreasen, S. A. Levin, W. M. Liu: Epidemiological models with age structure, proportionate mixing, and cross-immunity

L. Gardini, R. Lupini, M. G. Messia: Hopf bifurcation and transition to chaos in Lotka-Volterra equation

M. Mesterton-Gibbons: On compromise in foraging and an experiment by Krebs et al. (1977)

J. H. Swart: Viable controls in age-dependent population dynamics

J. B. T. M. Roerdink: The biennial life strategy in a random environment. Supplement

A. G. Pakes: A complementary note on the supercritical birth, death and catastrophe process

L. J. Cromme, I. E. Dammasch: Compensation type algorithms for neural nets: stability and convergence

O. Arino, M. Kimmel: Asymptotic behavior of a nonlinear functional-integral equation of cell kinetics with unequal division

T. Darden, N. L. Kaplan, R. R. Hudson: A numerical method for calculating moments of coalescent times in finite populations with selection

A. Scheib: Analysis of a model for random competition

Springer-Verlag Berlin
Heidelberg New York London
Paris Tokyo Hong Kong

Biomathematics

Managing Editor:
S. A. Levin

Editorial Board:
M. Arbib, J. Cowan,
C. DeLisi, M. Feldman,
J. Keller, K. Krickeberg,
R. M. May, J. D. Murray,
A. Perelson, T. Poggio,
L. A. Segel

Volume 18

S. A. Levin, Cornell University, Ithaca, NY; T. G. Hallam, L. J. Gross, University of Tennessee, Knoxville, TN, USA (Eds.)

Applied Mathematical Ecology

1989. XIV, 489 pp. 114 figs. ISBN 3-540-19465-7

Contents: Introduction. – Resource Management. – Epidemiology: Fundamental Aspects of Epidemiology Case Studies. – Ecotoxicology. – Demography and Population Biology. – Author Index. – Subject Index.

This book builds on the basic framework developed in the earlier volume – "Mathematical Ecology", edited by T. G. Hallam and S. A. Levin, Springer 1986, which lays out the essentials of the subject. In the present book, the applications of mathematical ecology in ecotoxicology, in resource management, and epidemiology are illustrated in detail. The most important features are the case studies, and the interrelatedness of theory and application. There is no comparable text in the literature so far. The reader of the two-volume set will gain an appreciation of the broad scope of mathematical ecology.

Volume 19

J. D. Murray, Oxford University, UK

Mathematical Biology

1989. XIV, 767 pp. 262 figs. ISBN 3-540-19460-6

This textbook gives an in-depth account of the practical use of mathematical modelling in several important and diverse areas in the biomedical sciences.
The emphasis is on what is required to solve the real biological problem.
The subject matter is drawn, for example, from population biology, reaction kinetics, biological oscillators and switches, Belousov-Zhabotinskii reaction, neural models, spread of epidemics.
The aim of the book is to provide a thorough training in practical mathematical biology and to show how exciting and novel mathematical challenges arise from a genuine interdisciplinary involvement with the biosciences. It also aims to show how mathematics can contribute to biology and how physical scientists must get involved.
The book also presents a broad view of the field of theoretical and mathematical biology and is a good starting place from which to start genuine interdisciplinary research.

In preparation

Volume 20

J. E. Cohen, Rockefeller University, New York, NY, USA; F. Briand, Gland, Switzerland; C. M. Newman, University of Arizona, Tucson, AZ, USA

Community Food Webs

Data and Theory

1989. Approx. 300 pp. 46 figs. ISBN 3-540-51129-6

Springer-Verlag Berlin
Heidelberg New York London
Paris Tokyo Hong Kong

Springer

Lecture Notes in Biomathematics